Cognitive Radio Networks

Cognitive Radio Networks

Professor Kwang-Cheng Chen
National Taiwan University, Taiwan

Professor Ramjee Prasad
Aalborg University, Denmark

A John Wiley and Sons, Ltd, Publication

This edition first published 2009
© 2009 by John Wiley & Sons Ltd

Registered office
John Wiley & Sons Ltd, The Atrium, Southern Gate, Chichester, West Sussex, PO19 8SQ, United Kingdom

For details of our global editorial offices, for customer services and for information about how to apply for permission to reuse the copyright material in this book please see our website at www.wiley.com.

Library of Congress Cataloging-in-Publication Data

Chen, Kwang-Cheng.
  Cognitive radio networks / Kwang-Cheng Chen, Ramjee Prasad.
      p. cm.
  Includes bibliographical references and index.
  ISBN 978-0-470-69689-7 (cloth)
1. Cognitive radio networks. I. Prasad, Ramjee. II. Title.
  TK5103.4815.C48 2009
  621.39′81–dc22

                                         2008055907

A catalogue record for this book is available from the British Library.

ISBN 978-0-470-69689-7 (H/B)

Set in 10/12pt Times by Thomson Digital, Noida, India.
Printed in Great Britain by CPI Antony Rowe, Chippenham, England

Contents

Preface   **xi**

1 Wireless Communications   **1**
    1.1   Wireless Communications Systems   1
    1.2   Orthogonal Frequency Division Multiplexing (OFDM)   3
       1.2.1   OFDM Concepts   4
       1.2.2   Mathematical Model of OFDM System   5
       1.2.3   OFDM Design Issues   9
       1.2.4   OFDMA   21
    1.3   MIMO   24
       1.3.1   Space-Time Codes   24
       1.3.2   Spatial Multiplexing Using Adaptive Multiple Antenna Techniques   27
       1.3.3   Open-loop MIMO Solutions   27
       1.3.4   Closed-loop MIMO Solutions   29
       1.3.5   MIMO Receiver Structure   31
    1.4   Multi-user Detection (MUD)   34
       1.4.1   Multi-user (CDMA) Receiver   34
       1.4.2   Suboptimum DS/CDMA Receivers   37
    References   40

2 Software Defined Radio   **41**
    2.1   Software Defined Radio Architecture   41
    2.2   Digital Signal Processor and SDR Baseband Architecture   43
    2.3   Reconfigurable Wireless Communication Systems   46
       2.3.1   Unified Communication Algorithm   46
       2.3.2   Reconfigurable OFDM Implementation   47
       2.3.3   Reconfigurable OFDM and CDMA   47
    2.4   Digital Radio Processing   48
       2.4.1   Conventional RF   48
       2.4.2   Digital Radio Processing (DRP) Based System Architecture   52
    References   58

3 Wireless Networks   **59**
    3.1   Multiple Access Communications and ALOHA   60
       3.1.1   ALOHA Systems and Slotted Multiple Access   61
       3.1.2   Slotted ALOHA   61

	3.1.3	Stabilised Slotted ALOHA	64
	3.1.4	Approximate Delay Analysis	65
	3.1.5	Unslotted ALOHA	66
3.2	Splitting Algorithms		66
	3.2.1	Tree Algorithms	67
	3.2.2	FCFS Splitting Algorithm	68
	3.2.3	Analysis of FCFS Splitting Algorithm	69
3.3	Carrier Sensing		71
	3.3.1	CSMA Slotted ALOHA	71
	3.3.2	Slotted CSMA	76
	3.3.3	Carrier Sense Multiple Access with Collision Detection (CSMA/CD)	79
3.4	Routing		82
	3.4.1	Flooding and Broadcasting	83
	3.4.2	Shortest Path Routing	83
	3.4.3	Optimal Routing	83
	3.4.4	Hot Potato (Reflection) Routing	84
	3.4.5	Cut-through Routing	84
	3.4.6	Interconnected Network Routing	84
	3.4.7	Shortest Path Routing Algorithms	84
3.5	Flow Control		89
	3.5.1	Window Flow Control	89
	3.5.2	Rate Control Schemes	91
	3.5.3	Queuing Analysis of the Leaky Bucket Scheme	92
References			93

4 Cooperative Communications and Networks — **95**

4.1	Information Theory for Cooperative Communications		96
	4.1.1	Fundamental Network Information Theory	96
	4.1.2	Multiple-access Channel with Cooperative Diversity	101
4.2	Cooperative Communications		102
	4.2.1	Three-Node Cooperative Communications	103
	4.2.2	Multiple-Node Relay Network	109
4.3	Cooperative Wireless Networks		113
	4.3.1	Benefits of Cooperation in Wireless Networks	114
	4.3.2	Cooperation in Cluster-Based Ad-hoc Networks	116
References			118

5 Cognitive Radio Communications — **121**

5.1	Cognitive Radios and Dynamic Spectrum Access		121
	5.1.1	The Capability of Cognitive Radios	122
	5.1.2	Spectrum Sharing Models of DSA	124
	5.1.3	Opportunistic Spectrum Access: Basic Components	126
	5.1.4	Networking The Cognitive Radios	126
5.2	Analytical Approach and Algorithms for Dynamic Spectrum Access		126
	5.2.1	Dynamic Spectrum Access in Open Spectrum	128
	5.2.2	Opportunistic Spectrum Access	130
	5.2.3	Opportunistic Power Control	131
5.3	Fundamental Limits of Cognitive Radios		132

5.4   Mathematical Models Toward Networking Cognitive Radios   136
       5.4.1   CR Link Model   136
       5.4.2   Overlay CR Systems   137
       5.4.3   Rate-Distance Nature   140
       References   142

6   Cognitive Radio Networks   **145**
6.1   Network Coding for Cognitive Radio Relay Networks   146
       6.1.1   System Model   147
       6.1.2   Network Capacity Analysis on Fundamental CRRN Topologies   150
       6.1.3   Link Allocation   154
       6.1.4   Numerical Results   156
6.2   Cognitive Radio Networks Architecture   159
       6.2.1   Network Architecture   159
       6.2.2   Links in CRN   161
       6.2.3   IP Mobility Management in CRN   163
6.3   Terminal Architecture of CRN   165
       6.3.1   Cognitive Radio Device Architecture   165
       6.3.2   Re-configurable MAC   168
       6.3.3   Radio Access Network Selection   169
6.4   QoS Provisional Diversity Radio Access Networks   171
       6.4.1   Cooperative/Collaborative Diversity and Efficient Protocols   172
       6.4.2   Statistical QoS Guarantees over Wireless Asymmetry
               Collaborative Relay Networks   174
6.5   Scaling Laws of Ad-hoc and Cognitive Radio Networks   177
       6.5.1   Network and Channel Models   177
       6.5.2   Ad-hoc Networks   178
       6.5.3   Cognitive Radio Networks   179
       References   180

7   Spectrum Sensing   **183**
7.1   Spectrum Sensing to Detect Specific Primary System   183
       7.1.1   Conventional Spectrum Sensing   183
       7.1.2   Power Control   187
       7.1.3   Power-Scaling Power Control   188
       7.1.4   Cooperative Spectrum Sensing   190
7.2   Spectrum Sensing for Cognitive OFDMA Systems   194
       7.2.1   Cognitive Cycle   195
       7.2.2   Discrimination of States of the Primary System   197
       7.2.3   Spectrum Sensing Procedure   203
7.3   Spectrum Sensing for Cognitive Multi-Radio Networks   206
       7.3.1   Multiple System Sensing   207
       7.3.2   Radio Resource Sensing   216
       References   228

8   Medium Access Control   **231**
8.1   MAC for Cognitive Radios   231

8.2   Multichannel MAC   232
        8.2.1   General Description of Multichannel MAC   235
        8.2.2   Multichannel MAC: Collision Avoidance/Resolution   238
        8.2.3   Multichannel MAC: Access Negotiation   242
8.3   Slotted-ALOHA with Rate-Distance Adaptability   251
        8.3.1   System Model   252
8.4   CSMA with AMC   259
        8.4.1   Carrier Sense Multiple Access with Spatial-Reuse
                Transmissions   261
        8.4.2   Analysis of CSMA-ST   263
        8.4.3   A Cross-Layer Power-Rate Control Scheme   268
        8.4.4   Performance Evaluations   270
        References   272

9   Network Layer Design   275
9.1   Routing in Mobile Ad-hoc Networks   275
        9.1.1   Routing in Mobile Ad-hoc Networks   275
        9.1.2   Features of Routing in CRN   276
        9.1.3   Dynamic Source Routing in MANET   278
        9.1.4   Ad-hoc On-demand Distance Vector (AODV)   283
9.2   Routing in Cognitive Radio Networks   286
        9.2.1   Trusted Cognitive Radio Networking   286
        9.2.2   Routing of Dynamic and Unidirectional CR Links in CRN   288
9.3   Control of CRN   291
        9.3.1   Flow Control of CRN   291
        9.3.2   End-to-End Error Control in CRN   292
        9.3.3   Numerical Examples   292
9.4   Network Tomography   296
9.5   Self-organisation in Mobile Communication Networks   298
        9.5.1   Self-organised Networks   298
        9.5.2   Self-organised Cooperative and Cognitive Networks   299
        References   304

10 Trusted Cognitive Radio Networks   307
10.1   Framework of Trust in CRN   308
        10.1.1 Mathematical Structure of Trust   308
        10.1.2 Trust Model   311
10.2   Trusted Association and Routing   311
        10.2.1 Trusted Association   312
        10.2.2 Trusted Routing   317
10.3   Trust with Learning   319
        10.3.1 Modified Bayesian Learning   319
        10.3.2 Learning Experiments for CRN   322
10.4   Security in CRN   328
        10.4.1 Security Properties in Cellular Data Networks   328
        10.4.2 Dilemma of CRN Security   330

10.4.3 Requirements and Challenges for Preserving User
                Privacy in CRNs   331
        10.4.4 Implementation of CRN Security   332
    References   334

11 Spectrum Management of Cognitive Radio Networks   **335**
    11.1 Spectrum Sharing   337
    11.2 Spectrum Pricing   339
    11.3 Mobility Management of Heterogeneous Wireless Networks   347
    11.4 Regulatory Issues and International Standards   350
        11.4.1 Regulatory Issues   351
        11.4.2 International Standards   354
    References   355

Index   **357**

Preface

Wireless communications and networks have experienced booming growth in the past few decades, with billions of new wireless devices in use each year. In the next decade we expect the exponential growth of wireless devices to result in a challenging shortage of spectrum suitable for wireless communications. Departing from the traditional approach to increase the spectral efficiency of physical layer transmission, Dr. Joe Mitola III's innovative cognitive radio technology derived from software defined radio will enhance spectrum utilization by leveraging spectrum "holes" or "white spaces". The Federal Communication Commission (FCC) in the US quickly identified the potential of cognitive radio and endorsed the applications of such technology. During the past couples of years, there now exist more than a thousand research papers regarding cognitive radio technology in the IEEE Xplore database, which illustrates the importance of this technology. However, researchers have gradually come to realize that cognitive radio technology, at the link level, is not sufficient to warrant the spectrum efficiency of wireless networks to transport packets, and networking these cognitive radios which coexist with primary/legacy radios through cooperative relay functions can further enhance spectrum utilization. Consequently, in light of this technology direction, we have developed this book on cognitive radio networks, to introduce state-of-the-art knowledge from cognitive radio to networking cognitive radios.

During the preparation of the manuscript for this book, we would like to thank the encouragement, discussion, and support from many international researchers and our students, including Mohsen, Guizani, Fleming Bjerge Frederiksen, Neeli Prasad, Ying-Chang Liang, Sumei Sun, Songyoung Lee, Albena Mihovska, Feng-Seng Chu, Chi-Cheng Tseng, Shimi Cheng, Lin-Hung Kung, Chung-Kai Yu, Shao-Yu Lien, Sheng-Yuan Tu, Bilge Kartal Cetin, Yu-Cheng Peng, Jin Wang, Peng-Yu Chen, Chu-Shiang Huang, Ching-Kai Liang, Hong-Bin Chang, Po-Yao Huang, Wei-Hong Liu, I-Han Chiang, Michael Eckl, Yo-Yu Lin, Weng Chon Ao, Dua Idris, and Joe Mitola III, the father of cognitive radio. Our thanks also to Inga, Susanne and Keiling who helped with so many aspects that the book could not have been completed without their support.

The first author (K.C. Chen) would especially like to thank *Irving T. Ho Foundation* who endowed the chair professorship to National Taiwan University which enabled him to dedicate his time to writing this book. For the readers' information, Dr. Irving T. Ho is the founder of Hsin-Chu Science Park in Taiwan. Our appreciation also goes to the National Science Council and CTiF Aalborg University who made it possible for KC and Ramjee to work together in Denmark. Last but not the least, KC would like to thank his wife Christine and his children Chloé and Danny for their support, especially during his absence from home in the summer of 2008 while he was completing the manuscript.

Kwang-Cheng Chen, Taipei, Taiwan
Ramjee Prasad, Aalborg, Denmark

1

Wireless Communications

Conventional wireless communication networks use circuit switching, such as the first generation cellular AMPS adopting Frequency Division Multiple Access (FDMA) and second generation cellular GSM adopting Time Division Multiple Access (TDMA) or the IS-95 pioneering Code Division Multiple Access (CDMA). The success of the Internet has caused a demand for wireless broadband communications and packet switching plays a key role, being adopted in almost every technology. From the third generation cellular and beyond, packet switching becomes a general consensus in the development of technology.

The International Standards Organisation (ISO) has defined a large amount of standards for computer networks, including the fundamental architecture of Open System Interconnection (OSI) to partition computer networks into seven layers. Such a seven-layer partition might not be ideal when optimising network efficiency, but it is of great value in the implementation of large scale networks via such a layered-structure. Engineers can implement a portion of software and hardware in a network independently, even plug-in networks, or replace a portion of network hardware and/or software, provided that the interfaces among layers and standards are well defined. Considering the nature of 'stochastic multiplexing' packet switching networks, the OSI layer structure may promote the quick progress of computer networks and the wireless broadband communications discussed in this book.

Figure 1.1 depicts the OSI seven-layer structure and its application to the general extension and interconnection to other portion of networks. The four upper layers are mainly 'logical' rather than 'physical' in concept in network operation, whereas physical signalling is transmitted, received and coordinated in the lower two layers: physical layer and data link layer. The physical layer of a wireless network thus transmits bits and receives bits correctly in the wireless medium, while medium access control (MAC) coordinates the packet transmission using the medium formed by a number of bits. When we talk about *wireless communications* in this book, we sometimes refer it as a physical layer and the likely MAC of wireless networks, although some people treat it with a larger scope. In this chapter, we will focus on introducing physical layer transmission of wireless communication systems, and several key technologies in the narrow-sense of wireless communications, namely *orthogonal frequency division multiplexing* (OFDM) and *multi-input-multi-output* (MIMO) processing.

1.1 Wireless Communications Systems

To support multimedia traffic in state-of-the-art wireless mobile communications networks, digital communication system engineering has been used for the physical layer transmission. To allow a smooth transition into later chapters, we shall briefly introduce here the fundamentals of digital communications,

Cognitive Radio Networks   Kwang-Cheng Chen and Ramjee Prasad
© 2009 John Wiley & Sons, Ltd

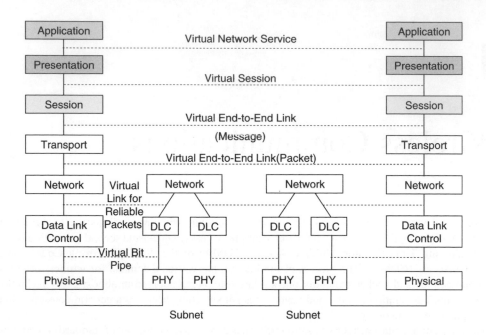

Figure 1.1   Seven-Layer OSI Network Architecture

assuming some knowledge of undergraduate-level communication systems and signalling. Interested readers will find references towards more advanced study throughout the chapter.

Following analogue AM and FM radio, digital communication systems have been widely studied for over half a century. Digital communications have advantages over their analogue counterparts due to better system performance in links, and digital technology can also make media transmission more reliable. In the past, most interest focused on conventional narrow-band transmission and it was assumed that telephone line modems might lead the pace and approach a theoretical limit. Wireless digital communications were led by major applications such as satellite communications and analogue cellular. In the last two decades, wireless broadband communications such as *code division multiple access* (CDMA) and a special form of narrowband transmission known as *orthogonal frequency division multiplexing* (OFDM) were generally adopted in state-of-the-art communication systems for high data rates and system capacity in complicated communication environments and harsh fading channels. A digital wireless communication system usually consists of the elements shown in Figure 1.2, where they are depicted as a block diagram.

Information sources can be either digital, to generate 1s and 0s, or an analogue waveform source. A source encoder then transforms the source into another stream of 1s and 0s with high entropy. Channel coding, which proceeds completely differently from source coding, amends extra bits to protect information from errors caused by the channel. To further randomise error for better information protection, channel coding usually works with interleaving. In this case, bits are properly modulated, which is usually a mapping of bits to the appropriate signal constellation. After proper filtering, in typical radio systems, such baseband signalling is mixed through RF (radio frequency) and likely IF (intermediate frequency) processing before transmission by antenna. The channel can inevitably introduce a lot of undesirable effects, including embedded noise, (nonlinear) distortion, multi-path fading and other impairments. The receiving antenna passes the waveform through RF/IF and an A/D converter translates the waveform into digital samples in state-of-the-art digital wireless communication systems. Instead of reversing the operation at the transmitter, synchronisation must proceed so that

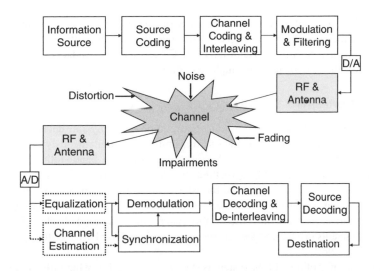

Figure 1.2   Block diagram of a typical digital wireless communication system

the right frequency, timing and phase can be recovered. To overcome various channel effects that disrupt reliable communication, equalisation of these channel distortions is usually adopted. For further reliable system design and possible pilot signalling, channel estimation to enhance receiver signal processing can be adopted in many modern systems.

To summarise, the physical layer of wireless networks in wireless digital communications systems is trying to deal with noise and channel impairments (nonlinear distortions by channel, fading, speed, etc.) in the form of *Inter Symbol Interference* (ISI). State-of-the-art digital communication systems are designed based on the implementation of these functions over hardware (such as integrated circuits) or software running on top of digital signal processor(s) or micro-processor(s).

In the next section of this chapter, we focus on OFDM and its multiple access, Orthogonal Frequency Division Multiple Access (OFDMA).

1.2 Orthogonal Frequency Division Multiplexing (OFDM)

In 1960, Chang [1] postulated the principle of transmitting messages simultaneously through a linear band limited channel without *Inter Channel Interference* (ICI) and *Inter Symbol Interference* (ISI). Shortly afterwards, Saltzberg [2] analysed the performance of such a system and concluded, 'The efficient parallel system needs to concentrate more on reducing crosstalk between the adjacent channels rather than perfecting the individual channel itself because imperfection due to crosstalk tends to dominate'. This was an important observation and was proven in later years in the case of baseband digital signal processing.

The major contribution to the OFDM technique came to fruition when Weinstein and Ebert [3] demonstrated the use of Discrete Fourier Transform (DFT) to perform baseband modulation and demodulation. The use of DFT immensely increased the efficiency of modulation and demodulation processing. The use of the guard space and raised-cosine filtering solve the problems of ISI to a great extent. Although the system envisioned as such did not attain the perfect orthogonality between subcarriers in a time dispersive channel, nonetheless it was still a major contribution to the evolution of the OFDM system.

To resolve the challenge of orthogonality over the dispersive (fading) channel, Peled and Ruiz [4] introduced the notion of the Cyclic Prefix (CP). They suggested filling the guard space with the cyclic extension of the OFDM symbol, which acts like performing the cyclic convolution by the channel as long as the channel impulse response is shorter than the length of the CP, thus preserving the orthogonality of subcarriers. Although addition of the CP causes a loss of data rate, this deficiency was compensated for by the ease of receiver implementation.

1.2.1 OFDM Concepts

The fundamental principle of the OFDM system is to decompose the high rate data stream (Bandwidth = W) into N lower rate data streams and then to transmit them simultaneously over a large number of subcarriers. A sufficiently high value of N makes the individual bandwidth (W/N) of subcarriers narrower than the coherence bandwidth (B_c) of the channel. The individual subcarriers as such experience flat fading only and this can be compensated for using a trivial frequency domain single tap equaliser. The choice of individual subcarrier is such that they are orthogonal to each other, which allows for the overlapping of subcarriers because the orthogonality ensures the separation of subcarriers at the receiver end. This approach results in a better spectral efficiency compared to FDMA systems, where no spectral overlap of carriers is allowed.

The spectral efficiency of an OFDM system is shown in Figure 1.3, which illustrates the difference between the conventional non-overlapping multicarrier technique (such as FDMA) and the overlapping multicarrier modulation technique (such as DMT, OFDM, etc.). As shown in Figure 1.3 (for illustration purposes only; a realistic multicarrier technique is shown in Figure 1.5), use of the overlapping multicarrier modulation technique can achieve superior bandwidth utilisation. Realising the benefits of the overlapping multicarrier technique, however, requires reduction of crosstalk between subcarriers, which translates into preserving orthogonality among the modulated subcarriers.

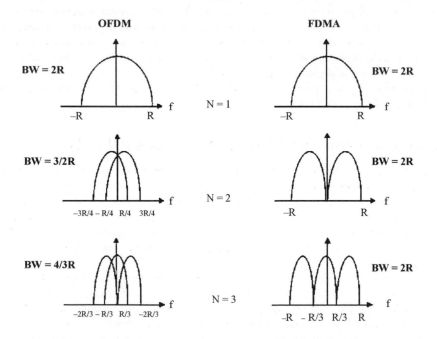

Figure 1.3 Orthogonal multicarrier versus conventional multicarrier

The 'orthogonal' dictates a precise mathematical relationship between frequencies of subcarriers in the OFDM based system. In a normal frequency division multiplex system, many carriers are spaced apart in such a way that the signals can be received using conventional filters and demodulators. In such receivers, guard bands are introduced between the different carriers in the frequency domain, which results in a waste of the spectrum efficiency. However, it is possible to arrange the carriers in an OFDM system such that the sidebands of the individual subcarriers overlap and the signals are still received without adjacent carrier interference. The OFDM receiver can therefore be constructed as a bank of demodulators, translating each subcarrier down to DC and then integrating over a symbol period to recover the transmitted data. If all subcarriers down-convert to frequencies that, in the time domain, have a whole number of cycles in a symbol period T, then the integration process results in zero ICI. These subcarriers can be made linearly independent (i.e., orthogonal) if the carrier spacing is a multiple of $1/T$, which will be proven later to be the case for OFDM based systems.

Figure 1.4 shows the spectrum of an individual data subcarrier and Figure 1.5 depicts the spectrum of an OFDM symbol. The OFDM signal multiplexes in the individual spectra with a frequency spacing equal to the transmission bandwidth of each subcarrier as shown in Figure 1.4. Figure 1.5 shows that at the centre frequency of each subcarrier there is no crosstalk from other channels. Therefore, if a receiver performs correlation with the centre frequency of each subcarrier, it can recover the transmitted data without any crosstalk. In addition, using the DFT based multicarrier technique, frequency-division multiplexing is achieved by baseband processing rather than the costlier bandpass processing.

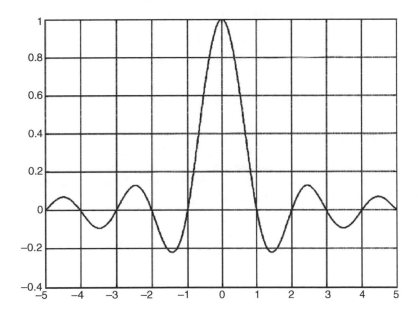

Figure 1.4 Spectra of OFDM individual subcarrier

The orthogonality of subcarriers is maintained even in the time-dispersive channel by adding the CP. The CP is the last part of an OFDM symbol, which is prefixed at the start of the transmitted OFDM symbol, which aids in mitigating the ICI related degradation. Simplified transmitter and receiver block diagrams of the OFDM system are shown in Figures 1.6 (a) and (b) respectively.

1.2.2 Mathematical Model of OFDM System

OFDM based communication systems transmit multiple data symbols simultaneously using orthogonal subcarriers as shown in Figure 1.7. A guard interval is added to mitigate the ISI, which is not shown

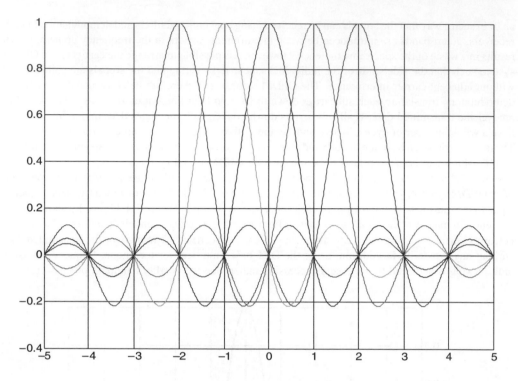

Figure 1.5 Spectra of OFDM symbol

in the figure for simplicity. The data symbols $(d_{n,k})$ are first assembled into a group of block size N and then modulated with complex orthonormal (exponential in this book) waveform $\{\phi_k(t)\}_{k=0}^{N}$ as shown in Equation (1.1). After modulation they are transmitted simultaneously as transmitter data stream. The modulator as shown in Figure 1.7 can be easily implemented using an Inverse Fast Frequency Transform (IFFT) block described by Equation (1.1):

$$x(t) = \sum_{n=-\infty}^{\infty} \left[\sum_{k=0}^{N-1} d_{n,k}\phi_k(t - nT_d) \right] \tag{1.1}$$

where

$$\phi_k(t) = \begin{cases} e^{j2\pi f_k t} & t\varepsilon[0, T_d] \\ 0 & otherwise \end{cases}$$

and

$$f_k = f_o + \frac{k}{T_d}, k = 0 \dots N-1$$

We use the following notation:

- $d_{n,k}$: symbol transmitted during nth timing interval using kth subcarrier;
- T_d: symbol duration;
- N: number of OFDM subcarriers;
- f_k: kth subcarrier frequency, with f_0 being the lowest.

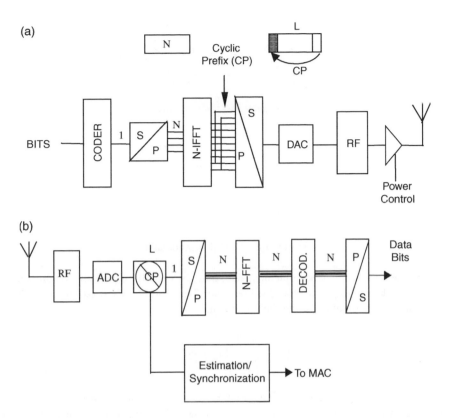

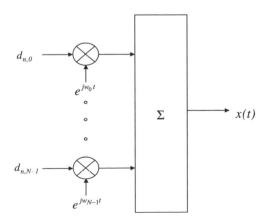

Figure 1.6   (a) Transmitter block diagram and (b) receiver block diagram

The simplified block diagram of an OFDM demodulator is shown in Figure 1.8. The demodulation process is based on the orthogonality of subcarriers $\{\phi_\kappa(\tau)\}$, namely:

$$\int_{\Re} \phi_k(t)\phi_l^*(t)dt = T_d\delta(k-l) = \begin{cases} T_d & k=l \\ 0 & otherwise \end{cases}$$

Figure 1.7   OFDM modulator

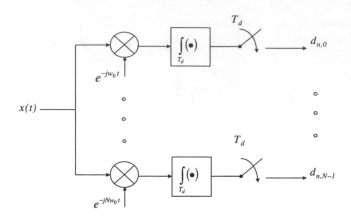

Figure 1.8   OFDM demodulator

Therefore, a demodulator can be implemented digitally by exploiting the orthogonality relationship of subcarriers yielding a simple Inverse Fast Frequency Transform (IFFT)/Fast Frequency Transform (FFT) modulation/demodulation of the OFDM signal:

$$d_{n,k} = \frac{1}{T_d} \int\limits_{nT_d}^{(n+1)T_d} x(t) * \phi_k^*(t) dt \tag{1.2}$$

Equation (1.2) can be implemented using the FFT block as shown in Figure 1.8.

The specified OFDM model can also be described as a 2-D lattice representation in time and frequency plane and this property can be exploited to compensate for channel related impairments issues. Looking into the modulator implementation of Figure 1.7, a model can be devised to represent the OFDM transmitted signal as shown in Equation (1.3). In addition, this characteristic may also be exploited in pulse shaping of the transmitted signal to combat ISI and multipath delay spread. This interpretation is detailed in Figure 1.8.

$$x(t) = \sum_{k,l} d_k \phi_{k,l}(t) \tag{1.3}$$

The operand $\phi k, l(t)$, represents the time and frequency displaced replica of basis function $\phi(t)$ by $l\tau_0$ and $k\nu_0$ in 2-D time and frequency lattice respectively and as shown in Figure 1.9. Mathematically it

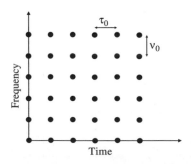

Figure 1.9   2-D lattice in time-frequency domain

can be shown that operand $\phi_{k,l}(t)$ is related to the basis function in Equation (1.4) as follows:

$$\phi_{k,l}(t) = \phi(t - l\tau_0)e^{j2\pi k v_0 t} \tag{1.4}$$

Usually the basis function $\phi(t)$ is chosen as a rectangular pulse of amplitude $1/\sqrt{\tau_0}$ and duration τ_0 and the frequency separation are set at $v_0 = 1/\tau_0$. Each transmitted signal in the lattice structure experiences the same flat fading during reception, which simplifies channel estimation and the equalisation process. The channel attenuations are estimated by correlating the received symbols with *a priori* known symbols at the lattice points. This technique is frequently used in OFDM based communication systems to provide the pilot assisted channel estimation.

1.2.3 OFDM Design Issues

Communication systems based on OFDM have advantages in spectral efficiency but at the price of being sensitive to environment impairments. To build upon the inherent spectral efficiency and simpler transceiver design factors, these impairment issues must be dealt with to garner potential benefits. In communication systems, a receiver needs to synchronise with a transmitter in frequency, phase and time (or frame/slot/packet boundary) to reproduce the transmitted signal faithfully. This is not a trivial task particularly in a mobile environment, where operating conditions and surroundings vary so frequently. For example, when a mobile is turned on, it may not have any knowledge of its surroundings and it must take few steps (based upon agreed protocol/standards) to establish communication with the base station/access point. This basic process in communication jargon is known as synchronisation and acquisition. The tasks of synchronisation and acquisition are complex issues anyway, but impairments make things even harder. Impairment issues are discussed in detail in the following sections.

1.2.3.1 Frequency Offset

Frequency offset in an OFDM system is introduced from two sources: mismatch between transmit and receive sampling clocks and misalignment between the reference frequency of transmit and receive stations. Both impairments and their effects on the performance are analysed.

The sampling epoch of the received signal is determined by the receiver A/D sampling clock, which seldom resumes the exact period matching the transmit sampling clock causing the receiver sampling instants slowly to drift relative to the transmitter. Many authors have analysed the effect of sampling clock drift on system performance. The sampling clock error manifests in two ways: first, a slow variation in the sampling time instant causes rotation of subcarriers and subsequent loss of the SNR due to ICI, and second, it causes the loss of orthogonality among subcarriers due to energy spread among adjacent subcarriers. Let us define the normalised sampling error as

$$t_\Delta = \frac{T' - T}{T}$$

where T and T' are transmit and receive sampling periods respectively. Then, the overall effect, after DFT, on the received subcarriers $R_{l,k}$ can be shown as:

$$R_{l,k} = e^{j2\pi k t_\Delta l \frac{T_s}{T_u}} X_{l,k} \, \mathrm{sin}\, c(\pi k t_\Delta) H_{l,k} + W_{l,k} + N_{t_\Delta}(l,k)$$

where l is the OFDM symbol index, k is the subcarrier index, T_s and T_u are the duration of the total and the useful duration of the symbol duration respectively, $W_{l,k}$ is additive white Gaussian noise and the last term N_{t_Δ} is the additional interference due to the sampling frequency offset. The power of the last term is approximated by $P_{t_\Delta} \approx \frac{\pi^2}{3}(k t_\Delta)^2$.

Hence, the degradation grows as the square of the product of offset t_Δ and the subcarrier index k. This means that the outermost subcarriers are most severely affected. The degradation can also be expressed as SNR loss in dB by following expression:

$$D_n \approx 10 \log_{10} \left[1 + \frac{\pi^2}{3} \frac{E_s}{N_0} (k t_\Delta)^2 \right]$$

In OFDM systems with a small number of subcarriers and quite small sampling error t_Δ such that $k t_\Delta \ll 1$, the degradation caused by the sampling frequency error can be ignored. The most significant issue is the different value of rotation experienced by the different subcarriers based on the subcarrier index k and symbol index l; this is evident from the term $\{e^{j2\pi k t_\Delta l \frac{T_s}{T_u}}\}$. Hence, the rotation angle is the largest for the outermost subcarrier and increases as a function of symbol index l. The term t_Δ is controlled by the timing loop and usually is very small, but as l increases the rotation eventually becomes so large that the correct demodulation is no longer possible and this necessitates the tracking of the sampling frequency in the OFDM receiver. The effect of sampling offset on the SNR degradation is shown in Figure 1.10.

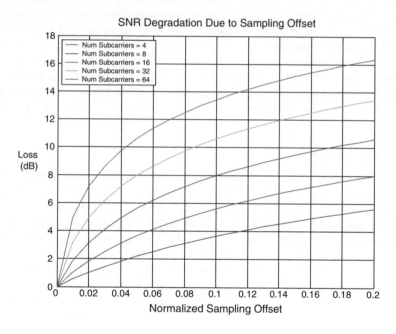

Figure 1.10 SNR degradation due to sampling mismatch

1.2.3.2 Carrier Frequency Offset

The OFDM systems are much more sensitive to frequency error compared to the single carrier frequency systems. The frequency offset is produced at the receiver because of local oscillator instability and operating condition variability at transmitter and receiver; Doppler shifts caused by the relative motion between the transmitter and receiver; or the phase noise introduced by other channel impairments. The degradation results from the reduction in the signal amplitude of the desired subcarrier and ICI caused by the neighbouring subcarriers. The amplitude loss occurs because the desired subcarrier is no longer sampled at the peak of the equivalent sinc-function of the DFT.

Adjacent subcarriers cause interference because they are not sampled at their zero crossings. The overall effect of carrier frequency offset effect on SNR is analysed by Pollet $et\ al$ [6] and for relatively small frequency error, the degradation in dB is approximated by

$$SNR_{loss}(dB) \approx \frac{10}{3 \ln 10}(\pi Tf_\Delta)^2 \frac{E_s}{N_0}$$

where f_Δ is the frequency offset and is a function of the subcarrier spacing and T is the sampling period. The performance of the system depends on modulation type. Naturally, the modulation scheme with large constellation points is more susceptible to the frequency offset than a small constellation modulation scheme, because the SNR requirements for the higher constellation modulation scheme are much higher for the same BER performance.

It is assumed that two subcarriers of an OFDM system can be represented using the orthogonal frequency tones at the output of the A/D converter at baseband as

$$\phi_k(t) = e^{j2\pi fkt/T} \quad \text{and} \quad \phi_{k+m}(t) = e^{j2\pi(k+m)t/T}$$

where T is the sampling period. Let us also assume that due to the frequency drift the receive station has a frequency offset of δ from kth tone to $(k+m)$th tone, i.e.,

$$\phi_{k+m}^\delta(t) = e^{j2\pi(k+m+\delta)t/T}$$

Due to this frequency offset there is an interference between kth and $(k+m)$th channels given by

$$I_m(\delta) = \int_0^T e^{jk2\pi t/T} e^{-j(k+m+\delta)2\pi t/T} dt = \frac{T(1-e^{-j2\pi\delta})}{j2\pi(m+\delta)}$$

$$|I_m(\delta)| = \frac{T|\sin(\pi\delta)|}{\pi|m+\delta|}$$

The aggregate loss (power) due to this interference from all N subcarriers can be approximated as following:

$$\sum_m I_m^2(\delta) \approx (T\delta)^2 \sum_{m=1}^{N-1} \frac{1}{m^2} \approx (T\delta)^2 \frac{23}{14} \quad \text{for } N \gg 1$$

1.2.3.3 Timing Offset

The symbol timing is very important to the receiver for correct demodulation and decoding of the incoming data sequence. The timing synchronisation is possible with the introduction of the training sequences in addition to the data symbols in the OFDM systems. The receiver may still not be able to recover the complete timing reference of the transmitted symbol because of the channel impairments causing the timing offset between the transmitter and the receiver. A time offset gives rise to the phase rotation of the subcarriers. The effect of the timing offset is negated with the use of a CP. If the channel response due to timing offset is limited within the length of the CP the orthogonality across the subcarriers are maintained. The timing offset can be represented by a phase shift introduced by the channel and can be estimated from the computation of the channel impulse response. When the receiver is not time synchronised to the incoming data stream, the SNR of the received symbol is degraded.

The degradation can be quantised in terms of the output SNR with respect to an optimal sampling time, $T_{optimal}$, as shown below:

$$\zeta = \frac{\Lambda(\tau)}{\Lambda(0)}$$

where $T_{optimal}$ is the autocorrelation function and τ is the delay between the optimal sampling instant $T_{optimal}$ and the received symbol time. The parameter τ is treated as a random variable since it is estimated in the presence of noise and is usually referred as the timing jitter. The two special cases of interest, baseband time-limited signals and band-limited signals with the normalised autocorrelation functions, are shown below in mathematical forms:

$$\Lambda(\tau) = \left[1 - \frac{|\tau|}{T_{symbol}}\right]$$

$$\Lambda(\tau) = \frac{1}{N}\left[\frac{\sin(\pi NW\tau)}{\sin(\pi W\tau)}\right]$$

where W is the bandwidth of the band-limited signal. The single carrier system is best described as the band-limited signal whereas the OFDM (multicarrier) system is best described as the time-limited signal. For single carrier systems, the timing jitter manifests as a noisy phase reference of the bandpass signal. In the case of OFDM systems, pilot tones are transmitted along with the data-bearing carrier to estimate residual phase errors.

Paez-Borrallo [7] has analysed the loss of orthogonality due to time shift and the result of this analysis is shown here to quantise its effect on ICI and the resulting loss in orthogonality. Let us assume the timing offset between the two consecutive symbols is denoted by τ, then the received stream at the receiver can be expressed as follows:

$$X_i = c_0 \int_{-T/2}^{-T/2+\tau} \phi_k(t)\phi_l^*(t-\tau)dt + c_1 \int_{-T/2+\tau}^{T/2} \phi_k(t)\phi_l^*(t-\tau)dt$$

where

$$\phi_k(t) = e^{j2\pi fkt/T}$$

Substitute $m = k - l$ and then the magnitude of the received symbol can be represented as

$$|X_i| = \begin{cases} 2T\left|\dfrac{\sin m\pi \dfrac{\tau}{T}}{m\pi}\right|, & c_0 \neq c_1 \\ 0, & c_0 = c_1 \end{cases}$$

This can be further simplified for simple analysis if $\tau \ll T$:

$$\frac{|X_i|}{T} \approx \frac{2m\pi\frac{\tau}{T}}{m\pi} = 2\frac{\tau}{T}$$

This is independent of m, for $\tau \ll T$.
We can compute the average interfering power as

$$E\left[\frac{|X_i|^2}{T^2}\right] = 4\left(\frac{\tau}{T}\right)^2\frac{1}{2} + 0\frac{1}{2} = 2\left(\frac{\tau}{T}\right)^2$$

The ICI loss in dB is computed as follows:

$$ICI_{dB} = 10 \log_{10}\left[2\left(\frac{\tau}{T}\right)^2\right]$$

1.2.3.4 Carrier Phase Noise

The carrier phase impairment is induced due to the imperfection in the transmitter and the receiver oscillators. The phase rotation could either be the result of the timing error or the carrier phase offset for a frequency selective channel. The analysis of the system performance due to carrier phase noise has been performed by Pollet et $al.$ [8] The carrier phase noise was modelled as the Wiener process θ (t) with $E\{\vartheta(t)\} = 0$ and $E[\{\vartheta(t_0 + t)^- - \vartheta(t_0)\}^2] = 4\pi\beta|t|$, where β (in Hz) denotes the single sided line width of the Lorentzian power spectral density of the free running carrier generator. Degradation in the SNR, i.e., the increase in the SNR needed to compensate for the error, can be approximated by

$$D(dB) \approx \frac{11}{6 \ln 10}\left(4\pi N\frac{\beta}{W}\right)\frac{E_s}{N_0}$$

where W is the bandwidth and E_s/N_0 is the SNR of the symbol. Note that the degradation increases with the increase in the number of subcarriers.

1.2.3.5 Multipath Issues

In mobile wireless communications, a receiver collects transmitted signals through various paths, some arriving directly and some from neighbouring objects because of reflection, and some even arriving because of diffraction from the nearby obstacles. These arriving paths arriving at the receiver may interfere with each other and cause distortion to the information-bearing signal. The impairments caused by multipath effects include delay spread, loss of signal strength and widening of frequency spectrum. The random nature of the time variation of the channel may be modelled as a narrowband statistical process. For a large number of signal reflections impinging on the receive antenna, the distribution of the arriving signal can be modelled as complex-valued Gaussian Random Processes based on central limit theory. The envelope of the received signal can be decomposed into fast varying fluctuations superimposed onto slow varying ones. When the average amplitude of envelope suffers a drastic degradation from the interfering phase from the individual path, the signal is regarded as fading. Multipath is a term used to describe the reception of multiple copies of the information-bearing signal by the receive antenna. Such a channel can be described statistically and can be characterised by the channel correlation function. The baseband-transmitted signal can be accurately modelled as a narrowband process as follows:

$$s(t) = x(t)e^{-2\pi f_c t}$$

Assuming the multipath propagation as Gaussian scatterers, the channel can be characterised by time varying propagation delays, loss factors and Doppler shifts. The time-varying impulse response of the channel is given by

$$c(\tau_n, t) = \sum_n \alpha_n(\tau_n, t)e^{-j2\pi f_{D_n}\tau_n(t)}\delta[t - \tau_n(t)]$$

where $c(\tau_n, t)$ is the response of the channel at time t due to an impulse applied at time $t - \tau_n(t)$; $\alpha_n(t)$ is the attenuation factor for the signal received on the nth path; $\tau_n(t)$ is the propagation delay for the nth path; and f_{D_n} is the Doppler shift for the signal received on the nth path.

The Doppler shift is introduced because of the relative motion between the transmitter and the receiver and can be expressed as

$$f_{D_n} = \frac{v \cos(\theta_n)}{\lambda}$$

where v is the relative velocity between transmitter and receiver, λ is the wavelength of the carrier and ϑ_n is the phase angle between the transmitter and the receiver.

The output of the transmitted signal propagating through channel is given as

$$z(t) = c(\tau_n, t) * s(t)$$

$$z(t) = \sum_n \alpha_n[\tau_n(t)]e^{-j2\pi(f_c + f_{D_n})\tau_n(t)} x(t - \tau_n(t))e^{-j2\pi f_c t}$$

where

$$\delta(t - \tau_n(t)) * x(t) = x(t - \tau_n(t))$$

$$\delta(t - \tau_n(t)) * e^{-j2\pi f_c t} = e^{-j2\pi f_c(t - \tau_n(t))}$$

$$\beta_n = \alpha_n[\tau_n(t)]e^{-j2\pi(f_c + f_{D_n})\tau_n(t)}$$

Alternately $z(t)$ can be written as

$$z(t) = \sum_n \beta_n x(t - \tau_n(t))e^{-j2\pi f_c t}$$

where β_n is the Gaussian random process. The envelope of the channel response function $c(\tau_n, t)$ has a Rayleigh distribution function because the channel response is the ensemble of the Gaussian random process. The density function of a Rayleigh faded channel is given by

$$f_z(z) = \frac{z}{\sigma^2} e^{-\left(\frac{z^2}{2\sigma^2}\right)}$$

A channel without a direct line of sight (LOS) path (i.e., only scattered paths) is typically termed a Rayleigh fading channel. A channel with a direct LOS path to the receiver is generally characterised by a Rician density function and is given by

$$f_z(z) = \frac{z}{\sigma^2} I_0\left(\frac{z\eta}{\sigma^2}\right) e^{-\left(\frac{z^2 + \eta^2}{2\sigma^2}\right)}$$

where I_0 is the modified Bessel function of the zeroth order and η and σ^2 are the mean and variance of the direct LOS paths respectively. Proakis [9] has shown the autocorrelation function of $c(\tau, t)$ as follows:

$$\Lambda_c(\tau, \Delta t) = E\{c(\tau, t)c^*(\tau, t + \Delta t)\}$$

In addition, it can be measured by transmitting very narrow pulses and cross correlating the received signal with a conjugate delayed version of itself. The average power of the channel can be found by setting $\Delta \tau = 0$, i.e., $\Lambda_c(\tau, \Delta t) = \Lambda_c(\tau)$. The quantity is known as the *power delay profile* or *multipath intensity profile*. The range of values of τ over which $\Lambda_c(\tau)$ is essentially nonzero is called the multipath delay spread of the channel, denoted by τ_m. The reciprocal of the multipath delay spread is a measure of the coherence bandwidth of the channel, i.e.,

$$B_m \approx \frac{1}{\tau_m}$$

The *coherence bandwidth* of a channel plays a prominent role in communication systems. If the desired signal bandwidth of a communication system is small compared to the coherence bandwidth of the channel, the system experiences flat fading (or frequency non-selective fading) and this eases signal processing requirements of the receiver system because the flat fading can be overcome by adding the extra margin in the system link budget. Conversely, if the desired signal bandwidth is large compared to the coherence bandwidth of the channel, the system experiences frequency selective fading and impairs the ability of the receiver to make the correct decision about the desired signal. The channels, whose statistics remain constant for more than one symbol interval, are considered a slow fading channel compared to the channels whose statistics change rapidly during a symbol interval. In general, broadband wireless channels are usually characterised as slow frequency selective fading.

1.2.3.6 Inter Symbol Interference (ISI) Issues

The output of the modulator as shown in Equation (1.1) is shown here for reference

$$x(t) = \sum_{n=-\infty}^{\infty} \left[\sum_{k=0}^{N-1} d_{n,k}\phi_k(t - nT_d) \right]$$

Equation (1.1) can be re-written in the discrete form for the nth OFDM symbol as follows:

$$x_n(k) = \sum_{k=0}^{N-1} d_{n,k}\phi_k(t - nT_d)$$

where $\phi_k(t) = e^{j2\pi fkt/T}$.

For the nth block of channel symbols, $d_{nP}, d_{nP+1} \ldots d_{nP+P-1}$, the ith subcarrier signal can be expressed as follows:

$$x_n^i(k) = \sum_{k=0}^{N-1} d_{nP+i,k}e^{j\frac{2\pi}{N}l_ik} \quad \text{For } i = 0, 1, 2 \ldots P-1; \; P = \text{number of subcarriers}$$

where l_i the index of time complex exponential of length N, i.e., $0 \leq l_i \leq N - -1$.

These are summed to form the nth OFDM symbol given as

$$x_n(k) \equiv \sum_{i=0}^{P-1} x_n'(k) = \sum_{i=0}^{P-1} d_{nP+i}e^{j\frac{2\pi}{N}l_ik} \tag{1.5}$$

The transmitted signal at the output of the digital-to-analogue converter can be represented as follows:

$$s(t) = \sum_{n} \left[\sum_{k=0}^{L-1} x_n(k)\delta(t - (nL+k)T_d) \right]$$

where, L is the length of data symbol larger than N (number of subchannels). Since the sequence length L is longer than N, only a subset of the OFDM received symbols are needed at the receiver to demodulate the subcarriers. The additional $Q = L - N$ symbols are not needed and we will see later that it could be used as a guard interval to add the CP to mitigate the ICI problem in OFDM systems. In multipath and additive noise environments, the received OFDM signal is given by

$$r_n(k) = \sum_{i=0}^{L-1} x_n(i)h(k-i) + \sum_{i=0}^{L-1} x_{n-1}(i)h(k+L-i) + v_n(k) \qquad (1.6)$$

The first term represents the desired information-bearing signal in a multipath environment, whereas the second part represents the interference from the preceding symbols. The length of the multipath channel, L_h, is assumed much smaller than the length of the OFDM symbol L. This assumption plus the assumption about the causality of the channel implies that the ISI is only from the preceding symbol. If we assume that the multipath channel is as long as the guard interval, i.e., $L_h \leq Q$, then the received signal can be divided into two time intervals. The first time interval contains the desired symbol plus the ISI from the preceding symbol. The second interval contains only the desired information-bearing symbol. Mathematically it can be written as follows:

$$r_n(k) = \begin{cases} \displaystyle\sum_{i=0}^{L-1} x_n(i)h(k-i) + \sum_{i=0}^{L-1} x_{n-1}(i)h(k+L-i) + v_n(k) & 0 \leq k \leq Q-1 \\ \displaystyle\sum_{i=0}^{L-1} x_n(i)h(k-i) + v_n(k) & Q \leq k \leq L-1 \end{cases} \qquad (1.7)$$

We are ready to explore the performance degradation due to ISI. ISI is the effect of the time dispersion of the information-bearing pulses, which causes symbols to spread out so that they disperse energy into the adjacent symbol slots. The *Nyquist criterion* paves the way to achieve ISI-free transmission with observation at the Nyquist rate samples in a band limited environment, to result in *zero-forcing equalisation*. The complexity of the equaliser depends on the severity of the channel distortion. Degradation occurs due to the receiver's inability to equalise the channel perfectly, and from the noise enhancement of the modified receiver structure in the process. The effect of the smearing of energy into the neighbouring symbol slots is represented by the second term in Equation (1.7). The effect of the ISI can be viewed in time and frequency domain.

One of the most important properties of the OFDM system is its robustness against multipath delay spread, *ISI mitigation*. This is achieved by using spreading bits into a number of parallel subcarriers to result in a long symbol period, which minimises the inter-symbol interference. The level of robustness against the multipath delay spread can be increased even further by addition of the guard period between transmitted symbols. The guard period allows enough time for multipath signals from the previous symbol to die away before the information from the current symbol is gathered. The most effective use of guard period is the cyclic extension of the symbol. The end part of the symbol is appended at the start of the symbol inside the guard period to effectively maintain the orthogonality among subcarriers. Using the cyclically extended symbol, the samples required for performing the FFT (to decode the symbol) can be obtained anywhere over the length of the symbol. This provides multipath immunity as well as symbol time synchronisation tolerance.

As long as the multipath delays stay within the guard period duration, there is strictly no limitation regarding the signal level of the multipath; they may even exceed the signal level of the shorter path. The signal energy from all paths just adds at the input of the receiver, and since the FFT is energy conservative, the total available power from all multipaths feeds the decoder. When the delay spread is larger than the guard interval, it causes the ISI. However, if the delayed path energies are sufficiently small then they may not cause any significant problems. This is true most of the time, because path delays longer than the guard period would have been reflected of very distant objects and thus have been diminished quite a lot before impinging on the receive antenna.

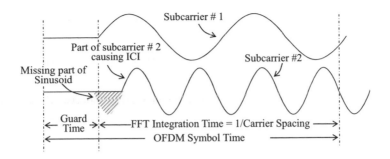

Figure 1.11   Effect of multipath on the ICI

The disaster of OFDM systems is ICI, which is introduced due to the loss of the orthogonality of subcarriers. The loss of orthogonality may be due to the frequency offset, the phase mismatch or excessive multipath dispersion. The effect of this is illustrated in Figure 1.11, where subcarrier-1 is aligned to the symbol integration boundary, whereas subcarrier-2 is delayed. In this case, the receiver will encounter interference because the number of cycles for the FFT duration is not the exact multiple of the cycles of subcarrier-2. Fortunately, ICI can be mitigated with intelligent exploitation of the guard period, which is required to combat the ISI. The frequency offset between the transmitter and the receiver generates residual frequency error in the received signal. The effect of the frequency offset can be analysed analytically by expanding upon Equation (1.7) as follows:

$$r_n(k) = \begin{cases} \sum_{i=0}^{L-1} x_n(i)h(k-i) + \sum_{i=0}^{L-1} x_{n-1}(i)h(k+L-i) + v_n(k) & 0 \le k \le Q-1 \\ \sum_{i=0}^{L-1} x_n(i)h(k-i) + v_n(k) & Q \le k \le L-1 \end{cases} \tag{1.8}$$

At the receiver the guard period is discarded and the remaining signal is defined for $k = 0, 1 \ldots N - -1$ as

$$r'_n(k) \equiv r_n(k+Q) \tag{1.9}$$

Substitute Equation (1.5) into Equation (1.9), which after simplification yields the following:

$$r'_n(k) = \sum_{\alpha} h(\alpha) \sum_i d_{nP+i} e^{j\frac{2\pi}{N}l_i(k+Q-\alpha)} + v_n(k)$$

or,

$$r'_n(k) = \sum_i d_{nP+i} e^{j\frac{2\pi}{N}l_i k} e^{j\frac{2\pi}{N}l_i Q} \sum_{\alpha} h(\alpha) e^{-j\frac{2\pi}{N}l_i \alpha} + v_n(k) \tag{1.10}$$

Equation (1.10) can be written in a simplified form as

$$r'_n(k) = \sum_i d_{nP+i} \phi_i H(l_i) e^{j\frac{2\pi}{N}l_i k} + v_n(k) \tag{1.11}$$

The (') is dropped from the equation without the loss of generality
where

$$\phi_i = e^{j\frac{2\pi}{N}l_i Q} \sim \text{Constant phase multiplier}$$

$$H(l_i) = \sum_{\alpha} h(\alpha) e^{-j\frac{2\pi}{N}l_i \alpha} \sim \text{Fourier Transform of the } h(n)$$

The received signal with frequency-offset Δf can be plugged into Equation (1.11) to yield the following:

$$r_n^{off}(k) \equiv r_n(k)e^{j2\pi\Delta fk} = \sum_i d_{nP+i}\phi_i H(l_i)e^{j\frac{2\pi}{N}k(l_i+\Delta fN)} + V_n(k) \tag{1.12}$$

It can be shown from Equation (1.12) that the frequency offset induces ICI as well the loss of orthogonality between subcarriers, which degrades performance by this ICI. In other words, the symbol estimate becomes

$$\hat{d}_{nP+i} = G_i\left[\{H(l_i)d_{nP+i}I_{\Delta f}(0)\} + \left\{\sum_{\substack{i=0 \\ i\neq m}}^{P-1} H(l_m)d_{nP+i}I_{\Delta f}(l_m-l_i)\} + V_n(l_i)\right]\right. \tag{1.13}$$

where the ICI term is

$$I_{\Delta f}(l_m-l_i) = e^{j\frac{2\pi}{N}k(l_m-l_i+\Delta fN)} \tag{1.14}$$

Starting from Equation (1.14) it can be shown that the SNR degradation due to small frequency offset is approximately

$$SNR_{loss}(dB) \approx \frac{10}{3\ln 10}(\pi\Delta fNT_s)^2\frac{E_s}{N_0} \tag{1.15}$$

where E_s/N_0 is the SNR in the absence of the frequency offset.

Please recall that ISI is eliminated by introducing a guard period for each OFDM symbol. The guard period is chosen larger than the expected delay spread such that multipath components from one symbol do not interfere with adjacent symbols. This guard period could be no signal at all but the problem of ICI would still exist. To eliminate ICI, the OFDM symbol is cyclically extended in the guard period as shown in Figure 1.12, by two intuitive approaches using cyclic prefix and/or cyclic suffix to facilitate the guard band. This ensures that the delayed replicas of the OFDM symbols due to multipath will always have the integer number of cycles within the FFT interval, as long as delay is smaller than the guard period. As a result, multipath signals with delays smaller than the guard period do not cause ICI.

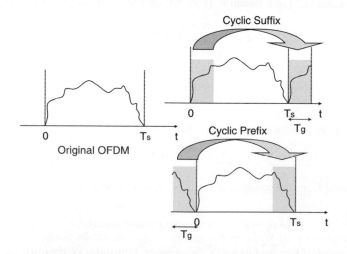

Figure 1.12 Cyclic prefix in the guard period

Mathematically it can be shown that the cyclic extension of the OFDM symbol in the guard period makes the OFDM symbol appear periodic at the receiver end even though there might be a delay because of the multipath environment. In OFDM system the N complex-valued frequency domain symbols $X(n), 0 < n < N - -1$, modulate N orthogonal carriers using the IDFT producing domain signal as follows:

$$x(k) = \sum_{n=0}^{N-1} X(n)e^{+j2\pi k\frac{n}{N}} = IDFT\{X(n)\} \tag{1.16}$$

The basic functions of the IDFT are orthogonal. By adding a cyclic prefix, the transmitted signal appears periodic:

$$s(k) = \begin{cases} x(k+N) & 0 \leq k < Q \\ x(k) & Q \leq k < L \end{cases}$$

where Q is the length of the guard period. The received signal now can be written as

$$y(k) = s(k)*h(k) + w(k) \quad 0 \leq k < L \tag{1.17}$$

If the cyclic prefix added is longer than the impulse response of the channel, the linear convolution with the channel will appear as a circular convolution from the receiver's point of view. This is shown below for any subcarrier $l, 0 \leq l < L$:

$$\begin{aligned} Y(n) &= DFT(y(k)) = DFT(IDFT(X(n)) \otimes h(k) + w(k)) \\ &= X(n)DFT(h(k)) + DFT(w(k)) = X(n)H(n) + W(n), \ 0 \leq k < N \end{aligned} \tag{1.18}$$

where $\otimes$ denotes circular convolution and $W(n) = DFT(w(k))$. Examining Equation (1.18) shows that there is no interference between subcarriers, i.e., zero ICI. Hence, by adding the cyclic prefix, the orthogonality is maintained through transmission. The obvious drawback of using the cyclic prefix is that the amount of data that has to be transmitted increases, thus reducing the usable throughput.

1.2.3.7 Peak to Average Power Ratio (PAPR)

Another challenge for OFDM systems (or multicarrier systems) is the accommodation of the large dynamic range of signal, caused by the *peak-to-average power ratio* due to the fact that the OFDM signal has a large variation between the average signal power and the maximum signal power. A large dynamic range is inherent to multicarrier modulations having essentially independent subcarriers. As a result, subcarriers can add constructively or destructively, which may contribute to large variation in signal power. In other words, it is possible for the data sequence to align all subcarriers constructively and accrue to a very large signal. It is also possible for the data sequence to make all subcarriers align destructively and diminish to a very small signal. This large variation creates problems for transmitter and receiver design requiring both to accommodate a large range of signal power with minimum distortion.

The large dynamic range of the OFDM systems presents a particular challenge for the Power Amplifier (PA) and the Low Noise Amplifier (LNA) design. The large output drives the PA to nonlinear regions (i.e., near saturation), which causes severe distortion. To minimise the amount of distortion and to reduce the amount of out-of-band energy radiation by the transmitter, the OFDM and other multicarrier modulations alike need to ensure that the operation of a PA is limited as much as possible in the linear amplification region. With an inherently large dynamic range, this means that the OFDM must keep its average power well below the nonlinear region of PA in order to accommodate the signal power fluctuations. However, lowering the average power hurts the efficiency and subsequently the range since it corresponds to a lower output power for the majority of the signal in order to accommodate the infrequent peaks. As a result, OFDM designers must make careful tradeoffs between allowable distortion and output power. That is, they must choose an average input level that generates sufficient output power and yet does not introduce too much interference or violate any spectral constraints.

To examine this tradeoff further, consider the IEEE802.11a version of an OFDM system that uses 52 subcarriers. In theory, all 52 subcarriers could add constructively and this would yield a peak power of $20*\log(52) = 34.4$ dB above the average power. However, this is an extremely rare event. Instead, most simulations show that for real PAs, accommodating a peak that is 3 to 6 dB above average is sufficient. The exact value is highly dependent on the PA characteristics and other distortions in the transmitter chain. In other words, the distortions caused by peaks above this range are infrequent enough to allow for low average error rates.

A simple method of handling PAPR is to limit the peak signals by *clipping* or replacing peaks with a smooth but lower amplitude pulse. Since this modifies the signal artificially, it does increase the distortion to some degree. However, if it is done in a controlled fashion then it generally limits the PA-induced distortion. As a result, it can in many cases improve the overall output power efficiency.

For packet-based networks the receiver can request a retransmission of any packet with error. A simple but effective technique may be to rely on a scramble sequence to control PAPR on retransmission. In other words, the data is scrambled prior to modulating the subcarriers for retransmission. This alone does not prevent large peaks and there may still be occasions when the transmitter introduces significant distortion due to a large peak power in the packet. However when the distortion is severe, the receiver will not correctly decode the packet and will request a retransmission. When the data is retransmitted, however, the scramble sequence is changed. If the first scramble sequence caused a large PAPR, the second sequence is extremely unlikely to do the same despite the fact that it contains the same data sequence. Since IEEE 802.11a/g/n networks use packet retransmissions already, this technique is used to mitigate some of problems with PAPR. The downside to this technique is that it does impact the network throughput because some of the data sequences must be transmitted more than once.

To minimise the OFDM system performance degradation due to PAPR, several techniques has been explored each with varying degrees of complexity and performance enhancements. These schemes can be divided into three general categories:

- Signal Distortion Technique:
  - Signal Clipping
  - Peak Windowing
  - Peak Cancellation
- Coding Technique
- Symbol Scrambling Technique

The simplest way to mitigate the peak-to-average power ratio problem is to limit (clip) the signal such that the peak level of the signal is always below the desired maximum level. However, this causes out of band radiation and signal distortion. The effect of this clipping is analogous to the rectangular windowing of the sample, which is equivalent to the spectrum of the desired signal being convolved by the sinc-function (spectrum of the rectangular window) causing the spectrum regrowth in the side bands and thus causing interference to the neighbouring channels. Simple clipping gives rise to spectral growth in side bands. Therefore, to tame the spectral growth in adjacent bands, other windowing functions with narrow bandwidth (such as Gaussian, Kaiser, Hamming and root raised cosine) have been applied.

The goal of the signal distortion techniques is to reduce the amplitude of the data samples, whose magnitude exceeds a certain threshold. The undesirable effect of signal distortion due to these can be avoided by using the peak cancellation technique. In this method, a time-shifted and scaled reference is subtracted from the signal such that each subtracted reference function reduces the peak power of at least one signal sample. A sinc-function could be used as a possible reference but this needs to be spectrally limited. The sinc-function can be spectrally limited by applying the raised cosine window function.

Researchers have looked into the applicability of coding techniques to mitigate the PAPR issue. To achieve a smaller PAPR level the achievable code rate also becomes smaller. Although there are a large number of code words available their implementation and properties for use as FEC codes (such as minimum distance) may not be suitable for implementation. However, Wilkinson et al. [10] observed that the largest portion of codes was Golay complementary sequences, which have a structured way of implementing the PAPR reduction codes. The Golay complementary sequences are sequence pairs for which the sum of autocorrelation function is zero for all delay shifts other than zero.

Another straightforward approach is the symbol scrambling technique. For each OFDM symbol, the input sequence is scrambled by a certain number of scrambling sequences. The output signal with the smallest PAPR is transmitted.

1.2.4 OFDMA

We note that the OFDM transmission uses all subcarriers for a packet time consisting of a number of symbol durations. However, we can introduce a more flexible approach to allow each user to use the partial frequency band (i.e., a number of subcarriers) and even partial packet time (i.e., a number of symbol durations), so that multiple users can share this OFDM transmission, which is known as *orthogonal frequency division multiple access* (OFDMA) or *multi-user OFDM*.

We have already discussed the OFDM as a multiplexing scheme, which provides better spectral efficiency and immunity to multipath fading especially in a wireless environment. The OFDM also lends itself to a simple implementation scheme based on highly optimised FFT/IFFT blocks. These advantages can also be extended for multiple access schemes by assigning a subset of tones (subcarriers) of OFDM to individual users. The allocation of subsets of tones to various users allows for simultaneous transmission of data from multiple users allowing the sharing the medium. In this way, it is equal to ordinary FDMA; however, OFDMA avoids the relatively large guard bands that are necessary in FDMA to separate different users. An example of an OFDMA time-frequency grid is shown in Figure 1.13, where seven users *a* to *g* each use a certain fraction – which may be different for each user – of the available subcarriers. The blank time-frequency grids may be unused or occupied

a		d		a		d		a		d	
a		d		a		d		a		d	
a	c	e		a	c	e		a	c	e	
a	c	e		a	c	e		a	c	e	
b		e	g	b		e	g	b		e	g
b		e	g	b		e	g	b		e	g
b		f	g	b		f	g	b		f	g
b		f	g	b		f	g	b		f	g

Frequency (vertical axis) — Time (horizontal axis)

Figure 1.13   OFDMA with users a, b, c, d, e, f, g to share time-frequency grids

by *pilot* signals. This particular example in fact is a mixture of OFDMA and TDMA, because each user only transmits in one out of every four timeslots, which may contain one or several OFDM symbols.

In the previous example of OFDMA, every user had a fixed set of subcarriers. It is a relatively easy change to allow hopping of the subcarriers per timeslot. Allowing hopping with different hopping patterns for each user actually transforms the OFDMA system into a frequency-hopping CDMA system. This has the benefit of increased frequency diversity, because each user uses all of the available bandwidth, as well as the interference averaging benefit that is common for all CDMA variants. By using forward-error correction coding over multiple hops, the system can correct for subcarriers in deep fades or subcarriers that are interfered with by other users. Because the interference and fading characteristics change for every hop, the system performance depends on the average received signal power and interference, rather than on the worst case fading and interference power. A major advantage of frequency-hopping CDMA systems over direct-sequence or *multicarrier CDMA* systems is that it is relatively easy to eliminate intra-cell interference by using orthogonal hopping patterns within a cell.

OFDMA has been popular in wireless communication systems, such as IEEE 802.16e (OFDMA version known as mobile WiMAX), IEEE 802.16m, 3GPP long-term evolution (LTE), etc. Hopping OFDM is also used in well known ultra-wide band (UWB) communications (i.e., WiMedia), which is going to be next phase system of the Bluetooth 3.0.

1.2.4.1 Radio Resource Allocation

A major difference of multiuser OFDM (OFDMA) from OFDM lies in radio resource allocation, which is a typical network layer problem but has a strong relationship with physical layer transmission. Since multiple users share the OFDM transmission in time domain (bits) and frequency domain (subcarriers), the radio resource allocation algorithm to dynamically exploit best-use of time-frequency grids can be viewed as the practical implementation of water-pouring to achieve Shannon capacity in the ISI channel.

We assume there are totally K users, and the kth user has the data rate R_k bit per OFDM symbol. Depending on the number of bits assigned to a subcarrier, the adaptive modulator uses a corresponding modulation and adjusts the transmit power level according to the combined subcarrier, bit, and power allocation algorithm. We define $c_{k,n}$ as the number of bits of the kth user that are assigned to the nth subcarrier. To model this problem, we do not allow more than one user to share one subcarrier or any symbol/bit. That is, if $c_{k',n} \neq 0$, $c_{k,n} = 0 \ \forall k \neq k'$, while $c_{k,n} \in \{0, 1, \cdots, M\} = D$ and M is the maximum number of bits per OFDM symbol for each subcarrier. We denote $f_k(c)$ as the required power in a subcarrier for reliable reception of c information bits/symbols when the channel gain is unity. In order to maintain the required Quality of Service (QoS) at the receiver, the transmit power allocated to the nth subcarrier by the kth user equals to

$$P_{k,n} = \frac{f_k(c_{k,n})}{\alpha_{k,n}^2}$$

Our goal is to find the best assignment of $c_{k,n}$ so that the overall transmit power is minimised under given transmission rates of users and given QoS through $f_k(\cdot)$. We further require $f_k(c)$ convex and increasing with $f_k(0) = 0$, which perfectly matches practical modulation and coding schemes. Mathematically,

$$P_T^* = \min_{c_{k,n} \in D} \sum_{n=1}^{N} \sum_{k=1}^{K} \frac{f_k(c_{k,n})}{\alpha_{k,n}^2}$$

subject to the following:

$$\text{Constraint 1} : \forall k \in \{1, 2, \cdots, K\}, \ R_k = \sum_{n=1}^{N} c_{k,n}$$

$$\text{Constraint 2} : \forall n \in \{1, 2, \cdots, N\}, \text{ if there exist } k' \text{ with } c_{k',n} \neq 0, \text{ then } c_{k,n} = 0 \ \forall k \neq k'$$

1.2.4.2 Time and Frequency Synchronisation for Multiuser OFDM

Accurate demodulation of OFDM signals requires subcarrier orthogonality, and thus good synchronisation to maintain such orthogonality. By assuming perfect sample clock (i.e., integer sample-duration misalignment in time), we try to estimate time and frequency offset of multiuser OFDM.

Statistical redundancy introduced by the cyclic prefix can provide the information about offset.

For one symbol received by the base station, we assume that N subcarriers constituting this symbol are subdivided into M bands of subcarriers (collectively as the set M_m). One transmitted OFDM symbol in the mth band of subcarrier is

$$S_m(t) = \sum_{n \in M_m} X_n e^{j2\pi nt/NT} \quad - T_g < t < NT$$

where NT is duration of OFDM symbol without cyclic prefix; T_g is length of cyclic prefix; θ_m is frequency offset relative to the receiver demodulation frequency; and ε_m is time offset relative to the receiver symbol clock.

The received signal is

$$r(k) = \sum_{m=0}^{M-1} r_m(k)$$
$$= \sum_{m=0}^{M-1} s_m(k-m) e^{j2\pi \theta_m k/N} + n_m(k)$$

where $s_m(k)$ is the transmitted signal.

We may use bandpass filters to separate subcarriers grouping if the estimator is good enough. For the outputs of the mth filter, we may obtain the one-shot estimator

$$\hat{\varepsilon}_m = \arg \max_{\varepsilon} \{|r_m(\varepsilon)| - \rho_m R_m(\varepsilon)\} \tag{1.19}$$

$$\hat{\vartheta}_m = \frac{-1}{2\pi} \angle r_m(\varepsilon_m)$$

where

$$r_m(\varepsilon) = \sum_{k=\varepsilon}^{\varepsilon+L-1} r_m^*(k) \cdot r_m(k+N)$$

$$R_m(\varepsilon) = \frac{1}{2} \sum_{k=\varepsilon}^{\varepsilon+L-1} |r_m(k)|^2 + | \cdot r_m(k+N)|^2$$

$$\rho_m = \frac{SNR_m}{SNR_m + 1} \qquad SNR_m = \frac{\sigma_{S_m}^2}{\sigma_{n_m}^2}$$

Such an estimator is shown to be the joint maximum likelihood (ML) estimate of ε and θ if the output of each filter

$$r(k) = \tilde{S}(k - \varepsilon)e^{j2\pi\theta k/N} + n(k) \tag{1.20}$$

where $\tilde{S}(k)$ are Gaussian distributed and uncorrelated except for the pairs of identical samples in the cyclic prefix.

1.3 MIMO

One of the major challenges in physical layer communication design is to increase the spectral efficiency to support broadband wireless communications over severe fading channels. From any standard textbook in fundamental communications, such as [9] [11] one of the best ways to combat fading is *diversity* to create independent communications channels. One straightforward way to create diversity is through antennas, such as the well known receive diversity by multiple receiving antennas with combining (where maximum ratio combining (MRC) is the optimal combining). Alamouti [13] pioneered optimal transmit diversity for two transmit antennas in addition to receiver diversity, which makes the communication channel into multiple-input and multiple-output (MIMO) by simultaneously using transmit diversity and receive diversity. In the following text, three basic types of MIMO technology will be introduced:

- beamforming;
- spatial multiplexing;
- space-time codes.

Depending on the geometry of the employed antenna array, two basic multi-antenna approaches can be considered: a *beamforming* approach for closely separated antenna elements (the inter-element separation is at most $\lambda/2$, where λ is the carrier wavelength) or a diversity approach for widely separated antenna elements (the typical inter-element spacing is at least a few λ). In this chapter, we explore the latter approach where the fading processes associated with any two possible transmit-receive antenna pair can be assumed to be independent. The fact that a MIMO system consists of a number of uncorrelated concurrent channels has been exploited from two different perspectives. First, from a pure diversity standpoint, one can enhance the fading statistics of the received signal by virtue of the multiple available replicas being affected by independent fading channels. By sending the same signal through parallel and independent fading channels, the effects of multipath fading can be greatly reduced, decreasing the outage probability and hence improving the reliability of the communication link.

In the second approach, referred to as *spatial multiplexing*, different information streams are transmitted on parallel spatial channels associated with the transmit antennas. This could be seen as a very effective method to increase spectral efficiency. In order to be able to separate the individual streams, the receiver has to be equipped with at least as many receive antennas as the number of parallel channels generated by the transmitter in general. For a given multiple antenna configuration, one may be interested in finding out which approach would provide the best performance.

1.3.1 Space-Time Codes

Space-Time Coding (STC) is a hybrid technique that uses both space and temporal diversity in a combined manner. There are two forms of STC namely Space-Time Block Code (STBC) and Space-Time Trellis Code (STTC). STBC efficiently exploits transmit diversity to combat multipath fading while keeping decoding complexity to a minimum. Tarokh *et al.* [19] show that no STBC can achieve full-rate and full-diversity for more than two transmit antennas, and proposed a 3/4 rate, full-diversity code for four transmit antennas. A full-rate quasi-orthogonal (QO) STBC was proposed by Jafarkhani [14] for four transmit antennas based on Alamouti's orthogonal STBC. In this case, the transmission matrix is given by

$$
\mathbf{C}_j = \begin{bmatrix} \mathbf{A}_{12} & \mathbf{A}_{34} \\ -\mathbf{A}_{34}^* & \mathbf{A}_{12}^* \end{bmatrix} = \begin{bmatrix} x_1 & x_2 & x_3 & x_4 \\ -x_2^* & x_1^* & -x_4^* & x_3^* \\ -x_3^* & -x_4^* & x_1^* & x_2^* \\ x_4 & -x_3 & -x_2^* & x_1 \end{bmatrix}
\tag{1.21}
$$

where $\mathbf{A}_{12}, \mathbf{A}_{34}$ are the Alamouti codes. It is noted here that since the channel matrix of the QO-STBC is not full-rank, full-diversity gain cannot be attained. To achieve the full-diversity and full-rate (FDFR) property, a new FDFR STC approach was recently proposed.

The Mixed (Hybrid Diversity and Spatial Multiplexing) mode combines diversity and spatial multiplexing by transmitting from four transmit antennas, each space-time block coded with the basic Alamouti scheme of order two. The transmission matrix of the space-time block coding for the ith data stream, $i = a, b$, is given by

$$
\mathbf{A}_i = \begin{bmatrix} x_1(i) & x_2(i) \\ -x_2(i)^* & x_1(i)^* \end{bmatrix}
\tag{1.22}
$$

To decode the data, Minimum Mean Square Error (MMSE) and Zero Forcing (ZF) receivers can be employed. For the MMSE receiver, we assume that the transmitted matrix is $[a_{2n}(k), a_{2n+1}(k), b_{2n}(k), b_{2n} + 1(k)]^T$, where a and b indicate different signal streams. First, the tap weight vector and decoding layer order are determined. If the first decoding layer is a the procedure can be represented by

$$
\begin{bmatrix} \hat{a}_{2n}(k) \\ \hat{a}_{2n+1}(k) \end{bmatrix} = decision \left\{ \begin{bmatrix} \mathbf{w}_1^H(k) \\ \mathbf{w}_2^H(k) \end{bmatrix} \mathbf{y}_n(k) \right\}
\tag{1.23}
$$

The interference from the original signal can be subtracted using $\hat{a}_{2n}(k)$ and $\hat{a}_{2n+1}(k)$, and accordingly, the other stream can be decoded as follows:

$$
\mathbf{y}_n'(k) = \mathbf{y}_n(k) - [\mathbf{h}_1(k)\mathbf{h}_2(k)] \begin{bmatrix} \hat{a}_{2n}(k) \\ \hat{a}_{2n+1}(k) \end{bmatrix}
$$

$$
\begin{bmatrix} \hat{b}_{2n}(k) \\ \hat{b}_{2n+1}(k) \end{bmatrix} = decision \left\{ \begin{bmatrix} \mathbf{h}_3(k) \\ \mathbf{h}_4(k) \end{bmatrix} \mathbf{y}_n'(k) \right\}.
\tag{1.24}
$$

Note that for comparative purposes we can also employ *Maximum Likelihood* (ML) decoding (explained in Section 1.3.1.1) to obtain the optimum performance, which was used as our baseline reference.

In the following, we briefly review the employed *spatial multiplexing* scheme. The V-BLAST architecture has been recently proposed for achieving high spectral efficiency over wireless channels characterised by rich scattering. In this approach, one way of detection is to use conventional *adaptive antenna array* (AAA) techniques, i.e., linear combining and nulling. Conceptually, each stream (i.e., layer) in turn is considered to be the desired signal, while regarding the remaining signals as interference. Nulling is performed by linearly weighting the received signals so as to satisfy some performance related criterion, such as ZF or MMSE. This linear nulling approach is viable, but superior performance is obtained if nonlinear techniques are used. One particularly attractive nonlinear alternative is to exploit symbol cancellation as well as linear nulling to perform detection. By using symbol cancellation, the interference from the already-detected components is subtracted from the received signal vector; reducing effectively the overall interference. Here we will consider ordered successive interference cancellation with ZF and MMSE. Also, a ML decoding receiver will be used as a reference.

We assume that $H_{ij}(k)$ is the channel coefficient from the j_{th} transmit antenna to the i_{th} receive antenna and $\mathbf{w}$ is white Gaussian noise with covariance matrix $\mathbf{C}_w = E[\mathbf{w}\mathbf{w}^H] = \sigma^2 \mathbf{I}R$. Then, the received signal vector can be written as follows:

$$\mathbf{y}_n(k) = \mathbf{H}(k)x_n(k) + \mathbf{w}(k) \tag{1.25}$$

where, the index k denotes the kth subcarrier, $y(k) = [y_1(k) \cdots y_{N_R}(k)]^T$, $\mathbf{x}(k) = [x_1(k) \cdots x_{N_R}(k)]^T$, and $\mathbf{w}(k)$ is the $(N_R \times 1)$ noise vector.

1.3.1.1 Maximum Likelihood Decoding (Optimal Solution)

The ML detection of $\mathbf{x}(k)$ can be found by maximising the conditional probability density function and this is equivalent to minimising the log-likelihood function:

$$\hat{\mathbf{x}}(k) = \min_{\mathbf{x}(k)} \{\mathbf{y}(k) - \mathbf{H}\mathbf{x}(k)\}^H \{\mathbf{y}(k) - \mathbf{H}\mathbf{x}(k)\} \tag{1.26}$$

where $\mathbf{x}(k) \in$ all possible constellation sets.

It is well known that ML decoding has a high complexity and thus suboptimal but practically implementable solutions are considered next.

1.3.1.2 Ordered Successive Interference Cancellation (OSIC)

Instead of ML decoding approach, linear detection techniques can be used, i.e., zero-forcing and MMSE. To improve the linear detection techniques, we try to decode according to received signal strength, and extract the decoded signal from the received signal. This approach is referred to as D-BLAST or V-BLAST according to the transmitted signal structure. For simplicity, we consider the OSIC. The receiving operation of OSIC can be summarised as follows:

✓ Step 1: Compute the tap weight matrix W.
✓ Step 2: Find the layer with maximum SNR.
✓ Step 3: Detection:

$$z_k(n) = \mathbf{W}_k^H \mathbf{y}(n)$$

$$\hat{x}_k(n) = decision[z_k(n)]$$

✓ Step 4: Interference cancellation:

$$\mathbf{y}(n) = \mathbf{y}(n) - \hat{\mathbf{h}}_k(n)$$

$$\mathbf{H} = [\mathbf{h}_1, \cdots, \mathbf{h}_{k-1}, 0, \mathbf{h}_{k+1}, \cdots, \mathbf{h}_T]$$

✓ Step 5: Repeat Step 1 until all symbols are detected.

Zero-Forcing (ZF): The cost function can be expressed as:

$$J_{ZF} = \{\mathbf{y}(k) - \mathbf{H}\hat{\mathbf{x}}(k)\}^H \{\mathbf{y}(k) - \mathbf{H}\hat{\mathbf{x}}(k)\} \tag{1.27}$$

Since J_{ZF} is a convex function over $\hat{x}(k)$, $\hat{x}(k)$ can be determined by using the minimum limit. Then, the tap weight vector is given by

$$\mathbf{W} = \{\mathbf{H}^H \mathbf{H}\}^{-1} \mathbf{H}^H \tag{1.28}$$

Minimum Mean Square Error(MMSE): To take into account the noise variance the cost function can be expressed as

$$J_{MMSE} = E[\{\mathbf{y}(k) - \mathbf{H\hat{x}}(k)\}^H \{\mathbf{y}(k) - \mathbf{H\hat{x}}(k)\}]$$ (1.29)

Using a similar method to the ZF detection method, the weight vector results in

$$\mathbf{W} = \{\mathbf{H}^H\mathbf{H} + \sigma^2\mathbf{I}\}^{-1}\mathbf{H}^H$$ (1.30)

Note that the noise variance has to be estimated in order to use the MMSE approach.

1.3.2 Spatial Multiplexing Using Adaptive Multiple Antenna Techniques

Several authors have considered the diversity-spatial multiplexing problem. The fundamental tradeoff between diversity and spatial multiplexing was explored by Zheng and Tse [15]. A scheme based on switching between diversity and spatial multiplexing was proposed by Heath and Paulraj [16]. The latter authors considered a fixed rate system in which the receiver adaptively selects one of the two transmission approaches based on the largest minimum Euclidean distance of the received constellation. The receiver informs its selection to the transmitting via a one-bit feedback channel. To ensure a fixed bit rate, the diversity scheme uses modulation with a higher order than that used by its counterpart spatial modulation case. Skjevling *et al.* [17] presented a hybrid method combining both diversity and spatial multiplexing. The proposed approach optimally assigns antennas to a given (fixed) transmission scheme combining diversity and spatial multiplexing. Antenna selection is based either on full channel feedback or long term statistics. Gorokhov *et al.* [18] studied the relationship between multiplexing gain and diversity gain in the context of antenna subset selection, thereby extending the recent results by Zheng and Tse [15].

1.3.3 Open-loop MIMO Solutions

Alamouti [13] developed a remarkable orthogonal full-diversity full-rate (FDFR) code for $N_T = 2$ transmit antennas, requiring a simple linear decoder at the receiver. Tarokh *et al.* [19] proved that a FDFR orthogonal code only exists for $N_T = 2$ and proposed some space-time block codes for $N_T > 2$ attaining full diversity but not full rate. A quasi-orthogonal full-rate code was proposed by Jafarkhani, although full diversity gain cannot be attained. Based on space-time constellation rotation, Xin *et al.* [20] and Ma *et al.* [21] proposed a FDFR encoder for an arbitrary number of transmit antennas. For an even number of transmit antennas, Jung *et al.* [22] obtained coding gain with a FDFR space-time block code by serially concatenating the Alamouti scheme with constellation rotation techniques. Although the Alamouti based space-time constellation rotation encoder (A-ST-CR) can effectively achieve full diversity and full rate, the decoding complexity is an issue and its practical implementation becomes prohibitive, even for a small number of transmit antennas, e.g., $N_T = 4$. This is due to the high computational complexity required by the maximum likelihood (ML) decoding algorithm.

In addressing the complexity problem, this chapter further extends these results by considering a system based on the serial concatenations of a new rotating precoding scheme with the basic Alamouti codes of order two. By a proper process of puncturing and shifting after the actual constellation-rotation operation, the encoding process can be conveniently decomposed into rotation operations carried out in a lower-order matrix space. The impact of this puncture and shift rotation coding scheme is very significant at the receiver, where due to the provided signal decoupling, the ML decoding is significantly reduced. It is shown in this chapter that the proposed method attains the same performance as the scheme presented with a substantial complexity reduction.

Researchers use a precoder based on the Vandermonde matrix to attain a FDFR system. After multiplying the received signal $\mathbf{x}$ by the Vandermonde matrix, each component of vector $\mathbf{r}$ combines all the symbols as can be observed in the next basic precoder equation:

$$
\mathbf{r} = \Theta\mathbf{x} = \frac{1}{\sqrt{4}}
\begin{bmatrix}
1 & \alpha_0^1 & \alpha_0^2 & \alpha_0^3 \\
1 & \alpha_1^1 & \alpha_1^2 & \alpha_1^3 \\
1 & \alpha_2^1 & \alpha_2^2 & \alpha_2^3 \\
1 & \alpha_3^1 & \alpha_3^2 & \alpha_3^3
\end{bmatrix}
\begin{bmatrix}
x_1 \\ x_2 \\ x_3 \\ x_4
\end{bmatrix}
=
\begin{bmatrix}
r_1 \\ r_2 \\ r_3 \\ r_4
\end{bmatrix}
\tag{1.31}
$$

where $\alpha_i = \exp(j2\pi(i+1/4)/N)$, $i = 0, 1, \cdots, N-1$.

Xin [20] and Ma [21] use a diagonal channel matrix after multiplying the information symbols by the Vandermonde matrix. This linear precoding is referred to as the constellation rotating operation. Notice that the coding advantage is not optimised, although the schemes successfully achieve FDFR. Jung [22] improves the coding advantages by concatenating the constellation rotating precoder with the basic Alamouti scheme, resulting in the following transmitted signals:

$$
\mathbf{S} =
\begin{bmatrix}
r_1 & r_2 & 0 & 0 \\
-r_2^* & r_1^* & 0 & 0 \\
0 & 0 & r_3 & r_4 \\
0 & 0 & -r_4^* & r_3^*
\end{bmatrix}
\tag{1.32}
$$

At the receiving end, the signal can be written as

$$
\mathbf{y} =
\begin{bmatrix}
y_1 \\ y_2^* \\ y_3 \\ y_4^*
\end{bmatrix}
=
\frac{1}{\sqrt{2}}
\begin{bmatrix}
h_1 & h_2 & 0 & 0 \\
h_2^* & -h_1^* & 0 & 0 \\
0 & 0 & h_3 & h_4 \\
0 & 0 & h_4^* & -h_3^*
\end{bmatrix}
\begin{bmatrix}
r_1 \\ r_2 \\ r_3 \\ r_4
\end{bmatrix}
+
\begin{bmatrix}
n_1 \\ n_2^* \\ n_3 \\ n_4^*
\end{bmatrix}
= \mathbf{Hr} + \mathbf{n}
\tag{1.33}
$$

Note that since r_1, r_2, r_3, r_4 already sums over $x_1 \sim x_4$ through the Vandermonde matrix, each symbol experiences the channel twice. At this point we can separate r_1, r_2, r_3, r_4 into two parts i.e., $(r_1, r_3$ and $r_2, r_4)$ or $(r_1, r_4$ and $r_2, r_3)$, consequently the Vandermonde matrix for the precoder needs not to be of size four, but smaller. Based on this observation, we can use a puncturing and shifting operation after the constellation rotation process resulting in a new precoder:

$$
\mathbf{r}_{1,3} =
\begin{bmatrix} r_1 \\ r_3 \end{bmatrix}
= \frac{1}{\sqrt{2}}
\begin{bmatrix} 1 & \alpha_0^1 \\ 1 & \alpha_2^1 \end{bmatrix}
\begin{bmatrix} x_1 \\ x_3 \end{bmatrix}
\qquad
\mathbf{r}_{1,4} =
\begin{bmatrix} r_1 \\ r_4 \end{bmatrix}
= \frac{1}{\sqrt{2}}
\begin{bmatrix} 1 & \alpha_0^1 \\ 1 & \alpha_3^1 \end{bmatrix}
\begin{bmatrix} x_1 \\ x_4 \end{bmatrix}
$$
$$
\text{or} \tag{1.34}
$$
$$
\mathbf{r}_{2,4} =
\begin{bmatrix} r_2 \\ r_4 \end{bmatrix}
= \frac{1}{\sqrt{2}}
\begin{bmatrix} 1 & \alpha_1^1 \\ 1 & \alpha_3^1 \end{bmatrix}
\begin{bmatrix} x_2 \\ x_4 \end{bmatrix}
\qquad
\mathbf{r}_{2,3} =
\begin{bmatrix} r_2 \\ r_3 \end{bmatrix}
= \frac{1}{\sqrt{2}}
\begin{bmatrix} 1 & \alpha_1^1 \\ 1 & \alpha_2^1 \end{bmatrix}
\begin{bmatrix} x_2 \\ x_3 \end{bmatrix}
$$

After puncturing and shifting, the encoder can be defined as

$$
\frac{1}{\sqrt{2}}
\begin{bmatrix}
1 & \alpha_0^1 & 0 & 0 \\
0 & 0 & 1 & \alpha_1^1 \\
1 & \alpha_2^1 & 0 & 0 \\
0 & 0 & 1 & \alpha_3^1
\end{bmatrix}
\quad \text{or} \quad
\frac{1}{\sqrt{2}}
\begin{bmatrix}
1 & \alpha_0^1 & 0 & 0 \\
0 & 0 & 1 & \alpha_1^1 \\
0 & 0 & 1 & \alpha_2^1 \\
1 & \alpha_3^1 & 0 & 0
\end{bmatrix}
\tag{1.35}
$$

Recently, Rajan et al. proposed a low decoding complexity (symbol by symbol decoding) improved space-time code with full diversity for three and four transmit antennas configurations. The following is the format obtained modification to the transmission matrix:

$$
s = \begin{bmatrix} x_1 + jy_3 & -x_2 + jy_4 & 0 & 0 \\ x_2 + jy_4 & x_1 - jy_3 & 0 & 0 \\ 0 & 0 & x_3 + jy_1 & -x_4 + jy_2 \\ 0 & 0 & x_4 + jy_2 & x_3 - jy_1 \end{bmatrix} \tag{1.36}
$$

where $x_i = s_{iI} \cos\theta - s_{iQ} \sin\theta$, $y_i = s_{iI} \cos\theta - s_{iQ} \sin\theta$ and $\theta = \tan^{-1}(\frac{1}{3})$. The complex symbols s_i take values from a QAM signal set.

1.3.4 Closed-loop MIMO Solutions

In this section we explain the widely used closed-loop MIMO solutions, which consist of two parts, i.e., antenna grouping and codebook based schemes using feedback information from the mobile station.

1.3.4.1 Antenna Grouping

The rate 1 transmission code for 4 Tx BS in the IEEE802.16e is

$$
A = \begin{bmatrix} s_1 & -s_2^* & 0 & 0 \\ s_2 & s_1^* & 0 & 0 \\ 0 & 0 & s_3 & -s_4^* \\ 0 & 0 & s_4 & s_3^* \end{bmatrix} \tag{1.37}
$$

Note that this scheme does not achieve full diversity. Using the equivalent model

$$
A^H A = \begin{bmatrix} \rho_1 & 0 & 0 & 0 \\ 0 & \rho_1 & 0 & 0 \\ 0 & 0 & \rho_2 & 0 \\ 0 & 0 & 0 & \rho_2 \end{bmatrix} \tag{1.38}
$$

If the BS can use channel state information, the performance of the existing matrix A approaches the performance of the full diversity full rate STC:

$$
\arg \min_{antenna\_pair} |\rho_1 - \rho_2| \tag{1.39}
$$

Let d_{min}^2 be the corresponding minimum distance of the normalised unit energy constellation. The 2^R-QAM Euclidean distance equation $d_{min}^2 = 12/(2^R - 1)$ will be used, corresponding to QAM modulation for diversity. Using this Euclidean distance equation, we can estimate the error probability as

$$
P_e \leq N_e Q\left(\sqrt{\frac{E_s}{N_0}} d_{MIN}^2\right) \tag{1.40}
$$

where d_{MIN}^2 is the squared Euclidean distance of the received signal and N_e is the number of nearest neighbours in the constellation, and can be found for each proposed mapping scheme based on the channel coefficient matrix $HQ(x) = \frac{1}{2} erfc(x/\sqrt{2})$, where $erfc$ is the complementary error function. For STC, the minimum distance of the diversity constellation at the receiver can be shown to be

$$
d_{MIN}^2(H) \leq \frac{\min\left(||H||_F^2(a,b), ||H||_F^2(c,d)\right)}{N_T} d_{min}^2 \tag{1.41}
$$

where (a, b) and (c, d) are the antenna grouping index and $\|\mathbf{H}\|_F$ is the Frobenius norm of matrix $\mathbf{H}$. The details for derivation follow the derivation procedure of the maximum SNR criterion for code design. Figure 1.14 shows the system block diagram, which makes use of a grouper to select the antenna pair based on feedback channel information from the MS.

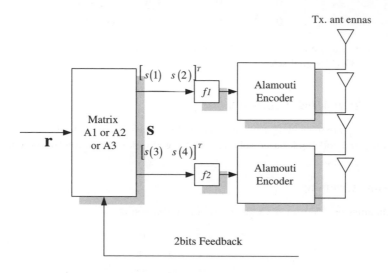

Figure 1.14 System block diagram

The performance of the proposed scheme can outperform the conventional STC without antenna grouping by 3.5 dB.

The rate 2 transmission code for four transmit antennas in the current standard is

$$
B = \begin{bmatrix}
s_1 & -s_2^* & s_5 & -s_7^* \\
s_2 & s_1^* & s_6 & -s_8^* \\
s_3 & -s_4^* & s_7 & s_5^* \\
s_4 & s_3^* & s_8 & s_6^*
\end{bmatrix}
\tag{1.42}
$$

At the mobile, the optimum transmission matrix is determined based on the following criteria. Let $y_{r,i}$ be the received signal at the ith symbol time at the rth receive antenna, and let $h_{i,r}$ denote the channel parameter between the tth transmit and rth receive antenna. When the number of receive antennas is two, the received signal can be represented as

$$
\mathbf{y} = \mathbf{X}(\mathbf{HW})\mathbf{s} + \mathbf{v}
\tag{1.43}
$$

where $\mathbf{y} = \begin{bmatrix} y_{1,1} & y_{1,2}^* & y_{2,1} & y_{2,2}^* \end{bmatrix}^T$, $\mathbf{s} = \begin{bmatrix} s_1 & s_2 & s_3 & s_4 \end{bmatrix}^T$, $\mathbf{v}$ is the noise vector,

$$
\mathbf{H} = \begin{bmatrix}
h_{1,1} & h_{2,1} & h_{3,1} & h_{4,1} \\
h_{1,2} & h_{2,2} & h_{3,2} & h_{4,2}
\end{bmatrix}
$$

and $\mathbf{X}(\cdot)$ is a function of a 2-by-4 input matrix which is defined as

$$
\mathbf{X}\left(\begin{bmatrix} a & b & c & d \\ e & f & g & h \end{bmatrix}\right) = \begin{bmatrix}
a & b & c & d \\
b^* & -a^* & d^* & -c^* \\
e & f & g & h \\
f^* & -e^* & h^* & -g^*
\end{bmatrix}
\tag{1.44}
$$

At the mobile station, the index of the transmission matrix **Bq** is determined based on the following criteria:

$$q = \arg \min_{l=1,\cdots,6} \left[abs(\det(\mathbf{H}_{l,1}) + \det(\mathbf{H}_{l,2})) \right] \qquad (1.45)$$

where $\mathbf{H}_{l,1}$ is the first two columns of $\mathbf{HW}_l$ and $\mathbf{H}_{l,2}$ is the last two columns of $\mathbf{HW}_l$. Note that the antenna grouping matrix selection rule in Equation (1.45) is equivalent to the following rule:

$$q = \arg \min_{l=1,\cdots,6} \left[trace\left(\left[(\mathbf{X}(\mathbf{HW}_l))^H \, \mathbf{X}(\mathbf{HW}_l) \right]^{-1} \right) \right] \qquad (1.46)$$

Alternative criteria for the antenna grouping can be applied to determine the antenna group index. For example, minimise BER, minimum mean square error, etc.

1.3.4.2 Codebook based Closed-Loop MIMO

The codebook is employed in the feedback from the MS (mobile station) to the BS (base station). The MS learns the channel state information from the downlink and selects a transmit beamforming matrix for the codebook. The index of the matrix in the codebook is then fed back to the BS. Each codebook corresponds to a combination of N_t, N_s and N_i, where N_t, N_s and N_i are the numbers of BS transmit antennas, available data streams and bits for the feedback index respectively. Once N_t, N_s and N_i are determined in the MS, the MS will feed back the codebook indexes, each of N_i bits. After receiving a N_i bit index, the BS will look up the corresponding codebook and select the matrix (or vector) according to the index. The selected matrix will be used as the beamforming matrix in MIMO precoding.

1.3.5 MIMO Receiver Structure

Let us consider uncoded MIMO transmission as

$$y = Hx + n \qquad (1.47)$$

where n is a spatio-temporally white circularly symmetric complex Gaussian noise vector with zero-mean and variance in each component, and H is the channel matrix.

ML detection of x gives

$$\hat{x} = \arg \min_x \|y - Hx\|^2$$

$$= \arg \min_x \sum_{i=1}^{M_R} |y_i - \sum_{j=1}^{M_T} h_{i,j} x_j|^2$$

where $h_{i,j}$ is the (i, j)th component of the matrix H. Such a complexity grows exponentially with M_T and Q (Q being bits per symbol in modulation).

1.3.5.1 Linear Receivers

The basic idea is to pre-process the received signal by linear transform

$$\tilde{y} = Ay = AHx + An$$

so that AH is close to a diagonal matrix. The immediate approach to select the pre-processing matrix is zero-forcing receiver (ZF) to remove the off-diagonal elements of AH.

$$\tilde{y}_{ZF} = H^+ y \quad H^+ : \text{Moore-Penrose pseudo-inverse of } H$$

1.3.5.2 LMMSE Receiver

Another approach is to minimise jointly the effects of off-diagonal elements of AH and of the filtered noise An, which is known as linear minimum mean-square error (LMMSE):

$$\tilde{y}_{LMMSE} = \left(H^H H + \frac{N_0}{E} I \right)^{-1} H^H y$$

where E is the average energy of one component of x, and I is the $M_T \times M_T$ identity matrix.

Linear receivers usually enjoy system complexity, at the price of poorer performance, especially $M_T = M_R$.

1.3.5.3 Decision-feedback Receivers

The pre-processing of decision-feedback detection involves the decomposition of the channel matrix H into the product form

$$H = QR$$

For the purpose of simplicity, we assume that $M_R \geq M_T$. Then, $Q^H Q = I_{M_T}$ and $R_{M_T \times M_T}$ is upper triangular. Using this decomposition, the transformed observation vector $\tilde{y} \equiv Q^H y$ has the form

$$\tilde{y} \equiv Rx + \tilde{n}$$

where $\tilde{n} \equiv Q^H n$ retains the statistical property of n. To minimise $m(x) \equiv ||y - Hx||_F^2$ is therefore to minimise

$$\tilde{m}(x) \equiv ||y - Rx||^2$$

From the structure of R, the detection algorithm is detect x_{M_T} by minimising $|\tilde{y}_{M_T} - r_{M_T,M_T} x_{M_T}|^2$ and then us x_{M_T} to detect $x_{M_T - 1}$ by minimising

$$|\tilde{y}_{M_T - 1} - r_{M_T, - 1, M_T - 1} - r_{M_T, - 1, M_T} \hat{x}_{M_T}|^2 + |\tilde{y}_{M_T} - r_{M_T,,M_T} \hat{x}_{M_T}|^2$$

This algorithm is clearly prone to error propagation.

1.3.5.4 Sphere Detection

The sphere detection algorithm (SDA) achieves performance close to ML, with lower complexity. The basic idea is to conduct the search for optimum x within a small subset of potential candidates. Typically, the search is constrained to a hyper-sphere centred at y of radius r

$$||y - Hx||^2 \leq r^2$$

If $r \to \infty$, it is just ML detection. The reduction of complexity is determined by appropriate selection of r.

SDA can be commonly implemented by considering a tree as shown in Figure 1.15, in which the bottom leaves correspond to all possible vectors x, with the components of x labelling branches from bottom to top.

Again, $\tilde{y} = Q^H y$, and ML detection is equivalent to traversing the tree by computing the metric $\tilde{m}(x) \equiv ||y - Rx||^2$ for all possible branches and retaining the minimum value. In other words, SDA reduces the number of branches to be checked by proper pruning of the tree. The operating algorithm can be summarised as follows:

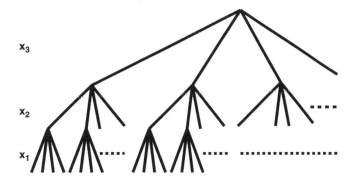

Figure 1.15   SDA for a MIMO with $M_T = 3$ and quaternary modulation

1. Use the decision feedback receiver to obtain a preliminary estimate $\hat{x}$ of x, and to compute the corresponding metric $\tilde{m}(\hat{x}) \equiv \|\hat{y} - Rx\|^2$.
2. Such a value is set as the squared radius of the sphere. The tree is now traversed, depth-first, from top to down, and the metric computed incrementally by adding one-by-one:

$$\|\ddot{y} - Rx\|^2 = \sum_{i-1}^{M_T} |\hat{y}_i - (Rx)_i|^2$$

3. Whenever the accumulated partial sum at a node is larger than (or equal to) $\tilde{m}(\hat{x})$, there is no checking of the leaves below the node (i.e., no further consideration).
4. If a new x is found with metric smaller than $\tilde{m}(\hat{x})$, then it replaces x in the rest of the algorithm (i.e., shrinking the hyper-sphere in ML search).

1.3.5.5 MIMO Receivers for Uncoded Signals

Detection of uncoded signals at the output of a MIMO channel can be viewed as a special case of equalisation for a general channel.

Consider a MIMO channel with M_T transmit and M_R receive antennas. The input-output relationship of the channel can be described by the conditional probability density function

$$f(y|x) \propto e^{\|y - Hx\|_F^2/N_0} = \prod_{j=1}^{M_R} f(y_j|x)$$

where $f(y_j|x) = e^{-|y_j - h_j x|^2/N_0}$ and h_j is the jth row of H.

$$f(y|x) \propto \prod_{i=1}^{M_T} f(x_i) \cdot f(y|x_1 \cdots x_{M_T}) = \prod_{i=1}^{M_T} f(x_i) \prod_{j=1}^{M_R} f(y_j|x)$$

To avoid the complexity of iterative algorithm, we may consider suboptimum MIMO receivers by

1. pre-processing the received signal to limit the effects of spatial interference:

$$\tilde{y} \equiv A(H)y = A(H)Hx + A(H)n$$

2. and exchanging the message in a simplified version;
3. and detection consists of a single sweep of a finite number of steps.

We now recall

$$\tilde{y}_{ZF} = H^+ y$$

to yield

$$\tilde{y} = x + H^+ n$$

with noise enhancement. Or, we may use LMMSE to minimise MSE

$$E\{\|Ay - x\|_F^2\}$$

$$A(H) = \left(H^H H + \frac{N_0}{E_s} I \right)^{-1} H^H$$

Since the off-diagonal terms in $A(H)H$ are small, the resulting detection algorithm is based on the approximation

$$f(y_i | x_1, x_2, \cdots x_{M_T}) \approx f(y_i | x_i)$$

Both ZF and LMMSE structures perform well if $M_R \gg M_T$.

1.3.5.6 Nonlinear Processing

The well-known V-BLAST falls into a class of suboptimum receivers based on the following operations:

- *Nulling of spatial interference:* This is achieved by modifying the pre-processing operation.
- *Cancellation of spatial interference:* This is obtained by simplifying

$$f(\tilde{y}_i | x_i, x_{i+1}, \cdots, x_{M_T}) \approx f(\tilde{y}_i | x_i, \hat{x}_{i+1}, \cdots, \hat{x}_{M_T})$$

in a sequential way for $i = M_T - 1, M_T - 2, \cdots, 1$ where $\hat{x}$ denotes a decision based on x.
- *Ordering:* Since V-BLAST is prone to error propagation, antenna ordering may significantly affect receiver performance.

1.4 Multi-user Detection (MUD)

Actually, the initial MIMO system can be a CDMA system as Figure 1.16 with multiple transmitters and multiple receivers. However, most CDMA systems adopt a single-user receiver structure and treat other users' signals as co-channel interference. Following Verdu's and Poor's [23] efforts to explore optimal reception of CDMA signals, a new branch of communication theory, known as *multi-user detection* (MUD), has been developed.

1.4.1 Multi-user (CDMA) Receiver

CDMA system performance is dominated by MAI conventional correlation receivers and cannot optimally detect signals in this situation.

Each user transmits digital information at the same data rate $1/T$. The bipolar data stream of the kth user is

$$\mathbf{b}_k^T = [b_k(-M), \cdots, b_k(0), \cdots, b_k(M)]$$

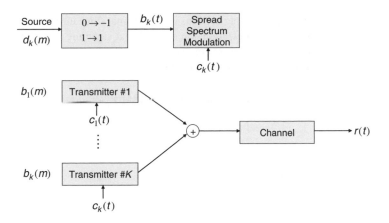

Figure 1.16   CDMA system

The transmitted signal of the kth user is given by

$$x_h(t) = \sum_{m=-M}^{M} b_k(m)c_k\,(t - mT - \tau_k)$$

where τ_k is time delay for the kth user.

Assume short code $N = G_p = \frac{T}{T_c}$ normalised condition

$$\frac{1}{T}\int_0^T c_h^2(t)dt = 1$$

The received energy per bit is

$$E_{c_k}\int_0^T b_k^2(m)c_k^2(t)dt = E_{c_k}$$

The received waveform is therefore

$$r(t) = \sum_{k=1}^{K}\sqrt{E_{ck}}\sum_{m=-M}^{M} b_k(m)c_k(t - mT - \tau_k) + n(t)$$

where $\tau_k < T$ and $\tau_k = 0$ for synchronous case.

For a one-shot signal ($m = 0$)

$$r(t) = \sum_{k=1}^{K}\sqrt{E_{c_k}}b_k(0)c_k(t) + n(t)$$

The cross correlation of signature is

$$\rho_{\ln}(j) = \frac{1}{T}\int_{T_l + mT}^{T_l + (m+1)T} c_l(t - \tau_l)c_m(t + jT - t_n)dt$$
$$l = 1, 2, \cdots, K$$
$$n = 1, 2, \cdots, K$$
$$j = -M, \cdots, 0, \cdots, M$$

A conventional receiver making one-shot decisions is not optimal since information on interference is ignored. While developing optimum detection, we assume that signature waveforms, time delays, phase shifts, amplitudes and numbers of users are available.

Optimum receivers could select the following bit sequence

$$\hat{\mathbf{b}} = \begin{bmatrix} \hat{b}_1(-M) & \cdots & \hat{b}_1(M) \\ \vdots & & \vdots \\ \hat{b}_k(-M) & \cdots & \hat{b}_k(M) \end{bmatrix}$$

which maximises the conditional probability

$$P[\hat{\mathbf{b}}|r(t)]$$

If we assume the transmitted bits are independent and equally probable, then it is equivalent to maximising

$$P\big[r(t)|\hat{\mathbf{b}}\big] = c \exp\left(-\frac{1}{2\sigma_n^2}\int_0^T \left[y_k(t) - \sum_{k=1}^K \hat{b}_k(0)\sqrt{E_{c_k}}c_k(t)\right]^2 dt\right), t \in [0,T]$$

This optimisation can be achieved by the Viterbi algorithm or in a form of maximum likelihood sequence detection (Figure 1.17). Unfortunately, it has exponential complexity (i.e., complexity growing exponentially with the number of users) as an NP-hard problem.

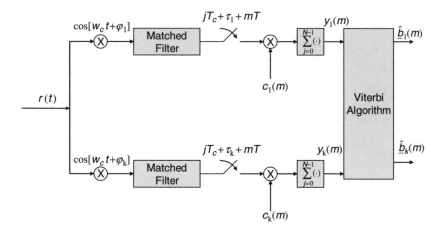

Figure 1.17 Optimal receiver

Let us look at the two-user case: $T_1 = 0, T_2 \le T$. The matched filter outputs at the mth sampling instant are given by

$$y_1(m) = \frac{1}{T}\int_{mT}^{(m+1)T} r(t)c_1(t)dt$$

$$y_2(m) = \frac{1}{T}\int_{mT+T_2}^{(m+1)T+T_2} r(t)c_2(t-T_2)dt$$

where

$$r(t) = \sqrt{E_{c_1}} \sum_{i=-M}^{M} b_1(i)c_1(t-iT_b) + \sqrt{E_{c_2}} \sum_{i=-M}^{M} b_2(i)c_2(t-iT_b-\tau_2)$$

$$y_1(m) = \sqrt{E_{c_1}}b_1(m) + \sqrt{E_{c_2}}b_2(m-1)\rho_{12}(1) + \sqrt{E_{c_2}}b_2(m)\rho_{12}(0) + \sqrt{E_{c_2}}b_2(m+1)\rho_{12}(-1) + n_1(m)$$
$$= z_1(\underline{\mathbf{b}}_1,\underline{\mathbf{b}}_2) + n_1(m)$$

$$y_2(m) = \sqrt{E_{c_2}}b_2(m) + \sqrt{E_{c_1}}b_1(m-1)\rho_{21}(1) + \sqrt{E_{c_1}}b_1(m)\rho_{21}(0) + \sqrt{E_{c_1}}b_1(m+1)\rho_{21}(-1) + n_2(m)$$
$$= z_2(\underline{\mathbf{b}}_1,\underline{\mathbf{b}}_2) + n_1(m)$$

$$\underline{\mathbf{b}}_1 = [b_1(m-1), b_1(m), b_1(m+1)]$$
$$\underline{\mathbf{b}}_2 = = [b_2(m-1), b_1(m), b_1(m+1)]$$

The resulting receiver is shown as Figure 1.18.

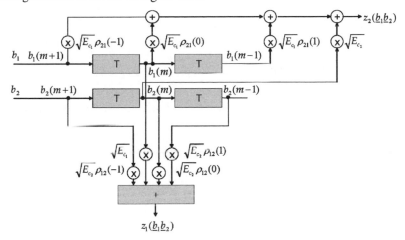

Figure 1.18   Tap delay line model for synchronous CDMA

Note: $\rho_{kh}(j) = \frac{1}{T}\int_{\tau_k}^{T+\tau_k} c_h(t-\tau_k)c_k(t+jT-\tau_c)dt = \delta_J$

1.4.2 Suboptimum DS/CDMA Receivers

Since the optimal multi-user receiver has complexity exponentially to the number of users, it is practically impossible to implement and suboptimal solutions are very much needed.

1.4.2.1 Decor-relating Receiver

The results for the two-user case can be easily generalised for a k-user system

$$y = RAB + n$$

where R is the cross correlation matrix $[\rho_{ij}]$

$$\mathbf{A} = \begin{bmatrix} \sqrt{E_{c_1}} & & 0 \\ & \ddots & \\ 0 & & \sqrt{E_{c_k}} \end{bmatrix}$$

$$\mathbf{b} = \begin{bmatrix} b_1(-M) & \cdots & b_1(M) \\ \vdots & & \vdots \\ b_k(-M) & \cdots & b_k(M) \end{bmatrix}$$

Then,

$$R^{-1}y = Ab + R^{-1}n$$

Ignoring the noise term t_0 results in a scale transmitted sequence at the right-hand side. The decorrelating receiver is based on the prior linear transformation, which consists of

- matched filter bank as the front-end processor producing **y**;
- decor relater computing R^{-1}.

The advantages for decorrelating the multi-user receiver are:

- low complexity (linear in most cases);
- no need of receiving amplitude and thus near-far resistant.

1.4.2.2 Interference Canceller (IC)

The operation of the interference canceller is based on successive cancellations of interference from the received waveform (Figure 1.19). It cancels the strongest detected signal and consists of

1. ranking the user signals $\sqrt{E_{c_1}} > \sqrt{E_{c_2}} > \sqrt{E_{c_3}} > \cdots > \sqrt{E_{c_k}}$;
2. detecting the strongest user by the conventional receiver;
3. regenerating the strongest user spread spectrum signal $\hat{x}_k(t) = \sqrt{E_{c_k}} b_k^{\wedge}(t) c_k(t)$;
4. cancelling the strongest interferer;
5. repeating until all user signals done.

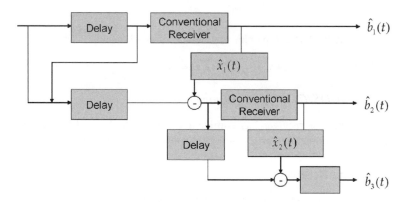

Figure 1.19   Interference canceller

If such cancellation proceeds serially, we call it 'successive interference cancellation' (SIC). It can also be done in parallel, and we call such schemes 'parallel interference cancellation' (PIC).

1.4.2.3 Adaptive MMSE Receiver

Adaptive minimum mean square error (MMSE) receivers do not require signature, timing and carrier phase information about interferes, but only the timing and carrier phase of the desired user (Figure 1.20).

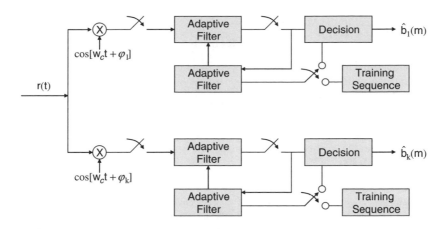

Figure 1.20   Adaptive MMSE receiver

The timing recovery of the receiver is greatly simplified by the fractional filter, which makes the detected desired signal insensitive to time differences. The adaptive FSE filter consists of $2L + 1$ taps. The tap coefficients a_l are determined by minimising the MSE between filter output and a known sequence. The sampled filter input is

$$r(mT_f) = \sum_{i=-M}^{M} \sum_{k=1}^{K} \sqrt{E_{c_k}} b_k c_k (mT_f - iT - \tau_k) + n(mT_f)$$

The filter output, which is only calculated at T seconds apart, is

$$y(mT) = \sum_{l=-L}^{L} a_l r(mT - lT_f)$$

Let us consider user 1 with no specific notation. The MSE is

$$
\begin{aligned}
\varepsilon &= E(l_m^2) = E\Big\{ [y(mT) - b(m)]^2 \Big\} \\
&= E\Big\{ [\boldsymbol{a}^T \boldsymbol{r} - b(m)]^2 \Big\} \\
\boldsymbol{a} &= (a_{-L}, \cdots, a_L)^T \\
\boldsymbol{r} &= [r(mT + LT_f), \cdots, r(mT), \cdots, r(mT - LT_f)]^T \\
\varepsilon &= \boldsymbol{a}^* \boldsymbol{D} \boldsymbol{a} - (\boldsymbol{a}^* \boldsymbol{f} + \boldsymbol{f}^* \boldsymbol{a}) - \sigma_b^2
\end{aligned}
$$

where $\boldsymbol{D} = E(\boldsymbol{r}\,\boldsymbol{r}^*), f = E[b^*(m)\boldsymbol{r}], \sigma_b^2 = E[|b(m)|^2]$
A direct calculation of the lth element, f_e, of the single-user channel vector f is

$$f_e = E[b(m)r(mT + lT_s)] = E[|b(m)|^2]c(lT_s)$$

The optimum tap setting is

$$\boldsymbol{a}_{opt} = 1 - \boldsymbol{f}^* \boldsymbol{D}^{-1} \boldsymbol{f}$$

Please note that f, $\boldsymbol{D}$ might not be available in practical cases. It makes LMMSE of great interest. As a matter of fact, the suboptimal MUD can simply share the same principle of MIMO receiver structure. Please recall that

$$\boldsymbol{y} = \boldsymbol{H}\boldsymbol{x} + \boldsymbol{n}$$

It is just as equivalent to the MUD problem as

$$\mathbf{y} = \mathbf{RAb} + \mathbf{n}$$

Therefore, suboptimal MUD and MIMO receivers share the same principle in the receiver structure if we treat H and $\mathbf{RA}$ equivalently.

References

[1] R.W. Chang, 'Synthesis of Band-Limited Orthogonal Signals for Multichannel Data Transmission', *Bell Systems Technical Journal*, **45**, 1960, 1775–1796.

[2] B.R. Saltzberg, 'Performance of an Efficient Parallel Data Transmission System', *IEEE Transactions on Communication*, **15**(6), 1967, 805–811.

[3] S.B. Weinstein, P.M. Ebert, 'Data Transmission of Frequency Division Multiplexing Using the Discrete Frequency Transform', *IEEE Transactions on Communication*, **19**(5), 1971, 623–634.

[4] A. Peled, A. Ruiz, 'Frequency Domain Data Transmission Using Reduced Computational Complexity Algorithms', *Proceedings of the IEEE International Conference on Acoustics, Speech, and Signal Processing (ICASSP'80), Denver, USA*, 1980, pp. 964–967.

[5] T. Pollet, M. Peeters, 'Synchronization with DMT Modulation', *IEEE Communications Magazine*, April 1999, 80–86.

[6] T. Pollet, P. Spruyt, M. Monenclaey, 'The BER Performance of OFDM Systems Using Non-Synchronized Sampling', *IEEE Global Telecommunications Conference*, 1994, pp. 253–257.

[7] J.M. Paez-Borrallo, 'Multicarrier Vs. Monocarrier Modulation Techniques: An Introduction to OFDM', *BWRC Retreat*, 2000.

[8] T. Pollet, M. Van Bladel, M. Monenclaey, 'BER Sensitivity of OFDM Systems to Carrier Frequency Offset and Wiener Phase Noise', *IEEE Transactions on Communication*, **43**(2), 1995, 191–193.

[9] J.G. Proakis, *Digital Communications*, 4th edition, McGraw-Hill, New York, 2000.

[10] T.A. Wilkinson, A.E. Jones, 'Minimization of the Peak-to-Mean Envelope Power Ratio of Multicarrier Transmission Schemes by Block Coding', *Proceedings of IEEE VTC '95*, July 1995, pp. 825–829.

[11] K.C. Chen, *Principles of Communications*, River, 2009.

[12] R. Van Nee, R. Prasad, *OFDM for Wireless Multimedia Communication*, Artech House Publishers, 2000.

[13] S.M. Alamouti, 'A Simple Transmit Diversity Technique for Wireless Communications', *IEEE Journal on Selected Areas in Communications*, **16**(8), 1998, 1451–1458.

[14] H. Jafarkhani, 'A Quasiorthogonal Space-Time Block Code', *IEEE Transactions on Communications*, **49**, 2001, 1–4.

[15] L. Zheng, D.N.C Tse, 'Diversity and Multiplexing: A Fundamental Tradeoff in Multiple-Antenna Channels', *IEEE Transactions on Information Theory*, **49**, 2003, 1073–1096.

[16] R.W. Heath Jr., A. Paulraj, 'Switching Between Spatial Multiplexing and Transmit Diversity Based on Constellation Distance', *Proceedings of Allerton Conference on Communication, Control and Computing*, October 2000.

[17] H. Skjevling, D. Gesbert, N. Christophersen, 'Combining Space Time Block Codes and Multiplexing in Correlated MIMO Channels: An Antenna Assignment Strategy', *Proceedings of Nordic Signal Processing Conference (NORSIG)*, June 2003.

[18] A. Gorokhov, D. Gore, A. Paulraj, 'Diversity Versus Multiplexing in MIMO System with Antenna Selection', *Allerton Conference*, October 2003.

[19] V. Tarokh, N. Seshadri, A. Calderbank, 'Space-Time Codes for High Data Rate Wireless Communications: Performance Criterion and Code Construction', *IEEE Transactions on Information Theory*, **44**, 1998, 744–765.

[20] Y. Xin, Z. Wang, G.B. Giannakis, 'Space-Time Diversity Systems Based on Linear Constellation Precoding', *IEEE Transactions on Communications*, **49**, 2001, 1–4.

[21] X. Ma, G.B. Giannakis, 'Complex Field Coded MIMO Systems: Performance, Rate, and Trade-offs', *Wireless Communication and Mobile Computing*, 2002, 693–717.

[22] T.J. Jung, K. Cheun, 'Design of Concatenated Space-Time Block Codes Using Signal Space Diversity and the Alamouti Scheme', *IEEE Communication Letters*, **7**, 2003, 329–331.

[23] S. Verdu, *Multiuser Detection*, Cambridge University Press, 1998.

2

Software Defined Radio

2.1 Software Defined Radio Architecture

Many different wireless communication systems exist and they are widely used for different purposes and application scenarios. Figure 2.1 illustrates some popular wireless communications international standards, ranging from body area networks, personal area networks, local area networks and metropolitan area networks, to wide area networks, with different application scenarios and optimised system parameters. As a matter of fact, for the cellular type system alone, we may have a variety of systems in use, such as legacy GSM, GPRS and EDGE, 3GPP wideband code division multiple access (WCDMA) with its update versions of HSDPA and HSUPA, and the upcoming 3GPP long-term evolution (LTE), and that is just considering air-interface technology. Figure 2.2 depicts possible cellular air-interface technology migration and we can easily observe a variety of systems in use or in future use, including time division multiple access (TDMA), code division multiple access (CDMA) and orthogonal frequency division multiple access (OFDMA). These air-interface technologies may co-exist in different geographical regions, and may co-exist simultaneously at the same location. A flexible realisation of terminal devices to allow users to appropriately use wireless communications is definitely essential.

Since the early days of electronic communications, however, typical (wireless) communication systems have been implemented by certain dedicated hardware and likely dedicated application specific integrated chips (ASIC) based on specific communication system parameters designed for use. It is quite different from the way that we use computers to run different applications and programs on the same computing facility. Software defined radio (SDR) that allows multiple systems realised by programs to run on a single hardware platform is therefore attractive under such scenarios (multiple co-existing systems in the same location or globally).

For a state-of-the-art terminal device, there are usually three types of processors for host featured applications and wireless communication/networking purposes:

- *Microprocessor:* We usually have a host processor for each terminal, such as in notebook computers, to serve the fundamental applications such as programming or execution of (multimedia) application software.
- *Embedded processor(s):* Embedded processors typically run networking functions of a device, and sometimes applications with/without the host processor.
- *Digital signal processor(s) (DSP):* When we execute wireless communications or extensive multimedia features such as audio and video, we usually adopt digital signal processors to facilitate

Cognitive Radio Networks   Kwang-Cheng Chen and Ramjee Prasad
© 2009 John Wiley & Sons, Ltd

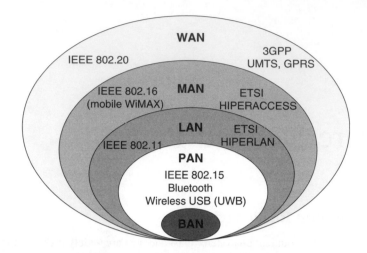

Figure 2.1 Global wireless communication standards

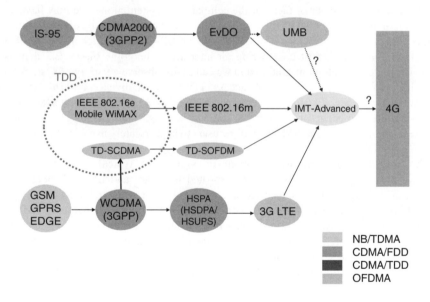

Figure 2.2 Air-interface technology migration of cellular systems

functions. DSP is a special purpose processor offering tremendous processing power for rather dedicated applications.

In other words, to execute wireless communications, we usually use DSPs for physical layer baseband implementation, and embedded processors for networking functions and control. SDR hardware structure can be summarised as shown in Figure 2.3. Antenna, PA and RF are just like conventional wireless communication hardware, with an analogue-to-digital converter (ADC) to transform analogue waveform into digital samples for the receiver, and DAC for the transmitter path. The flexibility of SDR primarily lies in a programmable baseband section. For SDR operating at multiple frequency bands, we may either use a set of dedicated RFs or a tunable wideband RF covering several frequency bands.

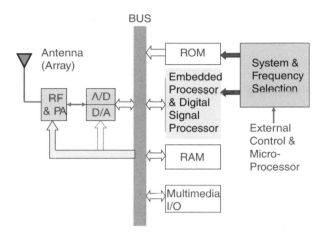

Figure 2.3   Hardware structure of SDR device

For physical-layer communication transceiver (transmitter and receiver) realisation, the RF segment and PA are handled by analogue circuits, and the baseband segment is implemented by DSP for SDR. The major design efforts are to transform each baseband communication block into algorithms computed by DSP programming. Higher layer networking functions are usually realised by embedded processor programs. Device (host) microprocessor and/or application processors handle multimedia and user interface applications (see Figure 2.4).

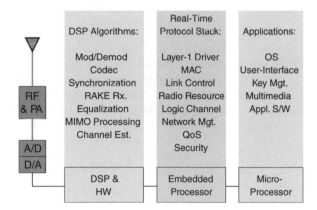

Figure 2.4   Software structure of SDR device

2.2 Digital Signal Processor and SDR Baseband Architecture

In contrast to typical general-purpose processors dealing with data manipulations, digital signal processors (DSP) target mathematical calculations for special-purpose applications such as control, media processing, communications, etc., operating at a much higher operation rate. Many state-of-the-art (wireless) communication products are implemented by DSPs.

Typical general-purpose processors use the well-known *Von Neumann machine* architecture in which memory is shared by instructions and data through a single data bus. DSP typically adopts the so-called *Harvard architecture* of separate program memory and data memory through a dedicated program memory bus and data memory bus. We may further add an instruction cache into the CPU and an I/O controller as a *super Harvard architecture* for DSP. Figure 2.5 illustrates these three architectures.

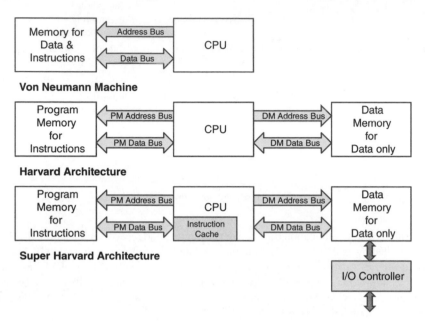

Figure 2.5   Processor architecture

Since the signal waveforms for multimedia or communications are usually analogue, analogue-to-digital conversion is required to encode the analogue waveform into digital samples for DSP to process. Digital signal processing using DSP has two categories: floating-point and fixed-point, based on the sample format to store and to manipulate. Fixed-point DSP represents each sample by a fixed number of bits, usually 16, 32 or 64 but 24 in some DSPs. For 16-bit fixed-point DSP, each sample has $2^{16} = 65,536$ possible 'unsigned integer' values. Floating-point DSP usually uses 32 bits or even 64 bits to represent each sample. In the common format of 32-bit floating-point, the largest and the smallest represented numbers are $\pm 3.4 \times 10^{38}$ and $\pm 1.3 \times 10^{-38}$ respectively.

DSP is supposed to be programmed by assembly, especially if we want to optimise DSP performance (i.e., faster execution). State-of-the-art DSPs usually support hundreds of instructions to execute. However, common development systems for DSP support the high-level C language, which is easy to program at the price of DSP efficiency. Fixed-point DSP processing power is typically measured by million instructions per second (MIPS) and floating-point DSP processing power is measured by million floating operations per second (MFLOPS). Modern DSPs can support several thousand MFLOPS, which is much more powerful than general purpose processors. To further meet the demand of digital signal processing power and multi-tasking, we may include multiple arithmetic logic units (ALUs) into one DSP and may combine several DSPs into a single system, known as multi-processing or parallel processing; Figure 2.6 depicts the common system architecture of paralleling DSPs. The bus to serve data exchange among DSPs and memory plays a central role, and the memory is shared with, and is likely to be controlled by, an arbitrator.

To implement the baseband portion of SDR using DSP, we may consider several system architectures to balance flexibility, power consumption, software complexity and cost (die size, system integration, etc.) as Figure 2.7 depicts.

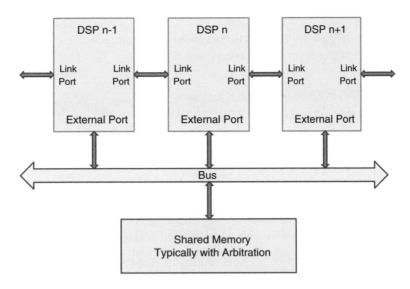

Figure 2.6   Multi-processing or DSP architecture

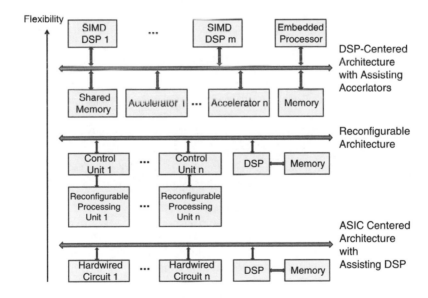

Figure 2.7   SDR baseband hardware system architecture

- ASIC centred architecture is a classic hardware realisation of multiple-standard devices, which may have DSP (or embedded processors) to assist processing. The software efforts would be entirely on layer-1 control or driver.
- Reconfigurable architecture supports more flexibility than the ASIC approach by controlling appropriately designed data processing units, in order, it is hoped, to maintain fully the advantages as against ASIC.
- DSP centred architecture provides the most flexibility. To enhance efficiency, single-instruction multiple-data (SIMD) DSP with special-purpose and dedicated-function accelerators are

usually applied. A general-purpose embedded processor is also usually used for the purpose of control and interfacing higher layers. The successful operation pretty much relies on software programming, in addition to the layer-1 control and driver.

2.3 Reconfigurable Wireless Communication Systems

We are now ready to introduce the methodology to implement the reconfigurable wireless communication systems over one platform.

2.3.1 Unified Communication Algorithm

Instead of direct mapping functional blocks into programming of any processor platform, reconfigurable wireless communication systems start from the unified algorithms over one programmable platform. A good example is the effort by researchers with the Rutgers University to implement various linear multi-user detection schemes (paper published in the *IEEE JSAC* 1999) [2]. As pointed out in the previous chapter (Section 1.4), we usually have several suboptimal Multi-user detection (MUD) receiver structures and here we consider the following three well-known linear versions:

- Matched filter (MF) receiver: the conventional receiver as a sort of degeneration of MUD;
- De-correlation (DC) receiver;
- Minimum mean-squared-error (MMSE) receiver.

Let us consider K users in a synchronous system. The output of the matched filter for the kth user can be written as

$$y_k = A_k b_k + \sum_{j \neq k} A_j b_j \rho_{jk} + n_k \tag{2.1}$$

where A_k is the amplitude of the kth user and $b_k \in \{-1, 1\}$ represents the bit of the kth user; $n_k = \sigma \int_0^T n(t) s_k(t) dt$ and $s_k(t)$ is the signature waveform of the kth user with unitary energy; ρ_{jk} represents a cross-correlation between signature waveforms $s_j(t)$ and $s_k(t)$, and is a component in covariance matrix $\mathbf{R}$. The MF outputs for users in the system can be represented by the vector form as

$$\mathbf{y} = \mathbf{RAb} + \mathbf{n} \tag{2.2}$$

where $\mathbf{y} = [y_1 \cdots y_K]^T$, $\mathbf{b} = [b_1 \cdots b_K]^T$, $\mathbf{A} = \mathbf{diag}\{A_1 \cdots A_K\}$, and $\mathbf{n}$ is the zero-mean Gaussian random vector with covariance matrix $\sigma^2 \mathbf{R}$. By such notation, linear MUD schemes can be realised as a linear transform of the received vector $\mathbf{y}$. The demodulation is simply as

$$\hat{b}_k = \text{sgn}(\mathbf{L y})_k$$

where L is the corresponding linear transform. For the linear MUD receivers we consider,

$$L_{\text{MF}} = \mathbf{I}$$
$$L_{\text{DC}} = \mathbf{R}^{-1}$$
$$L_{\text{MMSE}} = [\mathbf{R} + \sigma^2 (\mathbf{A}^T \mathbf{A})^{-1}]^{-1}$$

It is well known that the BER performance for these three receivers is $\text{BER}_{\text{MMSE}} \leq \text{BER}_{\text{DC}} \leq \text{BER}_{\text{MF}}$. The logical steps to organise one processing flow for these three receivers can be summarised as shown in Figure 2.8, which can reuse calculations to result in a multiple-system (i.e., multiple MUD receivers) configuration and algorithms based on communication theory.

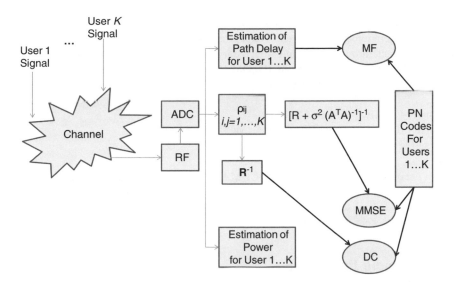

Figure 2.8 Logical functionality of reconfigurable linear MUD

2.3.2 Reconfigurable OFDM Implementation

Implementation of a reconfigurable communication system can be realised through (i) reconfigurable components such as tunable RF and field-programmable gate array (FPGA); or (ii) a compromise of different types of components typically processors, ASIC or hardwired circuits, memory and analogue devices. The first approach currently costs too much, which may mean power consumption, control overhead or size. Our immediate need is to build up a reconfigurable communication system through components as in the second approach. A good example is the RaPiD reconfigurable architecture, developed at the University of Washington especially for communication and signal processing by providing a reconfigurable pipelined data-path and reconfigurable control logic to reach a good compromise between programmable DSPs and ASIC. We can then implement an OFDM system over such a reconfigurable architecture as an example. The reconfigurable data-path contains a set of functional units including ALUs and multipliers, with surely a large number of registers for information required for functional units on every cycle, and a number of embedded (random access) memories for data repeatedly used through the computation.

2.3.3 Reconfigurable OFDM and CDMA

The next step of reconfigurable system architecture lies in demonstration of multiple wireless systems over a single platform. Lin et al. [4] proposed a SDR hardware platform consisting of one controller and a number of ultrawide SIMD processing elements (PEs), with data communication through direct memory access (DMA) instructions. This type of system structure was used in high-end general purpose computing such as the IBM Cell processor. Both WCDMA (3G cellular) and IEEE 802.11a (WLAN) are implemented over this programmable architecture for dual-mode demonstration. The implementation of wireless systems is done by first analysing wireless protocols. Four major functions are identified: filtering, modulation, channel estimation and error correction. Wireless protocols generally consist of multiple DSP algorithm kernels in feed-forward pipelines. Cache structure might provide some additional benefits. Computationally intensive algorithms shall have abundant data-level parallelism, such as searcher, LPF, FFT and the Viterbi decoding algorithm.

The control processor handles the system operations and manages the data processors through remote calls and DMA operations, while the set of data processors deals with heavy-duty data computations.

As a matter of fact, through communication theory, we can develop one programmable structure for OFDM, CDMA and their joint versions (OFDM-CDMA). There are three major types of OFDM-CDMA to merge OFDM and CDMA together:

- Multicarrier-CDMA (MC-CDMA) [9,10];
- Multicarrier direct sequence (DS)-CDMA (MC-DSCDMA) [11];
- Multitone (MT)-CDMA [12].

We [5] first need to develop a universal OF-CDMA structure to adjust parameters in order to realise any of above three systems, which can be degenerated into simple OFDM and CDMA. Then, we can adjust parameters to realise the programmable OFDM-CDMA platform that can serve as the basis for modern wireless communication systems.

2.4 Digital Radio Processing

Modern wireless communication systems are implemented by system-on-chip (SoC) technology, for both base stations and mobile stations. Owing to breakthroughs in the 1990s, even RF can be commercially implemented by CMOS and the entire transceiver can be put into a single SoC. Following the well-known Moore's law, density of gate-count in ICs doubles every 18 months. In other words, for a given wireless communication system, the SoC die size (i.e., usually proportional to cost and related to module size) can be reduced to half every 18 months. Unfortunately, such a conclusion is only true for a digital segment and not an analogue segment. Within a very few years, wireless communication SoC might be mainly occupied by analogue circuits for RF. A pioneering idea is to replace conventional RF circuits by digital processing/processors, which is known as *digital radio processing/processor* (DRP). Texas Instrument (TI) has brought the idea into commercial products recently [15–17].

2.4.1 Conventional RF

Super-heterodyne (SH) and direct conversion (DC) structures are two of most common RF system structures. The two step down-conversion model is easy to analyse and understand, but the need for more external components suggests a higher degree of complexity, which is inefficient and not suitable for SoC. On the other hand, the direct conversion structure shows a relative high degree of simplicity and is more suitable for the needs of SoC but has more practical implementation challenges.

2.4.1.1 Receiver

From a practical point of view, the most general description of an RF system is as presented in Figure 2.9.

In Figure 2.9(b), the incoming RF signal is amplified by a low-noise amplifier (LNA) and then directly to baseband in-phase (I) and quadrature (Q) signals. Channel selection and gain control are achieved by on-chip low-pass filters and a variable gain amplifier (VGA) at baseband. Channel selectivity in super-heterodyne receivers is achieved by down-converting the RF signal to fixed IF and passing through an IF surface acoustic wave (SAW) (or crystal) filters (see Figure 2.9(a)). Receivers operating in multiple bands/standards may have a different frequency plan and need different IF filters even on small form factor phones. Hence, there exist some advantages for the direct-conversion receiver structure.

(a)

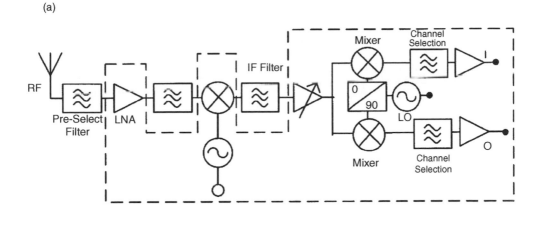

(b)

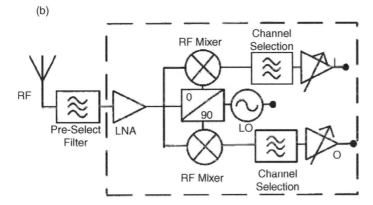

Figure 2.9   (a) Super-heterodyne and (b) direct-conversion structures

- Allowing channel filtering to be done at baseband not only makes the power-efficient on-chip filtering feasible, but also eliminates the need for external passive filters, contributing to the saving of board space and cost (see Figure 2.9(b)).
- Setting IF in a DCR is zero; no image frequency is present, as in the case of a super-heterodyne, thus eliminating the inter-stage image reject filter between the LNA and mixer in a DCR.
- Not needing a VHF voltage-controlled oscillator and phase-locked loops (PLL) that a conventional super-heterodyne would need saves die area and cost, as well as eliminating the effects of phase noise of the VHF VCO.

However, some of the key implementation challenges such as dc offset, second-order linearity, LO leakage, gain and phase imbalance are introduced in DCR.

2.4.1.2 Transmitter

A direct-conversion transmitter, as the name suggests, is one in which a baseband IQ signal directly modulates an RF carrier at the desired transmitter frequency. Variable RF gain stages controlled by the RF automatic gain control (AGC) signal amplify the modulated RF signal to the desired power output, as shown in Figure 2.10(b). RF SAW filters can be used to suppress noise floor in the receiver

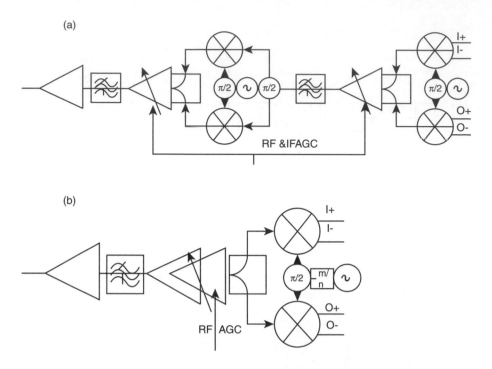

Figure 2.10   (a) Super-heterodyne and (b) DCT architectures

band before feeding the signal in to the power amplifier. This is different from the traditional dual frequency-conversion transmitter where dual LO sources are used and the required dynamic range is achieved using both IF and RF VGAs.

Current TX architectures in CDMA handsets are based on the dual-conversion transmit train as shown in Figure 2.10(a), wherein baseband IQ signals are modulated on an IF carrier followed by a VGA with > 80 dB dynamic range. Amplified IF is filtered using simple LC tuning elements and up-converted using image-reject mixers and variable gain driver blocks. Gain partitioning is made to achieve power control, noise floor reduction and linearity with optimal power efficiency. RFICs with 2 to 3 integrated PLLs and a VHF VCO are commercially available. Channel filtering and wave shaping of TX IQ signals is normally implemented in the baseband. IF filtering in a CDMA/WCDMA transmitter is only needed to suppress the receiver band noise generated in IF VGA stages. IF filters are typically simple LC parallel resonant tank circuits tuned at TX IF. The frequency rolloff offered by these 'filters' attenuate the RX band noise floor at TX IF + duplex spacing (e.g., Duplex spacing$_{WDMA}$ = 190 MHz; Duplex spacing$_{USPCS\ CDMA}$ = 80 MHz) to low enough levels.

DCT has some advantages compared to the dual-conversion transmit train:

- The saving component is achieved by eliminating on-chip IF VCO + PLL + tank circuit components and external LC IF filters normally used in a transmitter to reduce receive band noise.
- The DCT has frequency planning of multimode systems (e.g., GSM/WCDMA) with different duplex spacing and operating frequency ranges. This task becomes more complex for a dual-conversion transmitter. On-chip tuning and programmability opens up the potential for reusability of function blocks between different mobile-phone standards, saving die area and cost.
- Spurious emissions due to m*IF ∼ n*LO are entirely eliminated as IF is zero.

However, some of the key challenges for the DCT are shown below:

- IQ phase and gain imbalance;
- in-band noise floor;
- VCO pulling;
- dynamic range;
- power consumption.

The dynamic range is achieved by proper gain distribution in the baseband (analogue/digital) and RF sections with associated performance tradeoffs such as variable gain in digital (requires higher DAC resolution), variable gain in analogue baseband (causes increased LO leakage of mixer) and variable gain at RF (causes high power consumption).

2.4.1.3 Frequency Synthesiser

The most important organ in the RF frequency planning, which generates the required carrier frequency and sometimes controls the entire transceiver clock, is a LO based on a phase-locked loop (PLL). The most popular PLL in commercial products is illustrated in Figure 2.11.

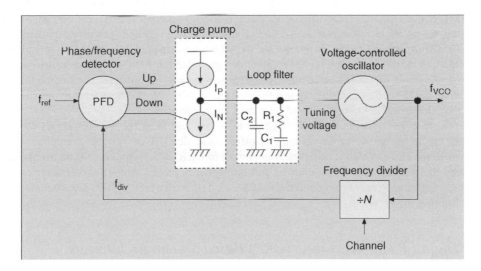

Figure 2.11 A typical local oscillator based on a charge-pump PLL

The phase/frequency detector (PFD) estimates the phase difference between the frequency reference f_{ref} input and the divided-by-N voltage controlled oscillator (VCO) clock f_{div} by measuring the time between their respective edges. It generates either an UP or a DOWN pulse whose width is proportional to the measured time difference. This signal, in turn, produces a current pulse I_P or I_N with the proportional duty cycle in a charge pump block. At the loop filter, this current is converted into a VCO tuning voltage. However, there are some drawbacks in charge-pump PLL.

- Period glitch, produced by the charge pump on every phase comparison, will modulate the VCO output frequency, thus giving rise to frequency spurs that degrade transmitter and receiver performance.
- Spur level attenuates by reducing loop bandwidth but degrades the transient response, and so a compromise is needed.

- Charge-pump PLL is analogue-intensive and ill-suited for integration in deep-submicron CMOS technology.
- The required precise matching of the charge-pump current sources and the required large external integrating capacitors of the loop filter are difficult in deep submicron CMOS due to relatively low transistor resistance and extremely thin gate oxide layer.
- The increasing nonlinearity of the voltage-to-frequency characteristics of the VCO make the PLL bandwidth subject to change and instability.

2.4.1.4 A/D Converter

The other most important role in the RF dynamic range is the A/D converter, which performs analogue-to digital conversion (ADC) of the RF input signal in the presence of large interferers. General commercial wireless standards specify operating conditions that dictate dynamic ranges that far exceed the reliable dynamic range of a single ADC. Whereas the levels of adjacent channel blockers and interferers, which contain no useful information, are much larger than the signal of interest, the SNR needed by the detector for the channel of interest is only 8 – 10 dB. In GSM, for example, the interferer at 3 MHz offset from the signal of interest is specified at − 23 dBm, while the sensitivity is specified at − 102 dBm. Although one could construct a GSM radio receiver with a low-noise amplifier (LNA) and a mixer providing a gain of about 20–30 dB followed by an ADC with a dynamic range of 90 dB in the 200 kHz band, it is neither practical nor desirable to do so. This is because typically with over 80 dB of dynamic range, the law of diminishing returns starts to play its role and each dB of dynamic range requires a disproportionate effort in improving the ADC performance (especially when addressing a multistandard radio). Therefore, typical receiver designs relax the ADC specifications by providing a series of gain stages followed by filtering stages. The gain reduces the impact of noise of the following stages, while the filter eliminates the extraneous information of the blocker/interferer so that more gain can be provided in the next stage without saturating the receiver.

Offsets reduce the available dynamic range because they can cause the received signal waveform to clip and create nonlinearities. One or more dc offset correction (DCOC) circuits are built to remove the offsets in the chain, which is particularly important in DCR. After removing the adjacent channel interferers, the residual offsets are removed and the signal is passed through the detector circuitry. Signal pre-conditioning may be required such as passing it through a match filter and an equaliser prior to hard limiting the decisions.

2.4.2 Digital Radio Processing (DRP) Based System Architecture

Together with the single on chip (SoC) integration trend and digital CMOS process technology advances (the deep-submicron CMOS process), a new paradigm facing analogue and RF designers of deep-submicron CMOS and SoC circuits is formulated below:

> In a deep-submicron CMOS process, time-domain resolution of a digital signal edge transition is superior to voltage resolution of analogue signals.

A successful design approach in this environment would exploit the paradigm by emphasising the following:

- fast switching characteristics or high f_T (40 ps and 100 GHz in this process, respectively) of MOS transistors: high-speed clock and/or fine control of timing transitions;
- high density of digital logic using state-of-the-art process makes digital functions extremely inexpensive;

- small device geometries and precise device matching made possible by fine lithography.

The design would, however, have to avoid the following:

- biasing currents that are commonly used in analogue designs;
- reliance on voltage resolution;
- nonstandard devices that are not needed for memory and digital circuits.

With the advanced process technology's rapid and precise advantages, Texas Instruments (TI) first delivered the almost digitalised SoC technology called digital radio processing (DRP), which migrates RF processing from the analogue to the digital domain and thus enables higher level of integration. This includes the following developments:

- Sampled-data processing techniques, which are used in the direct RF sampling, bring the discrete-time analogue domain processing concepts to replace the conventional continuous-time analogue domain mixer multiplication concepts, such as fast down-converting the signal carrier frequency and avoiding the conventional power-consuming active mixer.
- Switched-capacitor filters, where signal processing concepts are used to provide enough rejection of the noise and adjacent channel interferers, thus providing an RF filter of high selectivity and reducing the conventional external filter usage.
- Oversampling converters, whose oversampling technique not only relaxes the A/D resolution but also loosens the anti-aliasing filter constraint through the characteristics of noise shaping.
- All digital phase locked-loop (ADPLL), which is based on a digitally controlled oscillator (DCO) and a time-to-digital converter (TDC), thus avoiding the conventional VCO analogue tuning and providing a more precise digital phase detector to replace the analogue PFD in charge-pump PLL.

2.4.2.1 Receiver

In Figure 2.12, the received RF signal is amplified in the low-noise amplifier (LNA), split into I/Q paths and converted to current using a trans-conductance amplifier (TA) stage. The current is then down-converted to a programmable low-IF frequency (defaults to 100 kHz) and integrated on a sampling capacitor at the LO rate. Considering the plus and minus sides, the input signal is sampled at the Nyquist rate of the RF carrier. After initial decimation through a sinc filter response, a series of infinite impulse response (IIR) filtering follows RF sampling for close-in interferer rejection. These signal processing operations are performed in the multi-tap direct sampling mixer (MTDSM). A sigma-delta analogue-to-digital converter containing a front-end gain stage converts the analogue signal to a digital representation. A feedback control unit (FCU) provides a single-bit feedback to the MTDSM to establish the common mode voltage for the MTDSM while cancelling out differential offsets. The output of the I/Q ADCs are passed on to the digital receive (DRX) chain that performs decimation, down-conversion to 0-IF and adjacent channel interference removal. Thus, on the one hand, it is a low-IF architecture, which is similar to DCR and thus has the same implement challenges. On the other hand, there are some tradeoffs between this DRP system and the conventional analogue RF system.

- Using the direct RF sampling technique to down-convert the RF incoming signal brings more fleet computation into the digital-analogue concepts and thus eliminates traditional mixer challenges (e.g., image rejection problem or inter-modulation products). But subsequent following filtering is required to remove or alleviate the noise folding and the jitter noise problem shown in such sampling receivers.

(a)

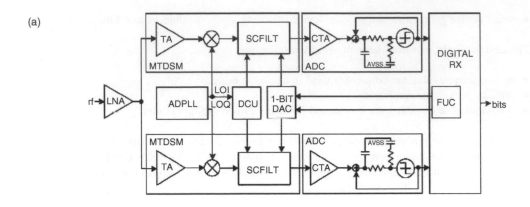

(b)

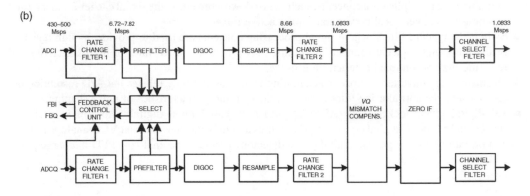

Figure 2.12   Simplified block diagram of (a) DRP receiver overview and (b) digital receive chain

■ Using IF digitalisation concepts to convert the analogue signal to the digital domain puts forward the ADC to close the antenna and shift a majority of function blocks (e.g., channel select filtering, 0-IF down-conversion, error correction and demodulation) to the digital domain and thus eliminate many sources of sensitivities of analogue solutions, such as device matching, phase noise, environmental sensitivity and performance variation over time. However, large interferers and blockers are present and the resolution and dynamic range performance of the higher speed ADC will be degraded and power dissipation increase with sampling rate, even impacting on other devices in the system.

■ Although it eliminates the IF filter and image rejection filter, due to direct RF sampling and IF digitalisation, there are a series of additional decimation filters needed to down-convert the data rate in Figure 2.12(b).

■ In addition, the external feedback loop, which includes feedback control unit (FCU) and 1-bit DAC in Figure 2.12(a), is needed in the digital domain to operate in different mode to cancel the dc offset, remove the blocker and interferer, and improve the linearity, continuously and respectively.

2.4.2.2 Transmitter

A transmitter that is well-suited for deep-submicron CMOS implementation is shown in Figure 2.13. It performs the quadrature modulation in the polar domain, in addition to the generation of the LO for

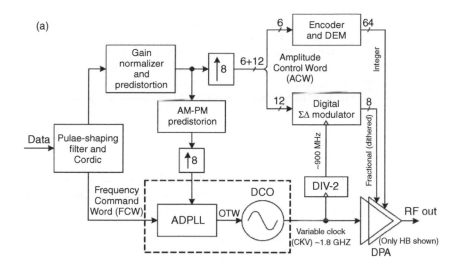

(a)

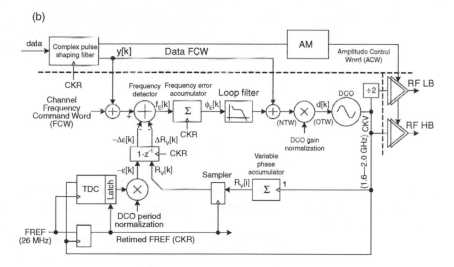

(b)

Figure 2.13 DRP polar transmitter: (a) amplitude modulation path polar transmitter; (b) polar transmitter based on an all-digital PLL

the receiver. All clocks in the system are derived directly from this source. The architecture is built using digital techniques that exploit the high speed and high density of the advanced CMOS, while avoiding problems related to voltage headroom. An ADPLL replaces the conventional RF synthesiser architecture, based on a voltage-controlled oscillator (VCO) and a phase/frequency detector (PFD) and charge-pump combination, with a digitally controlled oscillator (DCO) and a time-to-digital converter (TDC). All inputs and outputs are digital at multi-GHz frequency – the 40 ps rise time makes an almost perfect square wave. The full digital control of the RF frequency allows digital implementation of the PLL.

At the heart of the ADPLL lies a DCO. The oscillator core operates at twice the 1.6–2.0-GHz high-band frequency. The DCO tuning capacitance is split into a large number of tiny capacitors that are selected digitally. The advanced lithography allows creation of extremely fine variable capacitors (varactors) – about 40 attofarads of capacitance per step, which equates to the control of only

250 electrons entering or leaving the resonating LC tank. Despite the small capacitance step, the resulting frequency step at the 2 GHz RF output is 10–20 kHz, which is too coarse for wireless applications. Thus, the fast switching capability of the transistors is used by performing programmable high-speed (225–900 MHz) dithering of the 250 electrons in the finest varactors. The duty cycle of the high/low capacitive states establishes the time-averaged resonating frequency resolution, now less than 1 kHz. All the varactors are realised as n-poly/n-well MOSCAP devices that operate in the flat regions of their $C - V$ curves.

2.4.2.3 All-Digital PLL

The ADPLL shown in Figure 2.13(b) operates in a digitally synchronous fixed-point phase domain as follows. The variable phase $R_v[i]$ is determined by counting the number of rising clock transitions of the DCO oscillator clock CKV:

$$R_v[i] = \sum_{l=0}^{i} 1$$

The index i indicates the DCO edge activity. The FREF-sampled variable phase $R_v[k]$, where k is the index of the FREF edge activity, is fixed-point concatenated with the normalised TDC output $\varepsilon[k]$. The TDC measures and quantises the time differences between the FREF and DCO edges. The sampled differentiated variable phase is subtracted from FCW by the digital frequency detector. The frequency error $f_E[k]$ samples

$$f_E[k] = FCW - [(R_V[k] - \varepsilon[k]) - (R_V[k - 1] - \varepsilon[k - 1])] \tag{2.3}$$

are accumulated to create the phase error $\phi_E[k]$ samples

$$\phi_E[k] = \sum_{l=0}^{k} f_E[k]$$

which are then filtered by a fourth-order IIR filter and scaled by a proportional loop attenuator α. A parallel feed with coefficient ρ adds an integrated term to create type-II loop characteristics, which suppresses the DCO flicker noise.

The IIR filter is a cascade of four single stage filters, each satisfying the following equation:

$$y[k] = (1 - \lambda) \cdot y[k - 1] + \lambda \cdot x[k] \tag{2.4}$$

where $x[k]$ is the current input, $y[k]$ is the current output and λ is the configurable coefficient. The four-pole IIR filter attenuates the reference and TDC quantisation noise at the 80 dB/dec slope, primarily to meet the GSM spectral mask requirements at 400 kHz offset. The filtered and scaled phase error samples are then multiplied by the DCO gain normalisation factor $f_R/\hat{K}_{DCO}$, where f_R is the reference frequency and $\hat{K}_{DCO}$ is the DCO gain estimate, to make the loop characteristics and modulation independent from K_{DCO}. The modulating data is injected into two points of the ADPLL for the direct frequency modulation. A hitless gear-shifting mechanism for the dynamic loop bandwidth control serves to reduce the settling time. It changes the loop attenuator α several times during the frequency locking while adding the $(\alpha_1/\alpha_2 - 1)\phi_1$ DC-offset to the phase error, where indexes 1 and 2 stand for before and after the event, respectively. Of course, $\phi_1 = \phi_2$, since the phase is to be continuous. The FREF input is resampled by the RF oscillator clock, and the resulting retimed clock (CKR) is used throughout the system. This ensures that the massive digital logic is clocked after the quiet interval of the phase error detection by the TDC.

The transmitter architecture is fully digital and takes advantage of the wideband frequency modulation capability of the all digital PLL by adjusting its digital frequency command word. The modulation method is an exact digital two-point scheme, with one feed directly modulating the DCO frequency deviation while the other is compensating for the developed excess phase error. The DCO gain characteristics are constantly calibrated through digital logic to provide the lowest possible distortion of the transmitted waveform. The amplitude modulation path is built on a digitally controlled power amplifier (DPA) with a large array of MOS switches and operates in near-class-E mode. The RF amplitude is regulated by controlling the number of active switches. Fine amplitude resolution is achieved through high-speed transistor switch dithering. Despite the high speed of the digital logic operation, the overall power consumption of the transmitter architecture is lower than that of architectures to date.

In the transmitter, there also exist tradeoffs between the conventional transmitter and the DRP polar loop transmitter:

- Splitting the modulation path into AM and PM eliminates the mixer and filter and even the D/A converter, but the separate paths may delay each other and the high speed modulator, which may increase the power dissipation needed due to the need for a high resolution step.
- Although the sigma-delta technique is used to address fast fractional bit computation, relatively it needs more accurate and precise capacitor variation and linearity, and thus additional procedures such as retimed-clock and dynamic element matching (DEM) are required.

Up to this moment, realistic DRP has been realised by replacing analogue circuits with digital circuits directly. Developing a better digital design from communication theory with baseband communication design, as shown in Figure 2.14, remains an active research problem at this time, while digital SoC follows Moorse's law but analog portion of SoC does not.

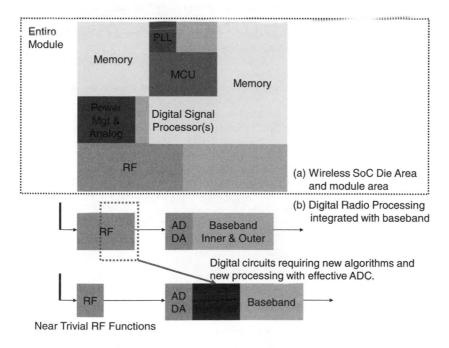

Figure 2.14   Digital radio processing

References

[1] J. Mitola, III, 'Software Radio Architecture: A Mathematical Perspective', *IEEE Journal on Selected Areas in Communications*, **17**(4), 1999, 514–538.

[2] J.E. Gunn, K.S. Barron, W. Ruczczyk, 'A Low-power DSP Core-based Software Radio Architecture', *IEEE Journal on Selected Areas in Communications*, **17**(4), 1999, 574–590.

[3] U. Ramacher, 'Software-Defined Radio Prospects for Multistandard Mobile Phones', *IEEE Computer*, October 2007, 62–69.

[4] Yuan Lin, *et al.*, 'SODA: A High-Performance DSP Architecture for Software-Defined Radio', *IEEE Micro*, January-February 2007, 114–123.

[5] K.C. Chen, S.T. Wu, 'A Programmable Architecture for OFDM-CDMA', *IEEE Communications Magazine*, feature subject on Software and DSP in Radio, November 1999, 76–78.

[6] G. Gerrari, G. Colavolpe, R. Raheli, *Detection Algorithms for Wireless Communications*, John Wiley & Sons, Ltd, Chichester, 2004.

[7] H. Meyr, M. Moeneclaey, S. Fechtel, *Digital Communication Receivers*, John Wiley & Sons, Inc., New Jersey, 1998.

[8] C. Ebeling, *et al.*, 'Implementation an OFDM Receiver on RaPiD Reconfigurable Architecture', *IEEE Transactions on Computers*, **53**(11), 2004, 1436–1448.

[9] N. Yee, J.P. Linnartr, G. Fettweis, 'Multicarrier CDMA in Indoor Wireless Radio Networks,' *Proc. IEEE PIMRC '93, Yokohama, Japan*, September 1993, pp. 109–113.

[10] K. Fazel, L. Papke, 'On the Performance of Convolutionally-coded CDMA/OFDM for Mobile Communication System', *Proc IEEE PJMRC '93, Yokohama, Japan*, September 1993, pp. 468–472.

[11] V.M. DaSilva, E.S. Sousa, 'Performance of Orthogonal CDMA Codes for Quasi-Synchronous Communication Systems', *Proc. IEEE ICUPC*, Ottawa, Canada, 1993.

[12] L. Vandendorpe, 'Multitone Direct Sequence CDMA System in an Indoor Wireless Environment', *Proc. 1st JEEE Symp. Commun.*

[13] A. Loke, F. Ali, 'Direct conversion radio for digital mobile phones – Design issues, status and trends', *IEEE Trans. Microwave Theory Tech.*, **50**, 2002, 2422–2435.

[14] K. Muhammad, R.B. Staszewski, D. Leipold, 'Digital RF Processing: Towards Low-cost Reconfigurable Radios', *IEEE Communications Magazine*, August 2005, 105–113.

[15] K. Muhammad *et al.*, 'The First Fully Integrated Quad-band GSM/GPRS Receiver in a 90-nm Digital CMOS Process', *IEEE J. of Solid-State Circuits*, August 2006,

[16] R. B. Staszewski *et al.*, 'All-digital PLL and GSM/EDGE Transmitter in 90nm CMOS', *Proc. IEEE Solid-State Circuits Conf.*, Sec. 17.5, 600, February 2005, pp. 316–317.

[17] T. Sowlati, D. Rozenblit, R. Pullela *et al.*, 'Quad-band GSM/GPRS/EDGE polar loop transmitter', *IEEE J. Solid-State Circuits*, **39**(12), 2004, 2179–2189.

[18] G. Fettweis *et al.*, 'Dirty RF: A New Paradigm,' *International Journal of Wireless Information Networks*, **14**(2), 2007.

[19] J. Smith, *Modern Communication Circuits*, McGraw-Hill, New York, 1986.

[20] I. Martinez G., 'Automatic Gain Control (AGC) Circuits: Theory and Design,' University of Toronto 2001.

[21] F. Chen, B. Leung, 'A 0.25-mW Low-pass Passive Sigma-delta Modulator with Built-in Mixer for a 10-MHz IF Input,' *IEEE J. Solid-State Circuits*, **32**, June 1997, 774–782.

[22] V. Jimenez, J. Garcia, F. Serrano *et al.*, 'Design and Implementation of Synchronization and AGC for OFDM-based WLAN Receivers,' *IEEE Transactions on Consumer Electronics*, **50**(4), 2004, 1016–1025.

[23] I.-Gu Lee, S.-Kyu, 'Efficient Automatic Gain Control Algorithm and Architecture for Wireless LAN Receiver', *Journal of Systems Architecture*, **53**, 2007, 379–385.

3

Wireless Networks

After introducing physical layer wireless communications and SDR, in this chapter we are going to orient the major concepts useful in higher layers of computer networks, as the foundation of wireless networks. Modern wireless networks or mobile networks can be categorised into two classes: *infrastructured* and *ad hoc*.

Each infrastructured wireless network (as shown in Figure 3.1(a)) has a high-speed backbone network (typically wired) to connect a number of base stations (or access points). The mobile stations communicate through the base stations then the backbone network and on toward the destination mobile station. The packet delivery relies on an infrastructure consisting of the backbone network and base stations. On the other hand, a number of mobile stations may establish an ad-hoc network without any infrastructure, as in Figure 3.1(b), where each link between two nodes (i.e., mobile stations) is plotted and these links build up the network topology of an ad-hoc network. Most modern commercial wireless

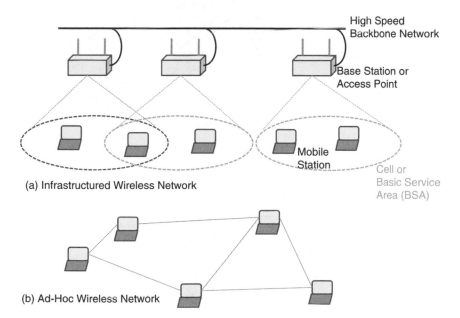

Figure 3.1 (a) Infrastructured and (b) ad-hoc wireless networks

Cognitive Radio Networks   Kwang-Cheng Chen and Ramjee Prasad
© 2009 John Wiley & Sons, Ltd

networks use the infrastructure network, although ad-hoc networking is allowed for occasional use. Ad-hoc networks have been adopted in military systems for a long time, as multi-hop packet radio networks. With a wide range of applications for wireless devices including sensors, ad-hoc networking will play a much more important role in future wireless networking and cognitive radio networking.

3.1 Multiple Access Communications and ALOHA

Recall the layering structure; we need a sub-layer called medium access control (MAC) between the DLC and physical layer. The purpose of this extra sub-layer is to allocate the multi-access medium various nodes. The method to coordinate physical transmission among various nodes in a computer/ communication network is known as a *multiple access* protocol, which also serves as the essential function in wireless networks.

The multiple access protocols operate in various network topologies and application scenarios. Figure 3.2(a) is an example of a satellite communication network with earth stations to communicate via the satellite, which has a star network topology. Figure 3.2(b) shows a bus network topology where nodes/stations are connected to a network through a bus (typically a cable or a fiber as physical medium). Another example in Figure 3.2(b) is a multi-hop packet radio network, a network topology widely considered in wireless ad-hoc networks (sometimes with a mesh network extension structure) and wireless sensor networks. The earliest widely known computer network was a satellite communication network to connect communications in the Hawaii islands, with the pioneering multiple access protocol ALOHA described in the next section.

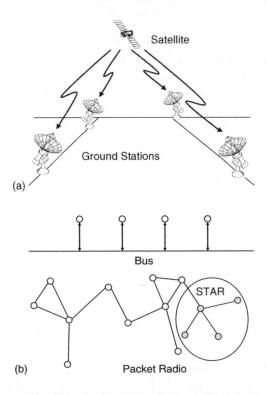

Figure 3.2   Examples of multiple access communication: (a) satellite and ground stations (star network topology as the simplest infrastructure); (b) bus, star and multi-hop packet radio

3.1.1 ALOHA Systems and Slotted Multiple Access

The pioneering multiple access system may well be the ALOHA for a satellite communication (actually information collection) system with a topology as in Figure 3.2(a). Nodes (earth stations) on the ground are trying to access the satellite to relay the packets and thus require a multiple access protocol to coordinate the transmission in a distributed way. When a set of nodes simultaneously share a communication channel, the reception is garbled if two or more nodes transmit simultaneously, which is known as *collision*. And, the channel is unused (or *idle*) if none transmit. The challenge of multiple accesses (also multi-access) is how to coordinate the use of such a channel through a distributed or centralised way. In this section, we focus on distributed multiple access protocol family, which is widely used in wireless data networks. Pure ALOHA is simply: (i) when a node has a packet to transmit, it transmits; and (ii) the node listens to the channel. If collision happens, the node re-schedules transmission of the packet by a (random) backlog algorithm. Otherwise, the node transmits the packet successfully.

To study the multiple access protocol, we usually consider the time axis to be slotted for node operation. Assumptions for the idealised slotted multi-access model are summarised as follows:

1. *Slotted system:* All transmitted packets have the same length and each packet requires one time unit (called a slot) for transmission. All transmitters' are synchronised so that the reception of each packet starts at an integer time and ends at the next integer time.
2. *Poisson arrival:* Packets arrive for transmission at each of the m transmitting nodes according to independent Poisson processes with λ/m arrival rate.
3. *Collision or perfect reception:* The packet is received either in a perfect way or in collision to lose information.
4. *{0,1,e} Immediate feedback:* The multiple access channel can provide feedback to distributed nodes with an alphabet set of size 3 {0,1,e}, where '1' stands for successful packet transmission and reception, '0' stands for channel idle with no packet transmission and 'e' stands for collision(s) in the multi-access channel.
5. *Retransmission of collisions:* Each packet involved in a collision must be retransmitted in some later slot, with further possible such retransmission until a successful transmission. A node with a retransmission packet is called backlogged.
6. (a) *No buffer in each node.*
   (b) *Infinite number of nodes in the systems.*

It is well known that pure ALOHA has *throughput* (i.e., average number of packets successfully transmitted per packet transmission time) of $1/(2e)$.

3.1.2 Slotted ALOHA

The basic idea of this approach is that each unbacklogged node simply transmits a newly arriving packet in the first slot after the packet arrival, thus risking occasional collisions but achieving a very small delay if collisions are rare. For m nodes, an arriving packet in TDM would have to wait for $m/2$ slots on average to transmit. Slotted ALOHA transmit packets almost immediately with occasional collisions, whereas TDM avoids collisions at the expense of larger delays. When a collision occurs in slotted ALOHA, each node sending one of the colliding packets discovers the collision at the end of the slot and becomes backlogged. Such nodes wait for some random number of slots before retransmitting. With infinite-node assumption, the number of new arrivals transmitted in a slot is a Poisson random variable with parameter λ. If the retransmission from backlogged nodes is sufficiently randomised, it is plausible to approximate the total number of retransmission and new transmissions in a given slot as a Poisson random variable with parameter $G > \lambda$. The parameter of a successful transmission in a slot is Ge^{-G} (see Figure 3.3).

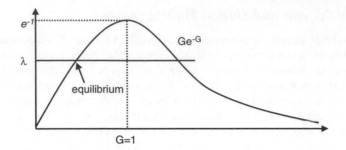

Figure 3.3 Transmission parameter

To construct a more precise model in order to understand system dynamics, assume that each backlogged node retransmits with same fixed parameter q_r in each successive slot until a successful transmission occurs. In other words, the number of slots from a collision until a given node involved in the collision retransmits is a geometric random variable $q_r(1 - q_r)^{i-1}, i \geq 1$. We use the assumption of no buffer (i.e., assumption 6(a) above). The behaviour of slotted ALOHA can be described as a node at the beginning of a given slot. Each of these nodes will transmit a packet in a given slot, independently of each other, with parameter q_r. Each of the $m - n$ other nodes will transmit a packet in a given slot if one (or more) such packets arrived during the previous slot. Since such arrivals are Poisson with mean λ/m, the parameter of no arrival is $e^{\lambda/m}$. Thus, the parameter with which an unbacklogged node transmits a packet in a given slot is $q_a = 1 - e^{-\lambda/m}$. Let $Q_a(I, n)$ be the parameter with which i unbacklogged nodes transmit packets in a given slot, and $Q_r(i, n)$ be the parameter which i backlogged nodes transmit:

$$Q_a(i, n) = \binom{m-n}{i}(1 - q_a)^{m-n-i}q_a^i \tag{3.1}$$

$$Q_r(i, n) = \binom{n}{i}(1 - q_r)^{n-i}q_r^i \tag{3.2}$$

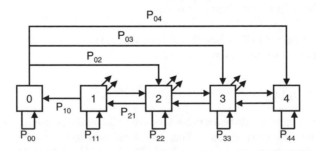

Figure 3.4 Markov chain model for slotted ALOHA

From one slot to the next, the state (the number of backlogged packets) increases by the number of new arrivals transmitted by unbacklogged nodes, reducing by one if a packet is transmitted successfully. A packet is transmitted successfully only if there is

- one new arrival and no backlogged packet;
- no new arrival and one backlogged packet.

The state transition parameter of going from state n to $n + i$ in Figure 3.4 is

$$
P_{n,n+i} = \begin{cases} Q_a(i,n), & 2 \leq i \leq (m-n) \\ Q_a(1,n)[1 - Q_r(0,n)], & i = 1 \\ Q_a(1,n)Q_r(0,n) + Q_a(0,n)[1 - Q_r(1,n)], & i = 0 \\ Q_a(0,n)Q_r(1,n), & i = -1 \end{cases}
\tag{3.3}
$$

We would like to choose the retransmission parameter q_r to be moderately large to avoid large delay. However, $q_r \cdot n \gg 1$ is possible, and then collisions occur in almost all successive slots and the system remains heavily backlogged for a long time.

Let $D_n(drift)$ be the expected change in backlog over one slot time, starting in state n, and P_{SUCC} the parameter of a successful transmission, the expected number of successful transmissions as follows:

$$
D_n = (m-n)q_a - P_{SUCC}
\tag{3.4}
$$

where

$$
P_{SUCC} = Q_a(1,n)Q_r(0,n) + Q_a(0,n)Q_r(1,n)
\tag{3.5}
$$

Define the attempt rate $G(n)$ as the expected number of attempted transmissions in a slot when the system is in state n:

$$
G(n) = (m-n)q_a + nq_r
\tag{3.6}
$$

If q_a, q_r are small,

$$
P_{SUCC} \approx G(n)e^{-G(n)}
\tag{3.7}
$$

The parameter of an idle slot is approximately $e^{-G(n)}$.

The drift is the difference between the curve and the straight line. Since the drift is the expected change in state from one slot to the next, though the system perhaps fluctuates, it tends to move in the direction of drift and consequently tends to cluster around the two stable points with rare excursions between the two.

We can have the following observations:

1. The departure rate (P_{SUCC}) is at most $1/e$ for large m.
2. The departure rate is almost zero for long periods whenever the system jumps to the undesirable stable point.

Consider the effect of changing q_r. As q_r is increased, the delay in retransmitting a collided packet decreases, but the linear relationship between n and the attempt rate $G(n) = (m-n)q_a + nq_r$ also changes. If the horizontal scale for n is held fixed, this change in attempt rate corresponds to a contraction of the horizontal $G(n)$ scale, and thus to a horizontal contraction of curve Ge^{-G}. This means that the number of backlogged packets required exceeding the unstable equilibrium point decreases. If q_r is decreased enough (still $> q_a$), the curve Ge^{-G} will expand enough that only one stable state will remain. At this stable point, when $q_r = q_a$, the backlog is an appreciable fraction of m (see Figure 3.5). This means both that an appreciable number of arriving packets is discarded and that the delay is excessively large.

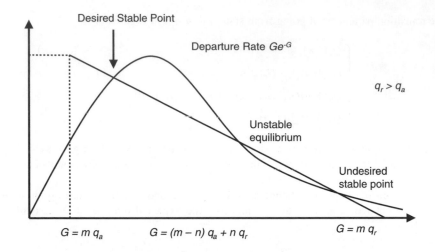

Figure 3.5   Stability in ALOHA throughput curve

3.1.3 Stabilised Slotted ALOHA

ALOHA (along with others described later in this family) suffers instability of random access protocols, which suggests zero throughput (or infinite large delay) if incoming traffic goes beyond the stable region to result in packet accumulation of total traffic.

To maintain good multi-access channel efficiency, it is desirable to change q_r dynamically to maintain $G(n)$ at 1 to maximise $P_{SUCC} \approx G(n)e^{-G(n)}$. The difficulty is that n is unknown to the nodes and can only be estimated from the feedback. All the strategies to estimate n or the appropriate q_r, in essence, increase q_r when an idle slot occurs and decrease q_r when a collision occurs.

With the infinite-node assumption (6b, on p.xxx), there is no steady-state distribution and the expected delay grows without bound as the system continues to run. With the no-buffer assumption (6a, also on p.xxx), the system discards a large number of arriving packets and has a very large but finite delay, whereas with the infinite-node assumption, no arrivals are discarded but the delay becomes infinite.

The maximum stable throughput of a multi-access system is defined as the least upper bound of arrival rates for which the system is stable. For example, for slotted ALOHA, it is zero.

When the estimate of backlog is perfect and $G(n)$ is maintained at the optimal value of 1, then idles occur with parameter $1/e \approx 0.368$, successes occur with parameter $1/e$ and collisions occur with parameter $1 - 2/e \approx 0.264$. Thus, the rule for changing q_r should allow fewer collisions than idles. The maximum stable throughput of such a system is at most $1/e$.

The Pseudo-Bayesian Algorithm may be the most well known algorithm to stabilise S-ALOHA and those protocols in the ALOHA family. This algorithm differs from slotted ALOHA in that new arrivals are regarded as backlogged immediately on arrival. Rather than being transmitted with certainty in the next slot, they are transmitted with parameter q_r in the same way as packets involved in previous collisions. If there are n backlogged packets (including new arrivals) at the beginning of a slot, the attempt rate is $G(n) = nq_r$ and the parameter of a successful transmission is $nq_r(1 - q_r)^{n-1}$. For unstabilised ALOHA, q_r has to be relatively small and new arrivals would be unnecessarily delayed. For stabilised ALOHA, q_r can be as large as 1 when the estimated backlog is negligible so that new arrivals are held up only when the system is already estimated to be congested.

The algorithm operates by maintaining an estimate $\hat{n}$ of the backlog n at the beginning of each slot. Each backlogged packet is then transmitted (independently) with parameter $q_r = \min\{1, 1/\hat{n}\}$. Subject to q_r at most 1 try to achieve an attempt rate $G = nq_r$.

$$\hat{n}_{k+1} \begin{cases} \max\{\lambda, \hat{n}_k + \lambda - 1\}, & \text{for idle or success} \\ \hat{n}_k + \lambda + (e-2)^{-1}, & \text{for collision} \end{cases} \tag{3.8}$$

3.1.4 Approximate Delay Analysis

Assume that λ is known and that the parameter of successful transmission P_{SUCC} is $1/e$ whenever n is 2 or more, and that $P_{SUCC} = 1$ for $n = 1$. This is a reasonable model for small λ since very few collisions occur and q_r is typically 1. It is also reasonable for large $\lambda < 1/e$ since typically n is large and $\hat{n} \approx n$.

Let W_i be the delay from the arrival of the ith packet until the beginning of the ith successful transmission. The average of W_i over i is equal to the expected queue delay W. Let n_i be the number of backlogged packets at the instant before i's arrival. n_i does not include any packet currently being successfully transmitted, but does include current unsuccessful transmissions:

$$W_i = R_i + \sum_{j=1}^{n_i} t_{j_i} + y_i \tag{3.9}$$

R_i is the residual time to the beginning of the next slot and t_1 (for $n_i > 0$) is the subsequent interval until the next successful transmission is completed. t_j, $1 < j \leq n_i$, is the interval from the end of the $(j-1)$th subsequent success to the end of the jth subsequent success. After those n_i successes, y_i is the remaining interval until the beginning of the next successful transmission.

Each slot is successful with parameter $1/e$ and the expected value of each t_j is e. By Little's theorem,

$$W = \frac{1}{2} + \lambda e W + E\{y\} \tag{3.10}$$

Consider the system at the first slot boundary at which both the $(i-1)$st departure has occurred and the ith arrival has occurred. If the backlog is 1, then $y_i = 0$. If $n > 1$, $E\{y_i\} = e - 1$. Let P_n be the steady-state parameter in which the backlog is n at a slot boundary. Since a packet is always successfully transmitted if the state is 1, P_1 is the fraction of slots in which the state is 1 and a packet is successfully transmitted. Since λ is the total fraction of slots with successfully transmissions, P_1/λ is the fraction of packets transmitted from state 1 and $1 - P_1/\lambda$ is the fraction transmitted from higher-numbered states:

$$E\{y\} = \frac{(e-1)(\lambda - P_1)}{\lambda} \tag{3.11}$$

$$\lambda = P_1 + (1 - P_0 - P_1)/e \tag{3.12}$$

State 0 can be entered at a slot boundary only if no arrivals occurred in previous slot and the previous state was 0 or 1:

$$P_0 = (P_0 + P_1)e^{-\lambda} \tag{3.13}$$

$$P_1 = \frac{(1 - \lambda e)(e^{\lambda} - 1)}{1 - (e-1)(e^{\lambda} - 1)} \tag{3.14}$$

$$W = \frac{e - \frac{1}{2}}{1 - \lambda e} - \frac{(e^{\lambda} - 1)(e - 1)}{\lambda[1 - (e-1)(e^{\lambda} - 1)]} \tag{3.15}$$

We therefore obtain the waiting time (i.e., delay) in an analytical way.

3.1.5 Unslotted ALOHA

There is no slotted concept in pure ALOHA that can be considered as the most primitive version of multiple access protocols. A station with a packet to transmit just transmits and listens to channel. If a packet is involved in a collision, it is retransmitted after a random delay.

Time

Figure 3.6   Packets in a collision

We may observe from Figure 3.6 that the collided duration is up to two times larger than that of slotted ALOHA. Therefore, as shown in Figure 3.7,

$$Throughput = G(n)e^{-2G(n)}$$

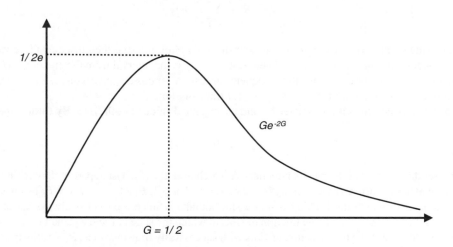

Figure 3.7   Unslotted ALOHA throughput

Regarding further analysis of delay in ALOHA family protocols, readers are directed to [1, 2].

3.2 Splitting Algorithms

We may observe a critical part in the ALOHA protocol, the random backoff algorithm, trying to resolve collisions. It generally results in more sophisticated collision resolution techniques that maintain stability without any complex estimation procedure and also increase the achievable throughput. To get an intuitive idea, with small attempted rates, it is most likely to be only between two packets that a collision occurs. If new arrivals could be inhibited from transmission until the collision was resolved, each of the colliding packets could be independently retransmitted in the following slot with parameter 1/2. It repeats until a successful transmission and is followed by the transmission of the remaining packet.

The expected number of slots for sending these two packets is 3, yielding a throughput of 2/3 for the period during which the collision is being resolved. There are various ways in which the nodes involved in a collision could choose whether or not to transmit in successive slots:

- flip a coin;
- use the arrival time of its collided packet;
- assume a finite set of nodes, each with a unique identifier represented by a string of bits, so that a node could use the successive bits of its identity to make the successive choices, an advantage of limiting the number of slots required to resolve a collision.

We call these types of algorithms splitting algorithms or collision resolution protocols (CRP).

3.2.1 Tree Algorithms

When a collision occurs in the kth slot, all nodes not involved in the collision go into a waiting mode, and all those involved in the collision split into two subsets (by flipping a coin, for example). The first subset transmits in slot $k + 1$ and if that slot is idle or successful, the 2nd subset transmits in slot $k + 2$. Alternatively, if another collision occurs in slot $k + 1$, the first of the two subsets splits again and the 2nd subset waits for the resolution of that collision (see Figure 3.8).

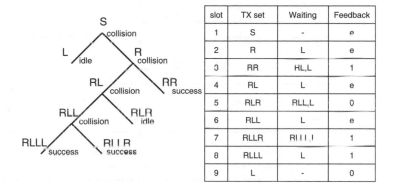

slot	TX set	Waiting	Feedback
1	S	-	e
2	R	L	e
3	RR	RL,L	1
4	RL	L	e
5	RLR	RLL,L	0
6	RLL	L	e
7	RLLR	RLLL,L	1
8	RLLL	L	1
9	L	-	0

Figure 3.8 Tree algorithm

One problem is what to do with the new packet arrivals that come in while a collision is being resolved. A collision resolution period (CRP) is completed when a success or an idle occurs and there are no remaining elements on the stack. At this time, a new CRP starts using the packets that arrived during the previous CRP. Then, there will be new waiting arrivals that collide and continue to collide until the subsets get small enough in this new CRP. The solution is: at the end of a CRP, the set of nodes with new arrivals is immediately split into j subsets, where j is chosen as that the expected number of packets per subset is slightly greater than 1. These new subsets are then placed on the stack and a new CRP starts. The maximum throughput is 0.43.

If a collision is followed by an idle slot, this means that all the packets involved in the collision were assigned to the second subset. The maximum throughput is 0.46.

In practice, this improvement has a slight problem in that if an idle slot is incorrectly perceived by the receiver as a collision, the algorithm continues splitting indefinitely, never making further successful transmissions. Thus, after some number h of idle slots followed by splits, the algorithm should be modified simply to transmit the next subset on the stack without first splitting it. If feedback is very reliable, h can be moderately large. Let x be the number of packets in the first collision, X_L, X_R be the number of packets in the resulting subsets. $X = X_L + X_R$. Assume *a priori* that X is a Poisson. Then X_L, X_R are independent Poisson random variables each with half the mean value of X. Given $X_L + X_R \geq 2$ and $X_R \geq 2$, rather than devoting a slot to this second subset that has an undesirable small expected number of packets, it is better to regard the second subset as just another part of the waiting new arrivals that have never been involved in a collision.

3.2.2 FCFS Splitting Algorithm

By splitting by arrival time, each subset consists of all packets that arrived in some given interval. When a collision occurs, that interval will be split into two smaller intervals. At each integer time k, the algorithm specifies the packets to be transmitted in slot k to be the set of all packets that arrived in some earlier interval. This interval is called the allocation interval for slot k, $T(k)$ to $T(k) + \alpha(k)$.

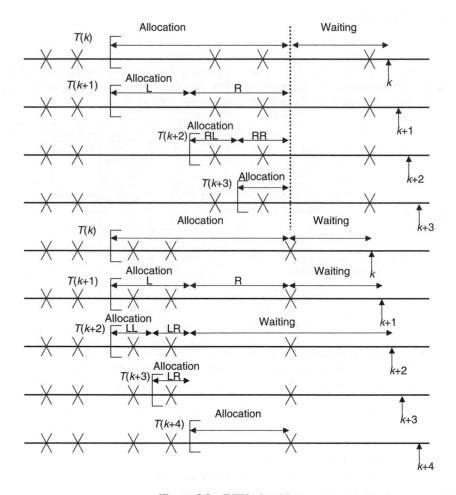

Figure 3.9 FCFS algorithm

Precise first-come-first-serve (FCFS) algorithm followed by each node is (see Figure 3.9):
If *feedback* = 0, then

$$T(k) = T(k-1)$$

$$\alpha(k) = \alpha(k-1)/2$$

$$\sigma(k) = \text{L}$$

If *feedback* $=1$ and $\sigma(k-1)=$ L, then

$$T(k) = T(k-1) + \alpha(k-1)$$

$$\alpha(k) = \alpha(k-1)$$

$$\sigma(k) = R$$

If *feedback* $=0$ and $\sigma(k-1)=$ L, then

$$T(k) = T(k-1) + \alpha(k-1)$$

$$\alpha(k) = \alpha(k-1)/2$$

$$\sigma(k) = R$$

If *feedback* $=0$ or 1 and $\sigma(k-1)=$ R, then

$$T(k) = T(k-1) + \alpha(k-1)$$

$$\alpha(k) = \min[\alpha_0, \, k - T(k)]$$

$$\sigma(k) = L$$

where α_0 can be chosen either to minimise delay for a given arrival rate or to maximise the stable throughput.

The algorithm gives the allocation interval (i.e., $T(k)$ $\alpha(k)$) and the status $\sigma = $ (L or R) for slot k in terms of the feedback, allocation interval and status from slot $k-1$.

3.2.3 Analysis of FCFS Splitting Algorithm

Node (R,0) which corresponds to the initial slot of a collision resolution period (CRP) is split in two as an artifice to visualise the beginning and end of a CRP.

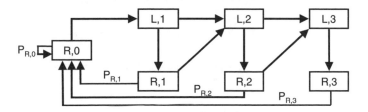

Figure 3.10   FCFS splitting algorithm

Assume the size of the initial allocation interval to be α_0. Each splitting of the allocation interval decreases by 2. (L,i) and (R,i) in Figure 3.10 correspond to allocation intervals of size $2^{-i}\lambda\alpha_0$. Refine G_i as the expected number of packets in an interval that has been split i times:

$$G_i = 2^{-i}\lambda\alpha_0 \tag{3.16}$$

To find transition probabilities and show Figure 3.10 as a Markov chain, we are only interested in one period of collision resolution period. The upper half of node (R,0) acts as the starting state and the lower half acts as the final state. $P_{(R,0)}$ is the parameter of an idle or success in the first slot. Since the number of packets in the initial allocation interval is Poisson with mean G_0

$$P_{(R,0)} = (1+G_0)e^{-G_0} \tag{3.17}$$

(L,1) is entered after a collision in (R,0) with parameter $1 - P_{(R,0)}$, and X_L, X_R are the number of packets in the new allocation interval L and R and independent Poisson random variables of mean G_1

$$P_{(L,1)} = \frac{P\{X_L\}P\{X_R \geq 1\}}{P\{X_L + X_R \geq 2\}} = \frac{G_1 e^{-G_1}(1 - e^{-G_1})}{1 - (1+G_0)e^{-G_0}} \tag{3.18}$$

(R,1) is entered if and only if the transition above takes place. Thus, the parameter of success $P_{R,1}$ is as follows:

$$P_{(R,1)} = \frac{P\{X_R = 1\}}{P\{X_R \geq 1\}} = \frac{G_1 e^{-G_1}}{1 - e^{-G_1}} \tag{3.19}$$

We can prove that

$$P_{(L,i)} = \frac{G_i e^{-G_i}(1 - e^{-G_i})}{1 - (1 + G_{i-1})e^{-G_{i-1}}} \tag{3.20}$$

$$P_{(R,i)} = \frac{G_i e^{-G_i}}{1 - e^{-G_i}} \tag{3.21}$$

The parameters $P(L,i)$ $P(R,i)$ that (L,i) (R,i) entered before returning to (R,0) can be calculated iteratively from (R,0)

$$P(L, 1) = 1 - P_{R,0} \tag{3.22}$$

$$p(R, i) = P_{(L,i)}p(L, i), I \geq 1 \tag{3.23}$$

$$P(L, I+1) = (1 - P_{L,i})\, p(L, i) + (1 - P_{(R,i)})\, p(R, i),\ I \geq 1 \tag{3.24}$$

Let k be the number of slots in a CRP. Thus, k is the number of states visited in the chain, including (R,0) before returning to (R,0)

$$E\{k\} = 1 + \sum_{i=1}^{\infty}[P(L, i) + P(R, i)] \tag{3.25}$$

For the assumed α_0, the change in $T(k)$ is at most α_0. But if left-hand intervals are returned to the waiting interval, the change is less than α_0. Let f be the fraction of α_0 returned in this way over a CRP so that $\alpha_0(1 - f)$ is the change in $T(k)$. The parameter of a collision in state (L, i) is the parameter that the left hand interval in (L, i) contains at least two packets given that the right and left intervals together contain at least two.

$$P\{e|(L, i)\} = \frac{1 - (1+G_i)e^{-G_i}}{1 - (1+G_{i-1})e^{-G_{i-1}}} \tag{3.26}$$

The fraction of the original interval returned on such a collision is 2^{-i}

$$E\{f\} = \sum_{i=1}^{\infty} p(L,i)P\{e|(L,i)\}2^{-i} \tag{3.27}$$

$E\{f\}$ and $E\{k\}$ are functions only of $G_i = \lambda\alpha_0 2^{-i}$, and hence are functions only of the product $\lambda\alpha_0$. For large i, $P_{L,i} \to 1/2$, $p(L,i)$ and $p(R,i)$ tend to zero.

Define drift D to be the expected change in the time backlog, $k - T(k)$, over a CRP. This is the expected number of slots in a CRP less the expected change in $T(k)$:

$$D = E\{k\} - \alpha_0(1 - E\{f\}) \tag{3.28}$$

The drift is negative if $E\{k\} < \alpha_0(1 - E\{f\})$ or equivalently

$$\lambda < \frac{\lambda\alpha_0(1 - E\{f\})}{E\{k\}} \tag{3.29}$$

$$E\{k\} = 0.4871 \text{ at } \lambda\alpha_0 = 1.266 \tag{3.30}$$

It means that the maximal achievable throughput of the FCFS CRP algorithm is 0.4871, which was later shown by S. Verdu with the precise maximisation reaching 0.48760.

3.3 Carrier Sensing

As a matter of fact, if nodes can collect some kind of information from the multiple access channel, it is intuitive that this results in better performance multi-access protocols. The most straightforward approach might be *carrier sensing*, in which a node listens to a possible transmission in the channel first, and then transmits. Consequently, carrier sensing can be called *listen-before-transmission* (LBT).

To model carrier sensing, let β be the normalised propagation and detection delay required for all sources to detect an idle channel after a transmission ends:

$$\beta = \frac{\tau c}{L} \tag{3.31}$$

where τ is this time in seconds; c is the raw channel bit rate; and L is the number of bits in a data packet.

3.3.1 CSMA Slotted ALOHA

The simple way to improve ALOHA is to introduce the concept of *carrier sensing multiple-access* (CSMA). The major differences between CSMA slotted ALOHA and ordinary slotted ALOHA are

- idle slots in CSMA have a duration β;
- if a packet arrives at a node while a transmission is in progress, the packet is regarded as backlogged and begins transmission with parameter q_r after each subsequent idle slot; packets arriving during an idle slot are transmitted in the next slot as usual.

There are generally three types of carrier sensing mechanisms:

- *non-persistent:* as per the above introduction;
- *Persistent (or 1-persistent):* all arrivals during a busy slot simply postpone transmission to the end of that slot;

- *p-Persistent:* collided packets and new packets waiting for the end of a busy period use different parameters for transmission.

In non-persistent versions of CSMA, a user that generated a packet and found the channel to be busy refrains from transmitting the packet and behaves exactly as if its packet collided, i.e., it schedules (randomly) the retransmission of the packet to some point in the future. Please note that non-persistent CSMA is equivalent to *p*-persistent with an appropriate value of *p*.

3.3.1.1 Throughput Analysis

Assume the following:

- An infinite population of users generating packets according to a Poisson process with parameter λ.
- All packets of the same length with T seconds of transmission. When observing the channel, packets (new and retransmitted) arrive according to a Poisson process with parameter g packets/sec.
- τ: maximum propagation delay.
  $a = \tau/T$: normalised propagation time.

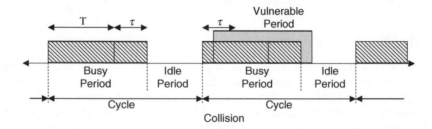

Figure 3.11   Non-persistent CSMA

From Figure 3.11, we observe along the time axis that a succession of cycles each consists of a transmission period followed by an idle period. $\tilde{B}$ is the duration of the busy period with mean B; $\tilde{U}$ is the duration within the transmission period in which a successful packet is being transmitted, with mean U; and $\tilde{I}$ is the duration of the idle period with mean I.

The cycle length is $\tilde{B} + \tilde{I}$. The throughput is $S = U/(B + I)$.

The duration of the idle period is the same as the duration between the end of transmission and the arrival of the next packet. Because packet scheduling is memoryless,

$$
\begin{aligned}
F_I(x) &= P_r[\tilde{I} \le x] = 1 - P_r[\tilde{I} > x] \\
&= 1 - P_r[\text{no packet scheduled during } x] \\
&= 1 - e^{-gx}
\end{aligned}
\tag{3.32}
$$

$\tilde{I}$ is exponentially distributed with mean $I = 1/g$.

To find U,

$$
\tilde{U} = \begin{cases} T, & \text{for successful period} \\ 0, & \text{for unsuccessful period} \end{cases}
\tag{3.33}
$$

If P_{SUCC} denotes the parameter that a transmitted packet is successful, then

$$U = E[\tilde{U}] = T \cdot P_{SUCC} \tag{3.34}$$

$$P_{SUCC} = P_r[\text{No arrival in } [t,\ t+\tau]] \tag{3.35}$$
$$= e^{-gt}$$

then,

$$U = Te^{-gt} \tag{3.36}$$

To compute B, let $\tilde{Y}$ be a random variable such that $t + \tilde{Y}$ denotes the time at which the last interfering packet was scheduled within a transmission period that started at time t (see Figure 13.12).

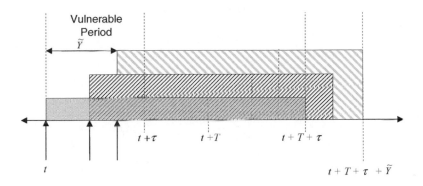

Figure 3.12 Illustration for computing B

Clearly, $\tilde{Y} < \tau$ and $\tilde{Y} = 0$ for a successful transmission period.

$$\tilde{B} = T + \tau + \tilde{Y} \tag{3.37}$$

The period $\tilde{Y}$ is characterised by the fact that no other packet is scheduled for transmission during $[t + \tilde{Y}, t + \tau]$. Otherwise, the packet that is transmitted at $t + \tilde{Y}$ would not have been the last packet to be transmitted in $[t + \tilde{Y}, t + \tau]$.

$$
\begin{aligned}
F_Y(y) \ &= P_r[\tilde{Y} \le y] \\
&= P_r[\text{no packet arrival during } \tau - y] \\
&= \begin{cases} e^{-g(\tau - y)}, & 0 \le y \le \tau, \\ 0, & y < 0 \\ 1, & y \ge \tau \end{cases}
\end{aligned} \tag{3.38}
$$

$$f_Y(y) = e^{-gt}\delta(y) + ge^{-g(\tau - y)} \tag{3.39}$$

$$E[\tilde{Y}] = \tau - \frac{1 - e^{-gt}}{g} \tag{3.40}$$

Therefore,

$$B = E[T+\tau+\tilde{Y}] = T+2\tau+\frac{1-e^{-gt}}{g} \tag{3.41}$$

$$S = \frac{gTe^{-g\tau}}{g(T+2\tau)+e^{-gt}} \tag{3.42}$$

We normalise the quantities with respect to the packet transmission time (Figure 3.13). Let G denote the average scheduling rate of packets measured in packets per packet transmission time, i.e.,

$$G = gT \tag{3.43}$$

$$S = \frac{Ge^{-aG}}{G(1+2a)+e^{-aG}} \tag{3.44}$$

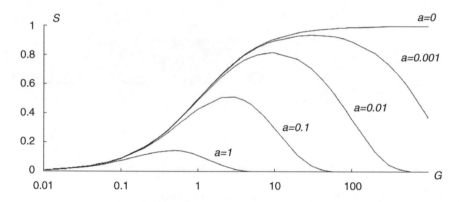

Figure 3.13 Normalisation of the quantities with respect to the packet transmission time

We note when $a \to 0$,

$$S \to \frac{G}{(1+G)}$$

which does not decrease to zero with increasing load. Finally, CSMA can be viewed as a member of the ALOHA family of protocols, and thus is accompanied by instability.

3.3.1.2 1-Persistent CSMA

A node/user, who senses the channel and finds it busy, persists in waiting and transmits as soon as the channel becomes idle. Consequently, the channel is always used if there is a user with a packet (see Figure 3.14).

A transmission period starts either with the transmission of a single packet (type 1) or with the transmission of at least two packets (type 2). A transmission period that follows an idle period is always a type 1 transmission period. An idle period is called type 0. Define the state of the system at the beginning of a transmission period to be the type of that transmission period. These states correspond to a three-state Markov chain embedded at the beginning of the transmission periods (Figure 3.15).

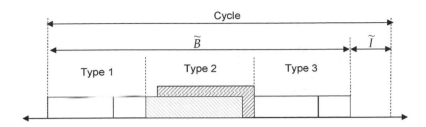

Figure 3.14   Operation of 1-persistent CSMA

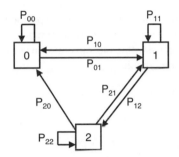

Figure 3.15   Three states Markov chain of 1-persistent CSMA

Only type 1 transmission periods may result in a successful transmission. However, for a type 1 transmission period to be successful, it is necessary that no packets arrive during its first τ seconds (vulnerable period). The parameter of this event is $e^{-g\tau}$.

Let π_i, $i = 0, 1, 2$, be the stationary parameter of being in state i. Let $\tilde{T}_i$, $i = 0, 1, 2$, be a random variable representing the length of type i transmission period and $E[\tilde{T}_i] = T_i$;

$$S = T \cdot \frac{\pi \cdot e^{-g\tau}}{2 \sum_{i=0}^{2} \pi_i T_i} \tag{3.45}$$

The idle period in 1-persistent CSMA is identical to that of non-persistent CSMA, i.e., exponentially distributed with mean $1/g$. $T_0 = 1/g$. $\tilde{T}_1$ and $\tilde{T}_2$ have the same distribution because the length of a transmission period with a single packet or two or more packets is determined only by the time of arrival of the last packet (if any) within the vulnerable period and does not depend at all on the type of transmission period. Let a transmission period start at time t and let $\tilde{Y}$ be a random variable representing the time (after t) of the last packet that arrived during the vulnerable period of a transmission period that started at time t ($\tilde{Y} = 0$ if no packet arrives [t, $t + \tau$]):

$$\tilde{T}_1 = \tilde{T}_2 = T + \tau + \tilde{Y} \tag{3.46}$$

$$f_{\tilde{Y}}(y) = e^{-g\tau}\delta(y) + ge^{-g(\tau - y)} \tag{3.47}$$

$$E[\tilde{Y}] = \tau - \frac{1 - e^{-g\tau}}{g} \tag{3.48}$$

$$T_1 = T_2 = T + \tau + E[\tilde{Y}] = T + \tau - \frac{1 - e^{-gy}}{g} \tag{3.49}$$

From the state diagram,

$$\pi_0 = \pi_1 p_{10} + \pi_2 p_{20} \tag{3.50}$$

$$\pi_1 p_{12} = \pi_2 (p_{21} + p_{20}) \tag{3.51}$$

$$\pi_0 + \pi_1 + \pi_2 = 1 \tag{3.52}$$

When a type 1 or 2 transmission period starts, the type of the next transmitter period is determined only by those packets scheduled for transmission after the transmission period begins. If no packet arrives within the transmission period, the next transmission period will be of type 0. If a single packet arrives within the transmission period, at least τ seconds after it begins, the next transmission period will be of type 1. If at least two packets arrive within the transmission period, at least τ seconds after it begins, the next transmission period will be type 2:

$$p_{1j} = p_{2j} \tag{3.53}$$

We have

$$\pi_0 = \frac{p_{10}}{1 + p_{10}} \quad \pi_1 = \frac{p_{10} + p_{11}}{1 + p_{10}} \quad \pi_2 = \frac{1 - p_{10} - p_{11}}{1 + p_{10}} \tag{3.54}$$

Assume that the type 1 transmission period starts at time t. Conditioning on $\tilde{Y} = y$, the next transmission period will be of type 0 only if no packet is scheduled in $[t + \tau, t + y + T + \tau]$. The parameter of this event is $e^{-g(T+y)}$. Unconditioning,

$$
\begin{aligned}
p_{10} &= \int_0^\tau e^{-g(T+y)} f_Y(y) \\
&= \int_0^\tau e^{-g(T+y)} [e^{-g\tau} \delta(y) + g e^{-g(\tau-y)}] dy \\
&= g e^{-g(T+\tau)} \left[T + g\tau \left(T + \frac{\tau}{2} \right) \right]
\end{aligned}
\tag{3.55}
$$

We can get

$$S = \frac{G e^{-G(1+2a)} \left[1 + G + aG(1 + G + \frac{aG}{2}) \right]}{G(1 + 2a) - (1 - e^{-aG}) + (1 + aG) e^{-G(1+a)}} \tag{3.56}$$

3.3.2 Slotted CSMA

Let the slot size be equal to τ, which means that any transmission starting at the beginning of a slot reaches (and could be sensed by) each and every user by the end of that slot. These slots maybe regarded as mini slots. Users are restricted to start transmissions only at the mini-slot boundary. Assume τ includes propagation delay and carrier sensing time. We also assume that T is a multiple of τ. ($1/a$ is an integer.)

When a packet is scheduled for transmission at a given time, the user waits to the beginning of the next mini-slot, at which time it senses the channel and transmits its packet if idle. (The packet occupies the channel $1/a + 1$ mini-slot before all other users have received it.) If the channel is sensed busy, the corresponding CSMA protocol is applied.

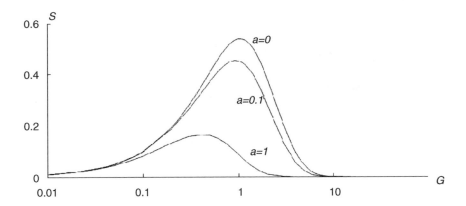

Figure 3.16 Performance of 1-persistent CSMA

3.3.2.1 Throughput of Slotted Non-persistent CSMA

The idle period has k mini-slots if there are no arrivals in the first $(k-1)$ mini-slot and at least one arrival in the kth mini-slot:

$$p\{\tilde{I} = k\tau\} = (e^{-g\tau})^{k-1}(1 - e^{-g\tau}) \quad k = 1, 2, \ldots \tag{3.57}$$

$$I = \frac{\tau}{1 - e^{-g\tau}} \tag{3.58}$$

Both successful and unsuccessful transmission periods last for $T + \tau$ seconds. A collision occurs if two or more packets arrive within the same mini-slot and are scheduled for transmission in the next mini-slot. A busy period will contain k transmission periods if there is at least one arrival in the last mini-slot of each of the first $k-1$ transmission periods and no arrival in the last mini-slot of the kth transmission period:

$$P_r\{\tilde{B} = k(T+\tau)\} = (1 - e^{-gt})^{k-1}e^{-gt}, k = 1, 2, \ldots \tag{3.59}$$

$$B = \frac{T+\tau}{e^{-gt}} \tag{3.60}$$

Similar to earlier work,

$$U = E\{\tilde{U}\} = T \cdot \frac{B}{T+\tau} \cdot P_{succ} \tag{3.61}$$

$$
\begin{aligned}
P_{succ} &= P_r\{\text{successful transmission period}\} \\
&= P_r\left\{ \begin{array}{l} \text{single arrival in last mini-slot} \mid \text{source} \\ \text{before the transmission period} \mid \text{arrival} \end{array} \right\} \\
&= \frac{P_r\{\text{single arrival in a mini-slot}\}}{P_r\{\text{some arrivals in a mini-slot}\}} \\
&= \frac{g\tau e^{-g\tau}}{1 - e^{-g\tau}}
\end{aligned} \tag{3.62}
$$

$$S = \frac{U}{B+I} = \frac{\dfrac{T}{e^{-gt}} \dfrac{gte^{-gt}}{1-e^{-g\tau}}}{\dfrac{T+\tau}{e^{-gt}} + \dfrac{\tau}{1-e^{-g\tau}}} = \frac{Tg\tau e}{T+\tau - Te^{-gt}} \tag{3.63}$$

$$S = \frac{aGe^{-aG}}{1+a-e^{-aG}} \tag{3.64}$$

A final check can be done by considering the asymptotic case

$$S_{a \to 0} = \frac{G}{1+G} \text{ as unslotted case} \tag{3.65}$$

3.3.2.2 Throughput of Slotted 1-Persistent CSMA

For the 1-persistent case,

$$I = \frac{\tau}{1-e^{-gt}} \tag{3.66}$$

$$P_r\{\tilde{B} = k(T+\tau)\} = [1 - e^{-g(T+\tau)}]^{k-1}e^{-g(T+\tau)}, \ k = 1, 2, \ldots \tag{3.67}$$

Since a busy period will contain k transmission periods if at least one packet arrives in each of the first $k-1$ transmission periods and no packet arrives in the kth transmission period:

$$B = \frac{T+\tau}{e^{-g(T+\tau)}} \tag{3.68}$$

The parameter of success in the first transmission period in a busy period, $P_{SUCC,1}$, is different from the successful parameter in any other transmission period within the busy period, $P_{SUCC,2}$:

$$
\begin{aligned}
P_{SUCC,1} &= P_r\{\text{successful transmission in the first transmission period of a busy period}\} \\
&= P_r\{\text{single arrival in a (last) mini-slot}|\text{at least one arrival}\} \\
&= \frac{gte^{-gt}}{1-e^{-gt}}
\end{aligned} \tag{3.69}
$$

$$
\begin{aligned}
P_{SUCC,2} &= P_r\{\text{successful transmission in non-first period in a busy period}\} \\
&= P_r\{\text{single arrival in a transmission period}|\text{at least one arrival}\} \\
&= \frac{g(T+\tau)e^{-g(T+\tau)}}{1-e^{(T+\tau)}}
\end{aligned} \tag{3.70}
$$

We therefore have

$$U = T\left[P_{SUCC,1} + \left(\frac{B}{T+\tau} - 1\right) \cdot P_{SUCC,2}\right] \tag{3.71}$$

$$S = \frac{U}{B+I} = \frac{Ge^{-(1+a)G}(1+a-e^{-aG})}{(1+a)(1+e^{-aG}) + ae^{-(1+a)G}} \tag{3.72}$$

$$S_{a \to 0} = \frac{Ge^{-G}(1+G)}{G+e^G} \tag{3.73}$$

We compile the performance of several CSMA family protocols in Figure 3.17. We note that slotted systems provide a slightly better performance, although in practical terms this is negligible. Of course, slotted systems are generally easier to analyse.

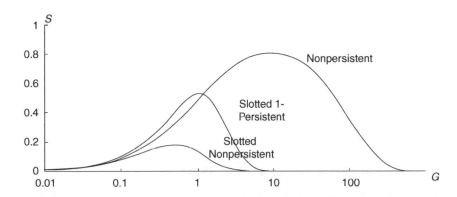

Figure 3.17   Comparison of multiple access protocols

3.3.3 Carrier Sense Multiple Access with Collision Detection (CSMA/CD)

To improve the performance of CSMA, we must reduce the cycle length. Shortening the idle period is possible by 1-persistent protocols but they perform poorly in most situations. Hence, shortening the duration of unsuccessful transmission periods is the only way to improve performance. In addition to the ability to sense carrier, users can detect interference among several transmissions while the transmission is in progress and abort transmission of their collided packets. If this can be done sufficiently fast, the duration of an unsuccessful transmission would be shorter than that of a successful one. This variation of CSMA is known as *carrier sense multiple access with collision detection* or *CSMA/CD* (Figure 3.18).

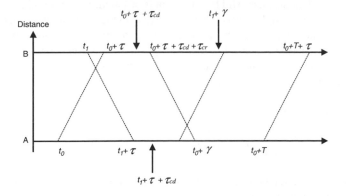

Figure 3.18   Illustration of CSMA/CD

Suppose that user A starts transmission at t_0 when the channel is idle, and then the transmission reaches user B at $t_0 + \tau$. Suppose B initiates a transmission at time $t_1 < t_0 + \tau$ (when B still senses idle). Since it takes τ_{cd} to detect the collision, user B positively determines the collision at $t_0 + \tau + \tau_{cd}$. Upon detection of a collision in many LANs, such as Ethernet, every user initiates a consensus reinforcement procedure, which is manifested by jamming the channel with a collision signal for duration τ_{cr} to ensure that all network users indeed determine that a collision takes place. At $t_0 + \tau + \tau_{cd} + \tau_{cr}$ user B completes the consensus reinforcement procedure, which reaches A at $t_0 + 2\tau + \tau_{cd} + \tau_{cr}$. From user A's standpoint, this transmission period lasted

$$\gamma = 2\tau + \tau_{cd} + \tau_{cr} \tag{3.74}$$

By a similar calculation, B completes this transmission period at time $t_1 + \gamma$. The channel is therefore busy for a period of $\gamma + t_1 - t_0$. In the worst case, in an unsuccessful transmission period, the channel remains busy for a duration of $\gamma + \tau$. Denote $\tilde{X}$ as the length of the transmission period:

$$\tilde{X} = \begin{cases} T + \tau, & \text{successful transmission period} \\ \gamma + \tau, & \text{unsuccessful transmission period} \end{cases} \tag{3.75}$$

3.3.3.1 Throughput of Slotted Non-persistent CSMA/CD

A cycle consists of a busy period (both successful and unsuccessful transmission) and an idle period. The distribution of the idle period is identical to that computed for slotted non-persistent CSMA:

$$P\{\tilde{I} = k\tau\} = (e^{-g\tau})^{k-1}(1 - e^{-g\tau}), \ k = 1, 2, \ldots \tag{3.76}$$

$$I = \frac{\tau}{1 - e^{-g\tau}} \tag{3.77}$$

We assume γ and T are integer multiples of τ.

$$P_{SUCC} = P_r\{\text{single transmisson}|\text{at least one transmission}\}$$
$$= \frac{g\tau e^{-g\tau}}{1 - e^{-gt}} \tag{3.78}$$

Each transmission period that contains a successful transmission is of length $T + \tau$ seconds while a transmission period with an unsuccessful transmission is of length $\gamma + \tau$ seconds. A busy period will contain l transmission periods if there was at least one arrival in the last mini-slot of each of the first $l - 1$ transmission periods and no arrival in the last mini-slot of the lth transmission period. Therefore, the parameter that the busy period contains in exactly l $(l \geq 1)$ transmission periods is $e^{-g\tau}(1 - e^{-g\tau})^{l-1}$ and the average number of transmission periods within the busy period is $1/e^{-g\tau}$. The parameter distribution of the length of the busy period is

$$P_r\{\tilde{B} = k(T+\tau) + (l-k)(\gamma+\tau)\}$$

$$= e^{-g\tau}(1 - e^{-g\tau})^{l-1} \binom{l}{k} P_{succ}^k (1 - P_{succ})^{l-k}, \text{ for } l = 1, 2, \ldots \text{ and } k = 0, 1, \ldots, l \tag{3.79}$$

When l corresponds to the total number of transmission periods in the busy period and k corresponds to the successful transmission periods:

$$B = \sum_{l=1}^{\infty} \sum_{k=0}^{l} [k(T+\tau) + (l-k)(\gamma+\tau)] P\{\tilde{B} = k(T+\tau) + (l-k)(\gamma+\tau)\}$$

$$= \frac{[P_{SUCC}(T+\tau) + (1 - P_{succ})(\gamma+\tau)]}{e^{-g\tau}} \tag{3.80}$$

$$P_r\{\hat{U} = kT\}$$

$$= P_r\{k \text{ successful transmission periods in a busy period}\}$$

$$= \sum_{l=k}^{\infty} P_r\{\tilde{B} = k(T + \tau) + (l - k)(\gamma + \tau)\}$$

(3.81)

$$U = \sum_{k=0}^{\infty} k \cdot T \cdot P_r\{\hat{U} = kT\} = \frac{T}{e^{-g\tau}} P_{succ}$$

(3.82)

Throughput in a normalised form is

$$S = \frac{U}{B + I} = \frac{aGe^{-aG}}{aGe^{-aG} + (1 - e^{-aG} - aGe^{-aG})\gamma' + a}$$

(3.83)

where $\gamma' = \gamma/T$. When $\gamma' = 1$, S is identical to slotted non-persistent CSMA.

3.3.3.2 Throughput of Slotted 1-Persistent CSMA/CD

Notice that the success or failure of a transmission period in the busy period depends only on the length of the preceding transmission period, except for the first transmission period that depends on arrivals during the preceding mini-slot. If $\hat{X}_i$ denotes the duration of the ith transmission period in the busy period, then the duration of the $(i + 1)$st transmission period depends only on $\hat{X}_i$ since the type of the ith transmission period (success or collision) is determined by the number of arrivals during the previous transmission period, which depends on the previous transmission period which in turn depends on (only) its duration. Hence, given that a transmission period is of length x, the length of the remainder of the busy period is a function of x whose average is denoted by $B(x)$. Similarly, the average time that the channel is carrying successful transmissions in the remainder of the busy period is denoted by $U(x)$. Let $A_i(x)$ be the parameter of i arrivals during a period of x. Under Poisson assumption, $a_i(x) = (gx)^i e^{-gx}/i!$.

$$B(x) = \frac{a_1(x)}{1 - a_0(x)}[T + \tau + [1 - a_0(T + \tau)]B(T + \tau)] \text{ successful}$$

$$+ \left[1 - \frac{a_1(x)}{1 - a_0(x)}\right][\gamma + \tau + [1 - a_0(\gamma + \tau)]B(\gamma + \tau)] \text{ unsuccessful}$$

(3.84)

By setting $x = T + \tau, \gamma + \tau$, we obtain two equations that can be solved easily. $B(x)|_{x=\tau}$ can yield the expected length of a busy period. In a similar manner,

$$U(x) = \frac{a_1(x)}{1 - a_0(x)}[T + [1 - a_0(T + \tau)]U(T + \tau)] + \left[1 - \frac{a_1(x)}{1 - a_0(x)}\right][[1 - a_0(\gamma + \tau)]U(\gamma + \tau)]$$

(3.85)

The average length of an idle period is $\tau/(1 - e^{-g\tau})$:

$$S = \frac{U}{B + I} = \frac{U(\tau)}{B(\tau) + \tau/(1 - e^{-g\tau})}$$

(3.86)

Though not in closed form, the computation is straightforward shown as Figure 3.19.

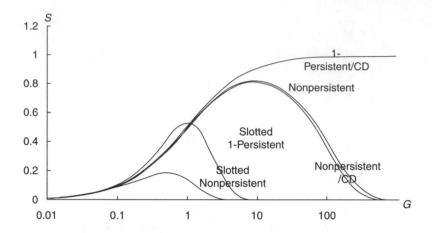

Figure 3.19   Comparison of CSMA protocols with CD

Please note that

- The 'gap' between the non-persistent and 1-persistent CSMA, when collision detection is used, has narrowed down.
- The good performance of 1-persistent/CD in light load is the reason to use it in LANs.

3.4 Routing

The *Routing* algorithm is typically referred to as the network layer protocol that guides packets through the communication subnet to their correct destination. In a datagram net, two successive packets of the same user pair may travel along different routes, and a routing decision is necessary for each individual packet. In a virtual circuit net, a routing decision is made when each virtual circuit is set up.

Two main functions performed by a routing algorithm are the selection of routes for various origin-destination pairs and the delivery of messages to their correct destination once the routes are selected. Two primary performance measures that are substantially affected by the routing algorithm are throughput and average packet delay.

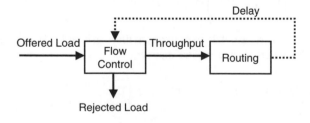

Figure 3.20   Interaction of routing and flow control

As the routing algorithm is more successful in keeping delay low, the flow control algorithm allows more traffic into the network (Figure 3.20).

We will now briefly discuss the routing methodology for wide area networks (WANs), which can be classified in two ways. If we consider the way to determine routing, we can classify routing algorithms as *centralised* or *distributed*. If we consider the routing paths, we can classify routing into *static* routing and *adaptive* routing, where adaptive routing searches good routes according to the most updated network information.

3.4.1 Flooding and Broadcasting

Broadcasting could be used as a primitive form of routing packets from a single transmitter to a single receiver or a subset of receivers. This is generally rather inefficient but may be sensible because it is simple or when the receivers' locations within the net are unknown. Flooding operates as follows. The origin node sends its information in the form of a packet to its neighbours. The neighbours relay it to their neighbours until the packet reaches all nodes in the network. Two additional rules limit the number of packet transmission:

- A node will not relay the packet back to the node from which the packet was obtained.
- A node will transmit the packet to its neighbours at most once.

This can be ensured by including the ID of the original node and a sequence number that is incremented with each new packet issued by the origin node, and by storing the highest sequence number received for each origin node and not delaying packets with sequence numbers that are less than or equal to the one stored.

Another broadcasting method is based on the use of a spanning tree. A spanning tree is a connected sub-graph of the network that includes all nodes and has no cycles. Broadcasting on a spanning tree is more efficient than flooding. It requires only $N - 1$ transmissions per packet broadcast, where N is the number of nodes.

3.4.2 Shortest Path Routing

More practical routing algorithms are based on the notion of a shortest path between two nodes. Here, each link is assigned a positive number called its length. A link may have different lengths in different directions. Each path between two nodes has a length equal to the sum of the lengths of its links. A shortest path routing algorithm routes each packet along a minimum length path between origin and destination nodes. More generally, the length of a link may depend on its transmission capacity and its projected traffic load. A shortest path should contain relatively few and uncongested links and therefore be desirable for routing.

A more sophisticated alternative is to allow the length of each link to change over time and to depend on the prevailing congestion level of the link. Then a shortest path may adapt to temporary overloads and route packets around points of congestion. Undesirable oscillations are possible.

An important distributed algorithm for calculating the shortest paths to a given destination, known as the Bellman-Ford method, has the form

$$D_i = \min_j [d_{ij} + D_j] \qquad (3.87)$$

where D_i is the estimated shortest distance of node i to the destination and d_{ij} is the length of link (i, j). In practice, the Bellman-Ford method can be implemented as an iterative process.

3.4.3 Optimal Routing

Shortest path routing has two drawbacks:

- It uses only one path per pair, thereby potentially limiting the throughput of the network.
- Its capability to adapt to changing traffic conditions is limited by its susceptibility to oscillations.

Optimal routing based on the optimisation of an average delay such as measure of performance can eliminate both by splitting any origin-destination pair traffic at strategic points and by shifting traffic gradually between alternative paths.

3.4.4 Hot Potato (Reflection) Routing

In networks where storage space at each node is limited, it may be important to modify the routing algorithm so as to minimise buffer overflow and the attendant loss of packets. The idea here is for nodes to get rid of their stored packets as quickly as possible, transmitting them on whatever link happens to be idle – not necessarily one that brings them closer to their destination.

3.4.5 Cut-through Routing

An incentive exists to split a long message into several smaller packets in order to reduce the delay of the message on multiple link paths, taking advantage of pipelining. This leads to cut-through routing, whereby a node can start relaying any portion of a packet to another node without waiting to receive the packet in its entirety. The delay can be reduced by as much as a factor on a path with n links. Error detection retransmission must be done on an end-to-end basis.

3.4.6 Interconnected Network Routing

As networks started to proliferate, it became necessary to interconnect them using various interface devices. In the case of WANs, the interfaces are called gateways and usually perform fairly complicated network layer tasks such as routing. The devices used for interconnection of LANs at the MAC layer to perform a primitive form of routing are known as bridges. LANs can also be connected with each other or with WANs using more sophisticated devices called routers (see Figure 3.21).

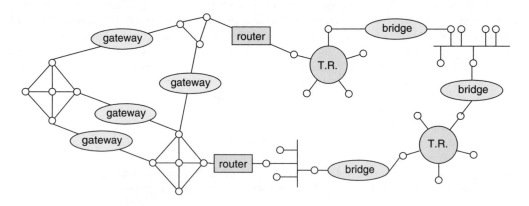

Figure 3.21   Illustration of interconnected network routing

3.4.7 Shortest Path Routing Algorithms

We will pay special attention to shortest path routing algorithms, which will be very useful in networking cognitive radios.

3.4.7.1 Bellman-Ford Algorithm

Suppose that node 1 is the destination node and consider the problem of finding a shortest path from every node to node 1. We assume that there exists at least one path from every node to the destination. Denote $d_{ij} = \infty$ if (i, j) is not an arc of the graph. A shortest walk from node i to node 1, subject to the constraint that the walk contains at most h arcs and goes through node 1 only once, is referred to as the shortest walk and its length is denoted by D_i^h $D_i^h = 0, \forall h$ (see Figure 3.22).

D_i^h can be generated by the iteration

$$D_i^{h+1} = \min_j[d_{ij} + D_j^h], \; \forall i \neq 1 \tag{3.88}$$

starting from the initial conditions

$$D_i^0 = \infty, \forall i \neq 1 \tag{3.89}$$

We claim that the Bellman-Ford algorithm first finds the one-arc shortest walk lengths, then two-arc ones, and so forth. We can also argue that the shortest walk lengths are equal to the shortest path lengths, under the additional assumption that all cycles not containing node 1 have nonnegative length. We say that the algorithm terminates after h iterations if

$$D_i^h = D_i^{h-1}, \forall i \tag{3.90}$$

Proposition 3.1: Consider the Bellman-Ford algorithm with the initial conditions $D_i^0 = \infty, \forall \neq 1$. Then:

1. D_i^h generated by the algorithm are equal to the shortest ($\leq h$) walk lengths from node i to node 1.
2. The algorithm terminates after a finite number of iterations if and only if all cycles not containing node 1 have nonnegative length. Furthermore, if the algorithm terminates, it does so after at most $h \leq N$ iterations, and at termination, D_i^h is the shortest path length from i to 1.

3.4.7.2 Dijkstra's Algorithm

This algorithm requires that all the arc lengths are nonnegative. Its worst-case computational requirements are considerably less than those of the Bellman-Ford algorithm (see the comparison in Figure 3.23). The general idea is to find the shortest paths in order of increasing path length. The shortest of the shortest paths to node 1 must be the single-arc path from the closest neighbour of node 1, because any multiple-arc path cannot be shorter than the first arc length because of the nonnegative-length assumption. The next shortest of the shortest paths must either be the single-arc path from the next closest neighbour of 1 or the shortest two-arc path through the previously chosen node, and so on.

We can view each node i as being labelled with an estimate D_i of the shortest path length to node 1. When the estimate becomes certain, we regard the node as being permanently labelled and keep track of this with a set of permanently labelled nodes. The node added to P at each step will be the closest to node 1 out of those that are not yet in P. The detailed algorithm is described as follows.

Initially, $P = \{1\}, D_1 = 0, D_j = d_{j1}, \forall j \neq 1$
Step 1: (Find the next closest node) Find $i \notin P$

$$D_i = \min_{j \notin P} D_j \tag{3.91}$$

Set $P: PU\{i\}$. If P contains all nodes, stop.

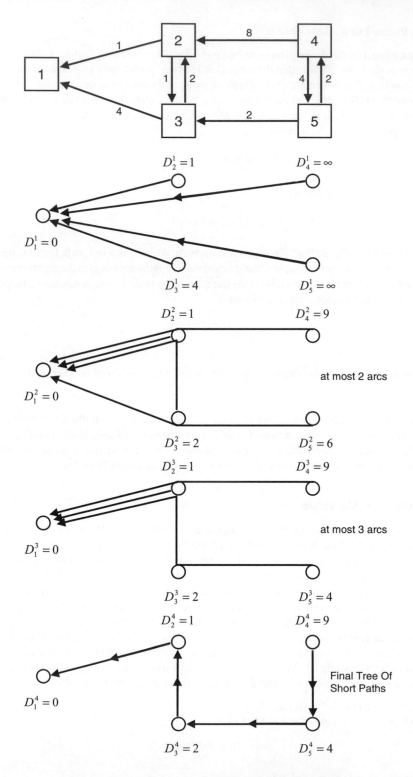

Figure 3.22   Illustration of the Bellman-Ford algorithm

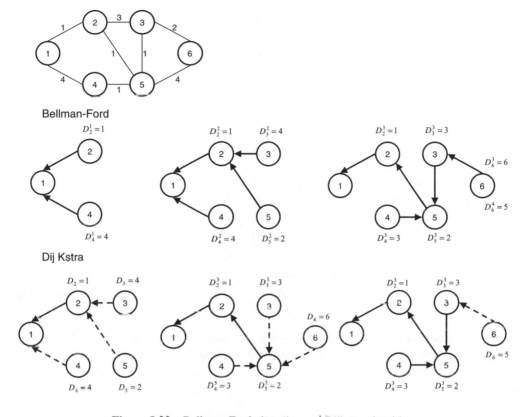

Figure 3.23 Bellman-Ford algorithm and Dijkstra algorithm

Step 2: (Updating of labels) For all $j \notin P$

$$D_j := \min[D_j, d_{ij} + D_i] \tag{3.92}$$

Go to step 1.

We claim that at the beginning of each step 1:

(a) $Di \leq Dj$, $\forall i \in P$ and $j \notin P$.
(b) D_j is, for each node j, the shortest distance from j to 1 using paths with all nodes except possibly j belonging to the set P.

3.4.7.3 Optimal Routing

The monotonically increasing function (cost function) is

$$\sum_{(i,j)} D_{ij}(F_{ij})$$

A frequently used formula is

$$D_{ij}(F_{ij}) = \frac{F_{ij}}{C_{ij} - F_{ij}} + d_{ij}F_{ij} \tag{3.93}$$

where C_{ij} is the transmission capacity of link (i, j) measured in the same unit as F_{ij}. F_{ij} is the total flow carried by link (i, j)

$$F_{ij} = \sum_{\substack{\text{all path } p \\ \text{containing}(i,j)}} x_p \qquad (3.94)$$

where x_p is the flow of path p. For every origin-destination pair w, there are the constraints

$$\sum_{p \in P_w} x_p = r_w, \quad x_p \geq 0, \forall p \in P_w \qquad (3.95)$$

where r_w is the given traffic input of the OD pair and P_w is the set of directed paths of w. In terms of the unknown path flow vector $x = \{x_p | p \in P_W, w \in W\}$, the problem is

$$\text{minimise} \sum_{(i,j)} D_{ij} \left[\sum_{\substack{\text{all paths } p \\ \text{containing}(i,j)}} x_p \right]$$

$$\text{subject to} \sum_{p \in P_W} x_p = r_w, \ \forall w \in W \qquad (3.96)$$

$$x_p \geq 0, \quad \forall p \in P_W, \ w \in W$$

We assume that each D_{ij} is a differentiable function of F_{ij} and is defined in an interval $[0, C_{ij}]$, where C_{ij} is a positive number or infinity. Let x be the vector of path flows x_p

$$D(x) = \sum_{(i,j)} D_{ij} \left[\sum_{\substack{\text{all paths } p \\ \text{contain}(i,j)}} x_p \right] \qquad (3.97)$$

Then,

$$\frac{\partial D(x)}{\partial x_p} = \sum_{\substack{\text{all links}(i,j) \\ \text{on path } p}} D'_{ij} \qquad (3.98)$$

where the first derivatives D'_{ij} are evaluated at the total flows corresponding to x. $\partial D(x)/\partial x_p$ is the length of path p when the length of each link (i, j) is taken to be the first derivative D'_{ij} evaluated at x. Consequently, in what follows, $\partial D(x)/\partial x_p$ is called the first derivative length of path p.

Let $x^* = \{x_p^*\}$ be an optimal path flow vector. Then, if $x_p^* > 0$ for some path p of an OD pair w, we must be able to shift a small amount $\delta > 0$ from path p to any other path p' of the same OD pair without improving the cost; otherwise, the optimality would be violated. To get x_p^*,

$$\delta \frac{\partial D(x^*)}{\partial x'_p} - \delta \frac{\partial D(x^*)}{\partial x_p} \geq 0 \qquad (3.99)$$

$$\frac{\partial D(x^*)}{\partial x'_p} \geq \frac{\partial D(x^*)}{\partial x_p}, \ \forall p' \in P_W \qquad (3.100)$$

Optimal path flow is positive only on paths with a minimum first derivative length. Furthermore, at the optimum, the paths along which the input flow r_w of OD pair w is split must have equal length (and less or equal length to that of all other paths of w). If D_{ij} is arc convex, the above is a necessary and sufficient condition for optimality.

In some datagram networks or in networks where detailed path descriptions are not known at the origin nodes, it may be necessary to maintain, in a routing table at each node i, a routing variable $\phi_{ik}(j)$ for each link (i, j) and destination j. The routing variable is defined as the fraction of all flow arriving at node i, destined for node j, and routed along link (i, j):

$$\phi(k,j) = \frac{f_{ik}(j)}{\sum_m f_{im}(j)}, \; \forall(i,k),j \tag{3.101}$$

where $f_{ik}(j)$ is the flow that travels on link (i, j) and is destined for node j. Given an optimal solution of the routing problem in terms of the path flow variables $\{x_p^*\}$, it is possible to determine $f_{ik}(j)$ and $\phi_{ik}(j)$.

3.5 Flow Control

In most networks, there are circumstances in which the externally offered band is larger than can be handled even with optimal routing. Then, if no measures are taken to restrict the entrance of traffic into the network, queue sizes at bottleneck links will grow and packet delays will increase similarly to a motorway traffic jam.

Generally, a need for flow control arises whenever there is a constraint on the communication rate between two points due to limited capacity of the communication lines or the processing hardware. Flow control may be required in the transport layer, network layer or Internet layer.

Approaches to flow control include the following:

- call blocking;
- packet discarding;
- packet blocking;
- packet scheduling.

The main objectives of flow control are to

- reach a good compromise between throttling sessions (subject to minimum data rate requirements) and keeping average delay and buffer overflow at a reasonable level;
- maintain fairness between sessions in providing the requisite quality of service.

3.5.1 Window Flow Control

A session between a transmitter A and a receiver B is said to be window flow controlled if there is an upper bound on the number of data units that have been transmitted by A and are not yet known by A to have been received by B. The upper bound (a positive integer) is called the window size.

The receiver B notifies the transmitter A that it has disposed of a data unit by sending a special message to A, which is called a permit. Upon receiving a permit, A is free to send one more data unit to B. The general idea in window strategy is that the input rate of the transmitter is reduced when permits return slowly. Therefore, if there is congestion along the communication path of the session, the attendant large delays of the permits cause a natural slowdown of the transmitter's data rate. However, the window strategy has an additional dimension, whereby the receiver may intentionally delay permits to restrict the transmission rate of the session, for example, to avoid buffer overflow.

The window flow control is typically executed end-to-end. In most common cases, the window size is αW, where α and W are some positive numbers. Each time a new batch of α data units is received at the destination node, a permit is sent back to the source allocating a new batch of α data units.

Usually, a numbering scheme for packets and permits is used so that permits can be associated with packets previously transmitted and loss of permits can use a protocol similar to those used for data link control, whereby a packet contains a sequence number and a request number. The latter number can serve as one or more permits for flow control purposes. For example, suppose that node A receives a packet from node B with request number k. Then A knows that B has disposed of all packets sent by A and numbered less than k, and A is free to send those packets up to number $k + W - 1$ that it has not sent yet (W is window size). In such a scheme, both the sequence number and the request number are represented modulo m ($\geq W + 1$) once packet ordering is preserved between transmitter and receiver.

It is assumed that the source node simply counts the number x of packets it has already transmitted but for which it has not yet received back a permit, and transmits new packets only as long as $x < W$.

Consider the case for a flow of packets where the round-trip delay d including round-trip propagation delay, packet transmission time and permit delay is smaller than the time required to transmit the full window of W packets; that is

$$d \leq WX \tag{3.102}$$

where X is the transmission time of a single packet. Then the source is capable of transmitting at full speed of $1/X$, and flow control is not active.

For $d > WX$ and the round-trip delay d is so large that full allocation of W packets can be transmitted before the first permit returns. Assuming that the source always has a packet waiting in queue, the rate of transmission is W/d.

The maximum rate of transmission corresponding to a round-trip delay d is given by

$$r = \min\left\{\frac{1}{x}, \frac{W}{d}\right\} \tag{3.103}$$

Please also note that the end-to-end windows cannot guarantee a minimum communication rate (Figure 3.24). In the mean time, we usually need to optimise the window size in wireless networks.

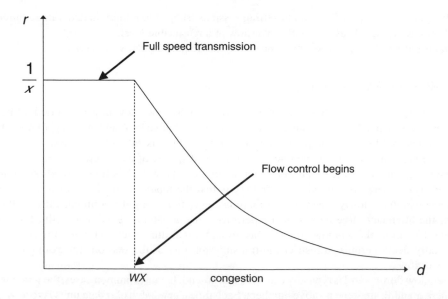

Figure 3.24   End-to-end windows

3.5.2 Rate Control Schemes

Window flow control is not very well suited to high-speed sessions in high speed wide area networks. An alternative form of flow control is based on giving each session a guaranteed data rate, which is commensurate to its needs. This rate should lie within certain limits that depend on the session type. The main considerations in setting input session rates are:

- Delay-throughput tradeoff.
- Fairness: If session rates must be reduced to accommodate some new sessions, the rate reduction must be done fairly, while obeying the minimum rate requirement of each session.

A more appropriate implementation is to admit as many as W packets every W/r seconds for the consideration of bursting traffic. An allocation of W packets is given to each session and a count x of the unused portion of this allocation is kept at the session origin. Packets from the session are admitted into the network as long as $x > 0$. Each time a packet is admitted, the count is decremented by 1, and W/r seconds later, the count is incremented by 1. This time window flow control is similar to window flow control with window size W except that the count is incremented W/r seconds after admitting a packet instead of after a round-trip delay when the corresponding permits return.

A related method that regulates the burstiness of the transmitted traffic somewhat better is the leaky bucket scheme. Here the count is incremented periodically, every $1/r$ seconds up to a maximum of W packets. Another way to view this scheme is to imagine that for each session, there is a queue of packets without a permit and a bucket of permits at the session's source. The packet at the head of the packets with permits is waiting to be transmitted. Permits are generated at the desired input rate r of the session (one permit each $1/r$ seconds) as long as the number in the permit bucket does not exceed a certain threshold W (Figure 3.25).

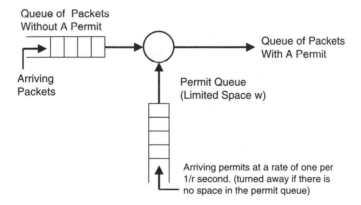

Figure 3.25   Procedure of packet permits

The leaky bucket scheme does not necessarily preclude buffer overflow and does not guarantee an upper bound on packet delay in the absence of additional mechanisms to choose and implement the session rates inside the network.

With proper choice of the bucket parameter, buffer overflow and maximum packet delay can be reduced. In particular, the bucket size W is an important parameter for the performance of the leaky bucket scheme. If W is small, bursty traffic is delayed waiting for permits to become available ($W = 1$ resembles time division multiplexing). If W is too large, long bursts of packets will be allowed into the net; these bursts may accumulate at a congested node downstream and cause buffer overflow. This leads to the idea of dynamic adjustment of the bucket size. It may be combined with the dynamic adjustment of permit rates and merged into a single algorithm. However, in high speed networks, the effectiveness of the feedback control messages from the congested nodes may be diminished because of relatively layer propagation delays. Therefore, some predictive mechanism may be needed to issue control messages before congestion sets in.

3.5.3 Queuing Analysis of the Leaky Bucket Scheme

To precede analysis, we assume the following:

- Packets arrive according to a Poisson process with rate λ.
- A permit arrives every $1/r$ seconds, but if the permit pool contains W permits, the arriving permit is discarded.

We view this system as a discrete time Markov chain with states $0, 1. \ldots$ The states $i = W + 1, W + 2 \ldots$ correspond to $1 - W$ packets without permits waiting and no permits available. The states $i = 0, 1, \ldots, W$ correspond to $W - i$ permits available and no packets without permits waiting. The state transitions occur at time $0, 1/r, 2/r \ldots$ just after permit arrival. The parameter of k packet arrivals in $1/r$ seconds is

$$a_k = \frac{e^{-\frac{\lambda}{r}}\left(\frac{\lambda}{r}\right)^k}{k!} \tag{3.104}$$

It can be seen that the transition parameters of the chain are

$$P_{0i} = \begin{cases} a_{i+1}, & i \geq 1 \\ a_0 + a_1, & i = 0 \end{cases} \tag{3.105}$$

for $j \geq 1$

$$P_{ji} = \begin{cases} a_{i-j+1}, & j \leq i+1, \\ 0 & \text{otherwise} \end{cases} \tag{3.106}$$

The global balance equations yield

$$P_0 = a_0 P_1 + (a_0 + a_1) P_0 \tag{3.107}$$

$$P_i = \sum_{j=0}^{i+1} a_{i-j+1} P_j, \ i \geq 1 \tag{3.108}$$

These equations can be solved recursively. The steady-state parameter can be obtained as

$$P_0 = \frac{r - \lambda}{r a_0} \tag{3.109}$$

To see this, note that the permit generation rate averaged over all states is $(1 - P_0 a_0)r$ while the packet arrival rate is λ. The system is stable, that is, the packet queue stays bounded, if $\lambda < r$. The average delay for a packet to obtain a permit is

$$T = \frac{1}{r}\sum_{j=0}^{\infty} P_j \max\{0, j - w\}$$
$$= \frac{1}{r}\sum_{j=w+1}^{\infty} P_j(j - w) \tag{3.110}$$

To obtain a closed-form expression for the average delay needed by a packet to get a permit and also to obtain a better model of the practical system, we modify the leaky bucket model slightly so that permits are generated on a per bit basis. This approximates the real situation where messages are broken up into small packets upon arrival at the source node. Particularly, we assume the following:

- Credit for admission into the network is generated at rate r bits/sec for transmission and the size of the bucket is w bits.
- Messages arrive according to a Poisson process with rate λ and the storage space for messages is infinite. Message lengths are independent and exponentially distributed with mean L bits.

Let $\mu = r/L$ so that $1/\mu$ is the mean time to transmit a message at the credit rate r. Let $c = w/r$ be the time over which credit can be saved up. The state of the system can be described by the number of bits in the queue and the available amount of credit. At time t, let $X(t)$ be either the number of bits in the queue (if it is nonempty) or minus the available credit.

Thus, whenever a message consisting of x bits arrives, $X(t)$ increases by x. (One of these three things happens: the credit decreases by x; the queue increases by x; or the credit decreases to 0 and the queue increases to $X(t) + x$ if $X(t) < 0$ and $X(t) + x > 0$.)

Letting $Y(t) = X(t) + c$, it can be seen that $Y(t)$ is the unfinished work in a fictitious $M/M/1$ queue with arrival rate λ messages/sec and service rate μ. An incoming bit is transmitted immediately if the size of the fictitious queue ahead of it is less than c and is transmitted L seconds earlier than in the fictitious queue otherwise. Focusing on the last bits of messages, we see that if T_i is the system delay in the fictitious queue for the ith message, $\max\{0, T_i - c\}$ is the delay in real queue. The steady-state distribution of the system time T_i at the fictitious queue is

$$P\{T_i \geq \tau\} = e^{-\tau(\mu - \lambda)} \tag{3.111}$$

Let $T'_i = \max\{0, T_i - c\}$ be the delay of the ith packet in the real queue:

$$P\{T'_i \geq \tau\} = \begin{cases} 1, & \tau \leq 0 \\ e^{-(c+\tau)(\mu - \lambda)}, & \tau > 0 \end{cases} \tag{3.112}$$

The average delay of a packet in the real queue is

$$T = \int_0^\infty P\{T'_i \geq \tau\} d\tau = \frac{1}{\mu - \lambda} e^{-c(\mu - \lambda)} \tag{3.113}$$

For more advanced study of computer networks, we recommend good textbooks such as *Data Networks* by Bertsekas and Gallager [1]. The descriptions in this chapter can suffice, however, as the basis for the many concepts of ad-hoc networks used in later chapters of this book.

References

[1] D. Bertsekas, R. Gallager, *Data Networks*, Prentice-Hall, New Jersey, 1992.
[2] N. Abramson,'The ALOHA System – Another Alternative for Computer Communications', *AFIPS. Conf. Proc.* **37**. Montvale. N.J. (ed.), AFIPS Press, in *1970 Fall Joint Computer Conf.*, 1970, pp. 281–285.
[3] F.A. Tobagi, V.B. Hunt, 'Performance Analysis of Carrier Sense Multiple Access with Collision Detection', *Computer Networks*, **4**,November 1980.
[4] G. Frenkel, 'The Grade of Service in Multiple Access Satellite Communication with Remand Assignment' *IEEE Tr. on Communications (Technology)*, **12**(10), 1972, 1681–1685.
[5] L. Kleinrock, F. Tobagi, 'Packet Switching in Radio Channels, Part I – Carrier Sense Multiple Access and Their Throughput – Delay Characteristics', *IEEE Tr. on Communications*, **23**(12), 1975, 1400–1416.

[6] L. Kleinrock, F. Tobagi, 'Packet Switching in Radio Channels, Part II – The Hidden Terminal Problem and Busy Tone Solution', *IEEE Tr. on Communications*, **23**(12), 1975, 1400–1416.

[7] N. Abramson, 'The Throughput of Packet Broadcasting Channels', *IEEE Tr. on Communications*, **25**(1), 1977, 117–128.

[8] J. Hayes, 'An adaptive Technique for Local Distribution', *IEEE Tr. on Communications*, **26**(8), 1978, 1178–1186.

[9] W. Bux, 'Local Area Subnetworks: A Performance Comparison', *IEEE Tr. on Communications*, **29**(10), 1981.

[10] E. Cole, B. Liu, 'Finite Population CSMA-CD Networks', *IEEE Tr. on Communications*, **31**(11), 1983, 1247–1251.

[11] J. Meditch, C.T. Lea, 'Switching and Optimization of the CSMA and CSMA/CD Channels', *IEEE Tr. on Communications*, **31**(6), 1983, 763–774.

[12] H. Takaqi, 'Queuing Analysis of Polling Models', *ACM Computing Surveys*, **20**,1988, 5–28.

[13] T. Kample, 'Optimal Scheduling of Jobs with Exponential Service Times on Identical Parallel Processors', *Operation Research*, **37**,1988, 126–133.

[14] F. Baccelli, M. Malcowski, 'Queuing Models for Systems with Synchronization Constraint', *IEEE Proc.*, **39**(1), 1989, 138–161.

[15] J. Hui, 'Resource Allocation for Broadband Networks', *IEEE Journal on Selected Areas in Communications*, **6**(9), 1989, 1598–1618.

[16] N. Abramson, 'Fundamental of Packet Multiple Access for Satellite Networks' *IEEE Journal on selected Areas in Communications*, **10**(1), 1992, 309–316.

[17] A. Ephremides, B. Hajek, 'Information Theory and Communications Networks: An Unconsummated Union', *IEEE Transactions on Information Theory*, **44**(6), 1998, 2416–2434.

[18] K.C. Chen, 'Medium Access Control of Wireless LANs for Mobile Computing', *IEEE Networks*, 1994.

4

Cooperative Communications and Networks

MIMO communications have successfully improved system capacity and/or system coverage under severe fading, as described in Chapter 1, by using a large number of transmit antennas and diversity antennas to generate good diversity order. However, a good number of antennas may not be available at one terminal device due to its size, and it is obvious that we can use relay node(s) to increase system coverage. Researchers also note that we can use cooperative diversity to exploit spatial diversity among distributed single-antenna nodes, which opens the door towards cooperative communications and thus cooperative networks.

The basic idea of cooperative diversity is to create the transmit diversity via the spatial diversity by transmitting and relaying the same signals through independent channels from a single antenna as shown in Figure 4.1. The source node A transmits the source signal X to both destination node C and relay node B, and relay node B transmits signal X_1 that includes the information about X, to destination node C, to form the simplest three-node (or three-terminal) relay network. The destination node C receives two signals, Y and Y_1. Both received signals contain information of X through different paths to create spatial diversity, which forms *cooperative communication* or *cooperative diversity*.

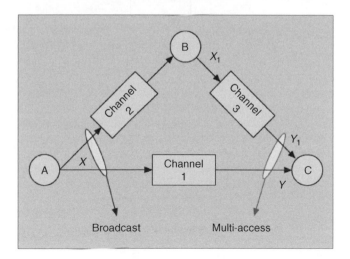

Figure 4.1   Relay channel [20]

Cognitive Radio Networks   Kwang-Cheng Chen and Ramjee Prasad
© 2009 John Wiley & Sons, Ltd

However, in addition to the increase of diversity order against fading, there exists another benefit for cooperative communications and cooperative networking (since three nodes or more are involved), and that is to increase the aggregated bandwidth via cooperative relay.

4.1 Information Theory for Cooperative Communications

Cooperative (diversity) communication is achieved by combining the relayed signal with the original direct signal in time, frequency or spatial domain, to decode the information rather than considering the relay signal as interference. In the following, we shall investigate the fundamentals of information theory towards cooperation. The original channel capacity analysis of relay channels was done by Cover and El Gamal [3], which considers a three-node/terminal relay network including source node, destination node and relay node. This work shows that the capacity of the cooperative scheme can be better than that of the non-cooperative scheme.

We start from the capacity region of simple point-to-point channel, multiple-access channel (uplink channel), broadcast channel (downlink channel), relay channel and interference channel. Then, we analyse the multiple-access channel with cooperative diversity, before proceeding to cooperative protocols.

4.1.1 Fundamental Network Information Theory

Let X be the random variable input to the channel and Y the random variable output of the channel. The simple point-to-point link has the capacity as $C = max_{p(x)} \, I(X; Y) = \log\left(1 + \frac{P}{N}\right)$, where P is the transmitter power constraint and N is the receiver Additive White Gaussian Noise (AWGN) noise power. The *multiple-access channel* with the simplest case of two senders is shown in Figure 4.2.

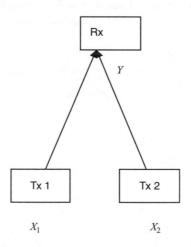

Figure 4.2   Simple multiple-access channel

Two senders, Tx1 and Tx2, communicate to the single receiver. The received signal is $Y = X_1 + X_2 + Z$, where Z is circular symmetric complex Gaussian noise with variance N_0. There is a power constraint P_j on sender j ($j = 1, 2$). It is well known that the capacity region that is to be the convex hull of the set of rate pairs satisfies:

$$R_1 \leq I(X_1; Y|X_2) = log\left(1 + \frac{P_1}{N_0}\right) \tag{4.1}$$

$$R_2 \leq I(X_2; Y|X_1) = log\left(1 + \frac{P_2}{N_0}\right) \tag{4.2}$$

$$R_1 + R_2 \leq I(X_1; X_2; Y) = log\left(1 + \frac{P_1 + P_2}{N_0}\right) \tag{4.3}$$

where $X_1 \sim N(0, P_1)$, $X_2 \sim N(0, P_2)$ meets the equality.

The capacity region of this simplest multiple-access channel is plotted in Figure 4.3.

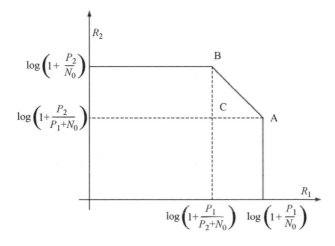

Figure 4.3 Capacity region of simple multiple-access channel

The scheme that achieves the corner point of the capacity region can be considered as a two-stage process. For example, we consider point A: In the first stage, the receiver decodes sender Tx2, considering sender Tx1 as part of the noise. This decoding has a low probability of error if $R_2 \leq I(X_2; Y) = log\left(1 + \frac{P_2}{P_1 + N_0}\right)$. After sender Tx2 has been decoded successfully, it can be subtracted out and sender Tx1 can be decoded correctly if $R_1 \leq I(X_1; Y|X_2) = log\left(1 + \frac{P_1}{N_0}\right)$. This idea is equivalent to the well known successive cancellation.

We now examine the *broadcast channel* with the topology shown in Figure 4.4.

There is a sender with power constraint P and two distant receivers Rx1, Rx2. The received signal is $Y_1 = X + Z_1$, $Y_2 = X + Z_2$. Please note that $X = X_1 + X_2$, where X_1 and X_2 are dedicated signals to Rx1 and Rx2 with power αP and $(1 - \alpha)P$, respectively. Z_1 and Z_2 are i.i.d. circular symmetric complex Gaussian noise with variance N_1 and N_2 (assume $N_1 < N_2$). The sender wishes to send independent messages X_1 and X_2 at rates R_1 and R_2 to receivers Rx1 and Rx2, respectively.

The capacity region of the Gaussian broadcast channel is determined by

$$R_1 \leq log\left(1 + \frac{\alpha P}{N_1}\right) R_2 \leq log\left(1 + \frac{(1 - \alpha)P}{\alpha P + N_2}\right) \tag{4.4}$$

where $0 \leq \alpha \leq 1$, α may be arbitrarily chosen to tradeoff rate R_1 for rate R_2.

The scheme that achieves the capacity region can be considered as follows. First, consider the 'bad' receiver Rx2 ($\because N_1 < N_2$), its effective signal to noise ratio is $\frac{(1 - \alpha)P}{\alpha P + N_2}$ because Rx1's message acts as

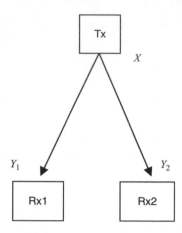

Figure 4.4   Broadcast channel

noise to Rx2. So, Rx2 transmits at rate $R_2 \leq log\left(1 + \frac{(1-\alpha)P}{\alpha P + N_2}\right)$. Second, consider the 'good' receiver Rx1; it first decodes Rx2's message, which it can accomplish because of its lower noise power N_1. Then it subtracts that message out and decodes its own message. So, Rx1 can transmit at rate $R_1 \leq log\left(1 + \frac{\alpha P}{N_1}\right)$.

The simplest *relay channel* consists of a source node, a destination node and one relay node as shown in Figure 4.5.

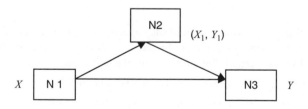

Figure 4.5   Relay channel

N1 is the source, N3 is the destination and N2 is the relay node. The relay channel combines a broadcast channel (N1 to N2 and N3) and a multiple-access channel (N1 and N2 to N3). Note that N2 receives Y_1 from N1 and transmits X_1 to N3.

The upper bound of the capacity of the relay channel is

$$C \leq max_{p(x,x_1)} min\{I(X, X_1; Y), I(X; Y, Y_1|X_1)\} \tag{4.5}$$

It is a consequence of a general *Max-flow min-cut* theorem described in the following.

Consider a general multi-terminal network where there are m nodes; each node i associates with a transmitted/received message pair $(X^{(i)}, Y^{(i)})$. The node i sends information at rate $R^{(ij)}$ to node j. Now, we divide the nodes into two sets, S and the complement S^c. The rate of flow of information from nodes in S to nodes in S^c is bounded by the following theorem:

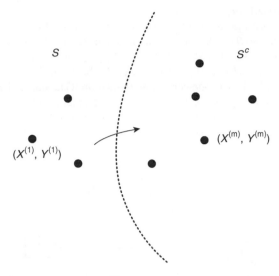

Theorem 4.1: If the information rates $\{R^{(ij)}\}$ are achievable, there exists some joint probability distribution $p(x^{(1)}, x^{(2)},\ldots, x^{(m)})$ such that

$$\sum_{i\varepsilon S, j\varepsilon S^c} R^{(ij)} \leq I\left(X^{(S)}; Y^{(S^c)}|X^{(S^c)}\right) \qquad (4.6)$$

for all $S \subset \{1, 2,\ldots, m\}$. Thus, the total rate of flow of information across cut sets is bounded by the conditional mutual information.

Remark: The above max-flow min-cut interpretation says that the rate of flow of information across any boundary is less than the mutual information between the inputs on one side of the boundary and the outputs on the other side, conditioned on the inputs on the other side. Please also note that these bounds are generally not achievable even for some simple channels.

Let us use the max-flow min-cut theorem to verify the bounds on capacity of the relay channel (Figure 4.6).

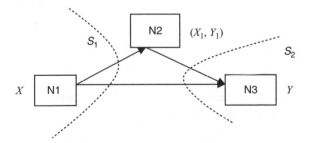

Figure 4.6   Cuts in the relay channel model

$$C \leq \max_{p(x,x_1)} min\{I(X, X_1; Y), I(X; Y, Y_1|X_1)\} \qquad (4.7)$$

The first term $I(X, X_1;Y)$ bounds the maximum rate of information transfer from senders N1 and N2 to receiver N3 (cut S_2). The second term $I(X;Y, Y_1|X_1)$ bounds the rate from N1 to N2 and N3 (cut S_1).

The capacity of a physically *degraded relay channel* (i.e., $I(X; Y, Y_1|X_1)) = I(X; Y_1|X_1)$ is achievable. This idea is known as *Block Markov Coding and Decode-Forward* by the following two schemes.

We encode information into B blocks each of length n channel uses. The same codebook of length n is used in each of the B blocks. $W_j[i] \in M_j = \{1, 2, \ldots, 2^{nRj}\}$ is the message of sender j $(j = 1)$ in block i $(i = 1, \ldots B)$. It would be encoded as $x_j[i'](w_j[i])$. (It can also be denoted as $x_{j'}^n[i'](w_j[i])$; we use the first vector notation here.) $x_{j'}[i'](\cdot)$. is a codeword from the random Gaussian codebook $C_{j.}$, i' is the block index and $j'(j' = 1, 2)$ is the sender index. Consider a relay channel with transmission between nodes 1 and 3 and with node 2 acting as relay. P_i is the power constraint on sender node i $(i = 1, 2)$. N_i is the AWGN noise power at node i $(i = 2, 3)$ (Figure 4.7).

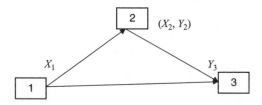

Figure 4.7   Scheme 1 (decode-forward + parallel channel decoding)

The procedure for this relay channel is therefore as follows:

1. Node 1 transmits the sequence of blocks $x_1[1]$ $(w_1[1]), \ldots, x_1[B]$ $(w_1[B])$.
2. The relay node 2 decodes $w_1[i]$ during block i, which is possible with zero error probability if $R_1 \leq \log\left(1 + \frac{P_1}{N_2}\right)$. It then transmits $w_1[i]$ during block $i + 1$, i.e., it transmits the sequence $0, x_2[2]$ $(w_1[1]), \ldots, x_2[B + 1](w_1[B])$. This signalling method is called 'decode-forward'.
3. At receiver node 3, the decoding is forward in time. Suppose that block i has been received, and that $w_1[i - 2]$ has been decoded error-free, so $x_2[i - 1](w_1[i - 2])$ is known. Based on the received signal at block i and $i - 1 - 1$:

$$\begin{bmatrix} y_3[i] \\ y_3'[i-1] \end{bmatrix} = \begin{bmatrix} y_3[i] \\ y_3[i-1] - x_2[i-1](w_1[i-2]) \end{bmatrix}$$

$$= \begin{bmatrix} x_1[i](w_1[i]) + x_2[i](w_1[i-1]) + z_3[i]) \\ x_1[i-1](w_1[i-1]) + z_3[i-1] \end{bmatrix}$$

Consider decoding of $w_1[i - 1]$, the signal $x_1[i](w_1[i])$ can be considered part of the background noise. The two signals $(y_3[i], y_3'[i-1])$ then form a parallel Gaussian channel. The rate is

$$R_1 \leq \log\left(1 + \frac{P_2}{P_1 + N_3}\right) + \log\left(1 + \frac{P_1}{N_3}\right) = \log\left(1 + \frac{P_1 + P_2}{N_3}\right) (= I(X_1, X_2; Y_3)). \qquad (4.8)$$

This method is called *parallel channel decoding*.

Scheme 2 (decode-forward + backward decoding) is introduced from the concept of *multiplexed coding*. Suppose the transmitter wants to transmit two messages $w_1 \in \{1, \ldots 2^{nR}_1\}$ and $w_2 \in \{1, \ldots 2^{nR}_2\}$. The transmitter makes a table with 2^{nR}_1 rows and 2^{nR}_2 columns, and assigns a random Gaussian code to each entry in the table. A receiver can decode both w_1 and w_2 if the channel capacity $C > R_1 + R_2$. If it knows w_1, however, it can decode w_2 if $C > R_2$ simply by searching the row corresponding to w_1, as it knows w_2. The operation procedure is

1. The source node 1 transmits $x_1[1](0, w_1[1]), x_1[2](w_1[1]), w_1[2]). . ., x_1[B+1](w_1[B], 0)$, where forming $x_1[i](w_1[i-1], w_1[i])$ uses multiplexed coding.
2. The relay node 2 still decodes forward. Suppose that it has decoded $w_1[i-1]$ in block $i-1$. It can then decode $w_1[i]$ from $x_1[i](w_1[i-1], w_1[i])$ if $R_1 \le log\left(1 + \frac{P_1}{N_2}\right)$.
3. The destination node 3 receives $y_3[i] = x_1[i](w_1[i-1], w_1[i]) + x_2[i](w_1[i-1]), + z_3[i]$. The decoding starts from the last block and proceeds backward to the first block. Suppose $w_1[i]$ has been decoded; $w_1[i-1]$ can be decoded if $R_1 \le log\left(1 + \frac{P_1+P_2}{N_3}\right)$ by the multiplexed coding argument. This method is called backward decoding.

The interference channel plays an important role in cooperative communications and cognitive radio with the structure shown in Figure 4.8, although it has been an open problem for 30 years with the best known achievable region produced by Han & Kobayashi in 1981.

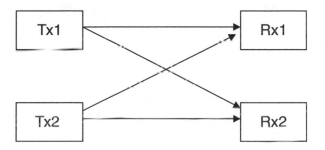

Figure 4.8   Interference channel

4.1.2 Multiple-access Channel with Cooperative Diversity

In the following, we want to analysis the model for a two-user multiple-access channel with cooperative diversity (the two sources cooperate). Two sources 1, 2 generate independent messages and these messages are encoded into codewords that can depend causally on one received message signal from the other. The decoder observes the channel output and estimates the source messages of the two sources, respectively (Figure 4.9).

Definition 4.1: P_i is the power constraint on sender i ($i = 1, 2$); N_i is the AWGN noise power at receiver i ($i = 1, 2, 3$); $a_{j,i}$ is the channel gain between node i and j; $s_{j,i} = ||a_{j,i}||^2 \frac{P_i}{N_j}$; and ρ is the correlation coefficient between X_1 and X_2.

To derive the outer bound on the capacity region, we use the max-flow min-cut strategy again. Considering the cut S_1, we obtain

$$R_1 \le I(X_1; Y_0, Y_2|X_2) \le log(1 + (1 - ||\rho||^2)(s_{0,1} + s_{2,1})). \tag{4.9}$$

Considering the cut S_2, we obtain

$$R_2 \le I(X_2; Y_0, Y_1|X_1) \le log(1 + (1 - ||\rho||^2)(s_{0,2} + s_{1,2})). \tag{4.10}$$

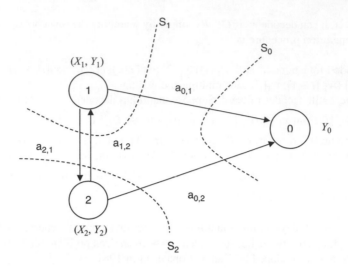

Figure 4.9   Model of multiple access with cooperation

Considering the cut S_3, we obtain

$$
\begin{aligned}
R_2 \leq I(X_1; X_2, Y_0) &\leq log(1 + s_{0,1} + s_{0,2} + 2\cos(\measuredangle\rho + \measuredangle a_{0,1} - \measuredangle a_{0,2})\|\rho\|\sqrt{s_{0,1}s_{0,2}}) \\
&\leq log(1 + s_{0,1} + s_{0,2} + 2\|\rho\|\sqrt{s_{0,1}s_{0,2}}).
\end{aligned}
\tag{4.11}
$$

where the last inequality assumes channel side information at transmitter.

To derive the *achievability* (inner bound on the capacity region), we consider two schemes here, non-cooperative transmission and decode-and-forward transmission.

■ *Non-cooperative transmission:* If the senders do not exploit their observations on each other, the system model reduces to the classical multiple-access channel, and thus the achievable rate region is

$$
\begin{aligned}
R_1 &\leq log(1 + s_{0,1}) \\
R_2 &\leq log(1 + s_{0,2}) \\
R_1 + R_2 &\leq log(1 + s_{0,1} + s_{0,2})
\end{aligned}
\tag{4.12}
$$

■ *Decode-and-forward transmission:* The idea of the coding scheme is similar to scheme 2 of the relay channel (i.e., decode-forward + backward decoding). We just show the results as follows:

$$
R_1 \leq log(1 + \alpha_1 s_{2,1}) \, R_2 \leq log(1 + \alpha_2 s_{1,2}) \, R_1 + R_2 \leq log(1 + s_{0,1} + s_{0,2} + 2\sqrt{(1 - \alpha_1)(1 - \alpha_2)s_{0,1}s_{0,2}}).
\tag{4.13}
$$

for some $0 \leq \alpha_1 \leq 1$, $i = 1, 2$.

The outer bound, the region of non-cooperative transmission and decode-and-forward transmission are plotted in Figure 4.10 with $s_{0,1} = s_{0,2} = 3$, $s_{2,1} = s_{1,2} = 15$.

4.2 Cooperative Communications

Cooperative schemes of the basic three-terminal relay network have been widely considered in the literature. Sendonaris *et al.* [8] proposed a simple cooperative protocol implemented by a CDMA system. Laneman *et al.* [9] proposed a cooperative diversity protocol in the time division channel. This time division assignment is applicable to different cooperative diversity protocols such as fixed relaying,

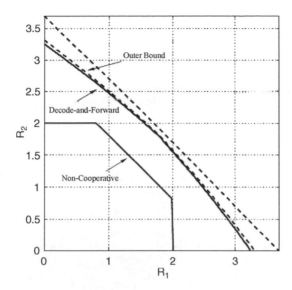

Figure 4.10 Achievability of multiple access with cooperation

selection relaying and incremental relaying. The cooperative schemes are then generalised to the multiple-terminal relay network, such as a space–time coded cooperative diversity protocol, in order to offer a signal design of the multiple relays by Laneman *et al.*, and an amplify and forward cooperative diversity protocol in multiple relays by Wittneben *et al.* [16, 17] Bletsas *et al.* [19] developed an opportunistic relaying such that the multiple-terminal relay network reduced to the three-terminal relay network by selecting the best relay terminal. The above concepts are very useful to the development of cognitive radio networking in co-existing multi-radio systems.

4.2.1 Three-Node Cooperative Communications

There are three fundamental cooperative (diversity) communication schemes in the three-terminal relay network: *decode and forward, amplify and forward* and *coded cooperation*, which are illustrated in Figure 4.11. The base-station (BS) can receive signals from two handsets (or nodes or mobile stations). Using the model in Figure 4.1, one handset is the source node and the other is the relay node. The BS receives those two handsets' signals to obtain the cooperative diversity. In the decode and forward method, the relay terminal decodes (or estimates) the received signal transmitted from the source terminal, and then forwards the decode signal to the destination terminal. It may cause error propagation if the relay terminal does not decode the source signal correctly. In the amplify and forward method, the relay terminal forwards the received source signal directly from the source terminal by amplifying a gain parameter to the destination without the decoding process. The gain parameter, called the *amplifying parameter*, is used to enhance the signal power. The coded cooperation method jointly considers the channel coding and cooperation protocol in the relay and source terminal. This methodology can offer not only cooperative diversity, but also coding gain.

4.2.1.1 Cooperative Communication with CDMA Implementation

In this model, we consider two mobiles communicating to a destination terminal as shown in Figure 4.11. Each mobile has its own information to be sent to the destination, denoted by W_i, $i = 1, 2$, with respect to mobile 1 and mobile 2. Moreover, each mobile also assists the other mobile to

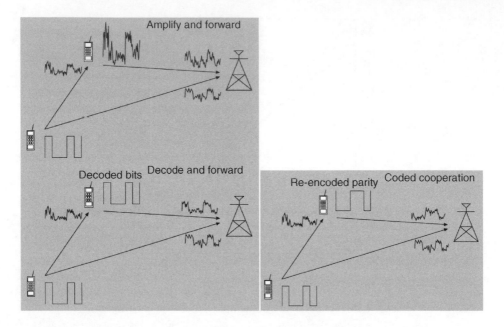

Figure 4.11 The three basic cooperative diversity protocols [22]

relay its signal, denoted by $Y_i, i = 1,2$ corresponding to mobile 2 and mobile 1, respectively, to the destination. Cleary, the received signal Y_0 can achieve cooperative diversity because the transmitted signal $X_i, i = 1,2$ in each mobile contains the information of both W_1 and W_2. To be specific, the mathematical baseband model during one symbol period is

$$
\begin{aligned}
Y_0 &= K_{10}X_1 + K_{20}X_2 + Z_0 \\
Y_1 &= K_{21}X_2 + Z_1 \\
Y_2 &= K_{12}X_1 + Z_2
\end{aligned}
\tag{4.14}
$$

where $\{K_{ij}\}$ denotes the channel fading coefficients which remain constant during one symbol period, and $Z_i, i = 0,1,2$ are the AWGN in destination, mobile 1 and mobile 2, respectively. The received signal Y_0 should be able to isolate the signals W_1 and W_2 for signal recovering. For this purpose, the use of two orthogonal channels may be useful: CDMA (Figure 4.12) and the time division scheme.

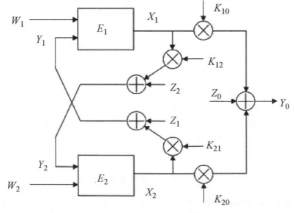

Figure 4.12 CDMA channel model

For example, we may consider using CDMA with both non-cooperative and cooperative implementations. Assume that each user has its own spreading code and those spreading codes are perfectly orthogonal. The transmission symbol periods are set to be $L=3$, and during one symbol period, each user transmits only one information bit. It also assumes that the coherence time of the channel is longer than the L symbol periods so that the channel fading coefficients are constant. In the non-cooperative case, the transmitted signals at the mobiles during three symbol periods are described by

$$X_1(t) = a_1 b_1^{(1)} c_1(t), \ a_1 b_1^{(2)} c_1(t), \ a_1 b_1^{(3)} c_1(t)$$
$$X_2(t) = a_2 b_2^{(1)} c_2(t), \ a_2 b_2^{(2)} c_2(t), \ a_2 b_2^{(3)} c_2(t)$$
(4.15)

where $b_j^{(i)}$ is i bits of mobile j's information, $c_j(t)$ is the spreading code of mobile j and a_j is the coefficient related to the power allocation in mobile j.

The three information bits are transmitted to the destination by their own spreading code without any cooperation. Obviously, this signal construction does not provide any diversity. Next, we consider a cooperative strategy in the cooperative case. The transmitted signals at the mobiles during the three symbol periods are described by

$$X_1(t) = a_{11} b_1^{(1)} c_1(t), \ a_{12} b_1^{(2)} c_1(t), \ a_{13} b_1^{(2)} c_1(t) + a_{14} \hat{b}_2^{(2)} c_2(t)$$
$$X_2(t) = a_{21} b_2^{(1)} c_2(t), \ a_{22} b_2^{(2)} c_2(t), \ a_{23} \hat{b}_1^{(2)} c_1(t) + a_{24} b_2^{(2)} c_2(t)$$
(4.16)

where $\hat{b}_j^{(i)}$ is the partner's estimate of i bits in mobile j and $\{a_{ji}\}$ is the coefficient related to the power allocation in mobile j. In the first symbol period, the first information bit is only sent to the destination. In the second symbol period, the information bit is sent to both destination and its partner. Therefore, in the third symbol period, each mobile can send its information bits and partner's information bits for generating the cooperative transmission. Clearly, in this strategy, we can achieve cooperative diversity and isolate the desire signal via the spreading code of each mobile. It should be noted that, in the non-cooperative case, each mobile transmits three information bits, but only two information bits in the cooperative case. However, it has been shown that the throughput of the cooperative case is better than the throughput of the non-cooperative case. Thus, the cooperative method not only gains the cooperative diversity but can also increase the throughput.

4.2.1.2 Cooperative Communication in Time Division Channel

To separate the source signal and relaying signal, we might need some means of orthogonal channel allocation. The time division multiplexing channel is therefore useful to serve the purpose of cooperative diversity.

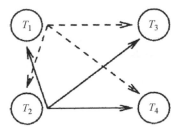

Figure 4.13   Terminal model of time division

The system model of time division cooperation is showed in Figure 4.13. The source, relay and destination terminals are denoted by T_s, T_r and T_d, respectively, where $s, r \in \{1,2\}$ and $d \in \{3,4\}$. We consider the scenario of narrow-band transmission under frequency nonselective fading channels, so that the channel fading coefficients are constants during the N time index. The time division channel allocation is illustrated in Figure 4.14 representing both non-cooperative and cooperative cases. In the non-cooperative case in Figure 4.14(b), with time index n, we represent the baseband signal as

$$y_d[n] = a_{s,d}[n]x_s[n] + z_d[n], \; i = 1, 2 \tag{4.17}$$

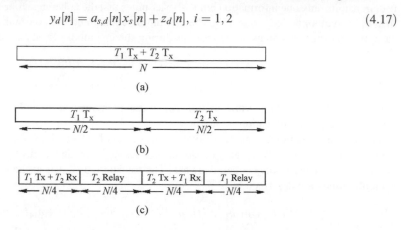

(a)

(b)

(c)

Figure 4.14 The time division channel allocations: (a) direct transmission with interference; (b) orthogonal direct transmission; and (c) orthogonal cooperative transmission [7].

where $x_s[n]$ and $y_d[n]$ are the source transmitted signal and the destination received signal, respectively, $\{a_{i,j}\}$ are the channel fading coefficients between node i and node j and $z_i[n]$ is the AWGN observed at node i. For $n = 1, 2, \ldots \frac{N}{2}$, the source node sends its source signal without cooperation, and then another source terminal sends its source signal for $n = \frac{N}{2} + 1, \ldots, N$. Thus, there is no diversity in this non-cooperative case. In the cooperative case in Figure 4.14(c), we divide two blocks of duration $\frac{N}{4}$ from the original block of duration $\frac{N}{2}$. In the first block, the source node transmits the source signal to both the destination and relay nodes, and the relay node listens for the transmitted source signal from the source node. In the second block, the relay node sends the relaying signal to the destination node. We can thus model the signals mathematically. For $n = 1, \ldots, \frac{N}{4}$,

$$\begin{aligned} y_r[n] &= a_{s,r}[n]x_s[n] + z_r[n] \\ y_d[n] &= a_{s,d}[n]x_s[n] + z_d[n] \end{aligned} \tag{4.18}$$

where $y_r[n]$ is the received signal in the relay terminal. And for $n = N/4 + 1, \ldots, N/2$,

$$y_d[n] = a_{r,d}[n]x_r[n] + z_d[n] \tag{4.19}$$

where $x_r[n]$ is the transmitted signal in the relay terminal for assisting the source terminal. Similarly, the rest of the symbol duration is in a revised manner. Clearly, if we use the relaying signal $x_r[n]$ well, cooperative diversity could be achieved. Next, we discuss these relaying methods for achieving cooperative diversity.

4.2.1.3 Fixed Relay

Regarding the fixed relay methodology, the basic relay mechanisms are generally (i) amplify and forward and (ii) decode and forward.

In the amplify and forward method, the relaying signal in the relay node is set by

$$x_r[n] = \beta y_r\left[n - \frac{N}{4}\right] \tag{4.20}$$

where β is the amplifying parameter. Inserting Equation (4.20) into Equation (4.19), we have

$$
\begin{aligned}
y_d[n] &= a_{r,d}x_r[n] + z_d[n] \\
&= a_{r,d}\beta y_r\left[n - \frac{N}{4}\right] + z_d[n] \\
&= a_{r,d}\beta\left\{a_{s,r}x_s\left[n - \frac{N}{4}\right] + z_d\left[n - \frac{N}{4}\right]\right\} + z_d[n], \quad n = \frac{N}{4} + 1, \ldots, \frac{N}{2}
\end{aligned}
\tag{4.21}
$$

Thus, from Equation (4.18), the received signal in the destination can be written as

$$
y_d[n] = \begin{cases} a_{s,d}[n]x_s[n] + z_d[n], \quad n = 1, \cdots, \dfrac{N}{4} \\ a_{r,d}\beta\left\{a_{s,r}x_s\left[n - \dfrac{N}{4}\right] + z_d\left[n - \dfrac{N}{4}\right]\right\} + z_d[n], \quad n = \dfrac{N}{4} + 1, \ldots, \dfrac{N}{2} \end{cases}
\tag{4.22}
$$

From Equation (4.22), we can observe that the cooperative diversity is obtained. Note that the noise term is also enhanced by the amplifying parameter.

In the decode and forward method, the relaying signal in the relay node is set by

$$
x_r[n] = \hat{x}_s\left[n - \frac{N}{4}\right], \quad n = \frac{N}{4} + 1, \ldots, \frac{N}{2}
\tag{4.23}
$$

Inserting Equation (4.23) into Equation (4.19), we have

$$
\begin{aligned}
y_d[n] &= a_{r,d}x_r[n] + z_d[n] \\
&= a_{r,d}\hat{x}_s\left[n - \frac{N}{4}\right] + z_d[n]
\end{aligned}
\tag{4.24}
$$

Thus, from Equation (4.18), the received signal in the destination can be written as

$$
y_d[n] = \begin{cases} a_{s,d}[n]x_s[n] + z_d[n], \quad n = 1, \ldots, \dfrac{N}{4} \\ a_{r,d}\hat{x}_s\left[n - \dfrac{N}{4}\right] + z_d[n], \quad n = \dfrac{N}{4} + 1, \ldots, \dfrac{N}{2} \end{cases}
\tag{4.25}
$$

Therefore, the cooperative diversity is obtained via the estimated source signal from the relay terminal. Note that if the estimation error is large, performance can be damaged.

4.2.1.4 Outage Behaviours of Fixed Relay

It is well known in mobile communications that we have to judge outage performance, in addition to link SNR, in the fading channels. We hereafter focus on slow fading, to capture scenarios in which delay constraints are on the order of the channel coherence time, and measure performance by outage probability and outage capacity. We use outage capacity instead of Shannon capacity because there is a conceptual difference between the AWGN channel and the slow fading channel. In the AWGN, one can send data at a positive rate while making the error probability as small as desired. This cannot be done for the slow fading channel as long as the probability that the channel is in deep fade is non-zero. Consequently, the Shannon capacity of the slow fading channel is zero in a strict sense, which is the reason why we define another measure of capacity. Outage probability $p_{out}(R) := P\{log(1 + |h|^2 SNR) < R\}$ is the probability that the channel cannot support transmission rate R. Outage capacity is the largest transmission rate R such that the outage probability $p_{out}(R)$ is less than ε (a predefined value). Furthermore, we assume only channel side information at receiver and employ half-duplex transmission, i.e., the node cannot transmit and receive simultaneously. The following discussion proceeds under the assumption of high-SNR approximations for orthogonal direct transmission and relaying transmission.

■ *Direct transmission:* From Equation (4.14), the maximum mutual information is given by $I_D = log$ $(1 + SNR\|a_{s,d}\|^2)$. The outage event for spectral efficiency R is given by $I_D < R$. For Rayleigh fading, i.e., $\|a_{s,d}\|^2$ exponentially distributed with parameter $\sigma_{s,d}^{-2}$, the outage probability is

$$p_D^{out}(SNR, R) := Pr[I_D < R] = Pr\left[\|a_{s,d}\|^2 < \frac{2^R - 1}{SNR}\right] \sim \frac{1}{\sigma_{s,d}^2} \cdot \frac{2^R - 1}{SNR}.$$

Please note that its diversity gain (the exponent of SNR in outage probability) is 1.

■ *Amplify-and-Forward:* The relay node amplifies what it receives:

$$x_r[n] = \beta y_r\left[n - \frac{N}{2}\right] \quad n = N/2 + 1 \ldots N$$

$$\beta \leq \sqrt{\frac{P}{\|a_{s,r}\|^2 + N_0}}.$$

Combining it with Equations (4.15) to (4.17), we have

$$\begin{bmatrix} y_d[n] \\ y_d[n+N/2] \end{bmatrix} = \begin{bmatrix} a_{s,d} \\ a_{r,d}\beta a_{s,r} \end{bmatrix} x_s[n] + \begin{bmatrix} 0 & 1 & 0 \\ a_{r,d}\beta & 0 & 1 \end{bmatrix} \begin{bmatrix} z_r[n] \\ z_d[n] \\ z_d[n+N/2] \end{bmatrix} \quad n = 1 \ldots N/2$$

It forms a vector Gaussian channel. The mutual information comes out to be

$$I_{AF} = log\left(1 + SNR\|a_{s,d}\|^2 + f\left(SNR\|a_{s,r}\|^2, \, SNR\|a_{r,d}\|^2\right)\right)$$

where $f(x, y) := \frac{xy}{x+y+1}$. The outage probability is therefore

$$p_{AF}^{out}(SNR, R) := Pr[I_{AF} < 2R] \sim \left(\frac{1}{2\sigma_{s,d}^2} \frac{\sigma_{s,r}^2 + \sigma_{r,d}^2}{\sigma_{s,r}^2 \sigma_{r,d}^2}\right)\left(\frac{2^{2R} - 1}{SNR}\right)^2$$

(R is scaled by 2 for consistency with the spectral efficiency of direct transmission.) Please note that its diversity gain is 2.

■ *Decode-and-Forward:* The relay node fully decodes $y_r[n]n = 1 \ldots N/2$, re-encodes (under a repetition-code scheme) and retransmits the source message: $x_r[n] = \hat{x}_s\left[n - \frac{N}{2}\right]$ $n = N/2 + 1 \ldots N$. The mutual information is

$$I_{DF} = min\left\{log\left(1 + SNR\|a_{s,r}\|^2\right), \, log\left(1 + SNR\|a_{s,d}\|^2 + SNR\|a_{r,d}\|^2\right)\right\}$$

The first term represents the maximum rate at which the relay can reliably decode the source message, while the second term represents the maximum rate at which the destination can reliably decode the source message given repeated transmissions. The outage probability is

$$p_{DF}^{out}(SNR, R) := P_r[I_{DF} < 2R] \sim \frac{1}{\sigma_{s,d}^2} \frac{2^{2R} - 1}{SNR}$$

We can observe that its diversity gain is 1. It indicates that fixed decode-and-forward does not offer diversity gain for large SNR, because it requires the relay to fully decode the source information which limits its performance.

4.2.1.5 Selection Relay and Incremental Relay

In the above fixed relaying method, the relay terminal relays the source signal from the source terminal to the destination all the time. However, when the channel condition between relay terminal and destination terminal is worse than the channel condition between source terminal and destination terminal, simply continuing the repetition of the source signal from source terminal to destination terminal may be better than relaying the signal from relay terminal to destination terminal in the relay duration. Hence, the selection relaying method decides whether to transmit the relaying signal to the destination terminal depending on the channel condition $|a_{s,r}|^2$. Specifically, if

$$|a_{s,r}|^2 < D \tag{4.26}$$

then the relay signal does not relay the signal to the destination terminal where D is certain threshold. The cooperative diversity is still held since the destination terminal always received the repetition of the source signal from the source terminal and/or the relaying signal during the second block.

Incremental relay tries to use the relay terminal more efficiently than the relaying method, based on the feedback information of the link condition. That is, the feedback information contains whether the transmission is a success or not, and the feedback information is obtained in the source terminal and relay terminal. Thus both the source terminal and relay terminal can decide reception from source terminal or relaying signal. Clearly, this method is more efficient than selection relaying since it has only one choice for reception or relaying so that the time of receptions can be reduced.

4.2.2 Multiple-Node Relay Network

In many communication scenarios such as cellular network, there can be more than three nodes in operation. Thus, we may need to extend the one relay node to multiple relay nodes that may offer more cooperative diversity. For interpreting this concept, there are three well-known multiple relay schemes, which we discuss in the following sections.

4.2.2.1 Space–Time Coded Cooperation

This multiple relay model is illustrated in Figure 4.15, with 2-phase operation. In phase I, the source node transmits the source signal to the destination and all the relay nodes. Thus, all the relay nodes listen

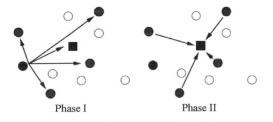

Phase I   Phase II

Figure 4.15   Multiple relay model

for the source signal in phase I. Then, in phase II, the relay nodes use the space–time code so that those relaying signals can be well-combined and transmitted to the destination node.

Again, in order to isolate these source signals and relaying signals, we divide the channel into a number of orthogonal sub-channels as in Figure 4.16. The set of cooperative nodes is denoted by $M = \{1,2,\ldots m\}$. For a source node s, the set of relay nodes assisting the source node s via the space–time code are denoted by $D(s)$, and the destination node corresponding to the source node s is denoted by $d(s)$.

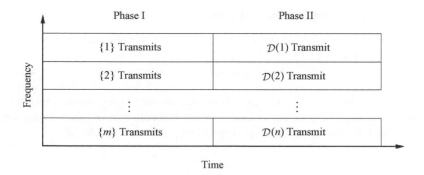

Figure 4.16 Channel allocations

We consider a baseband model of the time duration N during phase I and phase II. Each phase has half of the time duration N. For $n = 1,\ldots,\frac{N}{2}$ in the phase I, each relay node $r \in M - \{s\}$ receives the signal

$$y_r[n] = a_{s,r}[n]x_s[n] + z_r[n] \qquad (4.27)$$

where $x_s[n]$ and $y_r[n]$ are the source transmitted signal and the relay received signal, respectively; $\{a_{i,j}\}$ are the channel fading coefficients between node i and node j; and $z_i[n]$ is the AWGN in terminal i. The destination terminal $d(s)$ also receives the signal

$$y_{d(s)}[n] = a_{s,d(s)}[n]x_s[n] + z_{d(s)}[n] \qquad (4.28)$$

for $n = 1,\ldots,N/2$. For $n = N/2+1,\ldots,N$ in phase II, the destination node $d(s)$ receives the signal

$$y_{d(s)}[n] = \sum_{r\in D(s)} a_{r,d(s)}[n]x_r[n] + z_{d(s)}[n] \qquad (4.29)$$

where $x_r[n]$ denotes the relaying signal transmitted at the relay terminal r. These relaying signals are set by the space-–time codes to achieve the cooperative diversity.

4.2.2.2 Amplify and Forward Cooperation

This multiple-relay model is depicted in Figure 4.17, which can be considered for a general networking scenario such as ad-hoc networks. The source terminal and destination terminal are denoted by S_q and D_q, respectively, and we assume that the number of source terminals and destination terminals are equal to be N_a. The relay terminal is denoted by R_j, where the number of the relay terminals is N_r. This cooperative protocol follows the two phase scheme again. In phase I, the source nodes transmit the source signals to the relay nodes. In phase II, the relay nodes amplify and forward the received signal from source nodes to destination nodes.

The system model for this multiple-relay networking is illustrated in Figure 4.18. The transmitted signals of these source terminals are denoted by a signal vector $\vec{s} = [s_1, s_2, \ldots, s_{N_a}]^T$ whose element s_i

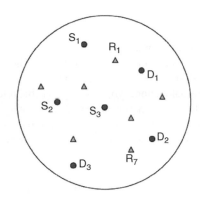

Figure 4.17   Multiple-relay cooperative model

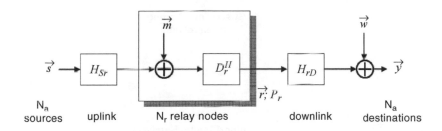

Figure 4.18   System model for multiple-relay networking

represents the source signal of the ith source node. The received signals corresponding to the destination nodes are denoted by $\vec{y} = [y_1, y_2, \ldots, y_{N_a}]^T$, whose element y_i is the received signal of the ith destination node. $\mathbf{H}_{sr}[j,q]$ is the channel fading coefficient between source terminal q to relay terminal j, and thus $\mathbf{H}_{sr}$ is the channel matrix between the source nodes to the relay nodes. $\vec{m}$ is the AWGN at the relay terminals. $\mathbf{D}_r$ is the diagonal gain matrix with elements d_j that represent the amplifying parameter for the amplify and forward scheme. $\mathbf{H}_{rD}$ is the channel matrix between the relay terminals and destination terminals. The signal model shown in Figure 4.18 is equivalent to Figure 4.19, where $\mathbf{H}_{SD}$ is the equivalent channel matrix between the source nodes and destination nodes. Obviously, if $\mathbf{H}_{SD}$ is a diagonal matrix, there is no interference in this equivalent model. Since this equivalent channel matrix is related to the diagonal gain matrix $\mathbf{D}_r$, we can design the amplifying parameter so that $\mathbf{H}_{SD}$ is a diagonal matrix. Consequently, the cooperative diversity is efficiently achieved.

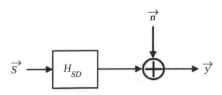

Figure 4.19   Equivalent system model

4.2.2.3 Opportunistic Relay

In the multiple-relay environment, all the relay nodes can assist the source nodes in relaying its signal. However, we can also select the best relay node to relay the signal. The basic idea of opportunistic relay is thus as illustrated in Figure 4.20, and we consider the two phase scheme again. Assume that there are only one source node and one destination node, with a set of possible relay nodes. In phase I, the source node transmits the source signal to the destination node and all the relay nodes. In phase II, all the relay nodes relay the received signals from source node to destination node in the original cooperative operation such as space–time codes. In order to reduce the overhead of those relay nodes, the

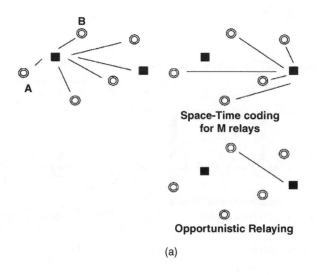

(a)

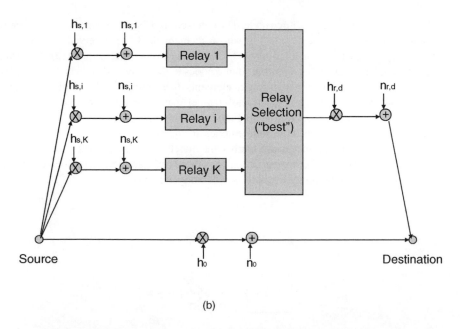

(b)

Figure 4.20   Opportunistic relay: (a) terminal link model; (b) relay selection

opportunistic relaying method selects the best relay path through only one relay node via the channel measurement to ensure the link performance.

The opportunistic relay concept can be generalised to cognitive radios in ad-hoc networks and co-existing multi-radio systems.

To be specific, denote $a_{s,i}$ and $a_{i,d}$ as the channel fading coefficient between the source node and relay node i and the channel fading coefficient between relay node i and destination node, respectively, where the channel is assumed to be a frequency nonselective fading channel. The method needs both the information $a_{s,i}$ and $a_{i,d}$ at the relay node i. There are two common policies to select the best relay node:

Relay Selection Policy A:

$$h_i = \min\{|a_{s,i}|^2, |a_{i,d}|^2\} \tag{4.30}$$

Relay Selection Policy B:

$$h_i = \frac{2}{\frac{1}{|a_{s,i}|^2} + \frac{1}{|a_{i,d}|^2}} \tag{4.31}$$

where h_i is the channel measurement (or channel gain) of the relay selection at relay node i. Clearly, Equations (4.30) and (4.31) provide a channel measurement of the source-relay-destination path at each relay node i. Thus, under those policies, the overhead from the multiple relay nodes is reduced, and the cooperative diversity is successfully obtained.

4.3 Cooperative Wireless Networks

Up to this moment, cooperative communications have been considered as a sort of spatial diversity at the physical layer since multiple independent faded copies of a transmitted signal can provide more information to the destination; even each node has a single-antenna arrangement. In the following, we focus on some networking scenarios from relay selection. The idea of selection relaying was originally proposed in [7], and exploited for distributed space time-coded (DSTC) protocols in [4]. However, the original approach becomes infeasible as the number of relays increases, hence relay selection was proposed [13]. Although both approaches achieve full diversity, we aim to provide more diversity gain with a reasonable cost. That is, our approach is equivalent to placing multiple antennas at the destination. However, via the concept of relay selection, it provides more meaning in ad-hoc networks as shown in Figure 4.21.

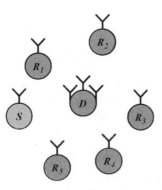

Figure 4.21   Topology of cooperative relay ad hoc network with $m=5$

4.3.1 Benefits of Cooperation in Wireless Networks

This ad-hoc network is composed of a single source, a single destination and $m=5$ relay nodes. The source and the destination can directly communicate with each other. There are a total of m fixed relay nodes, denoted by R_1, R_2,..., R_m. The relays operate a decode-and-forward (DF) scheme for relaying. The channel between two nodes, say i and j, is assumed to be slow-fading Rayleigh channel, with variance $\sigma_{i,j}^2$. With the assumption that the radios of all the devices are operating at half-duplex mode, the transmission of an information-bearing symbol can be split to two phases: broadcast phase and cooperation phase. During the broadcast phase, the source sends out its signal to the relays and its destination. After the broadcasting, each relay decodes the received signal. Once they can successfully decode, they will join to a decoding set, $D(S)$, where S denotes the source. The operation of the two phases is shown in Figure 4.22, in which we assume there are five relay nodes. In phase 1, all relays except R_2 can decode the transmitted signal from the source. Therefore R_2 fails to forward to the destination. In the next phase, the relays that have successfully decoded can forward. However, the channel between R_3 and the destination is not good enough, which results in a failed transmission.

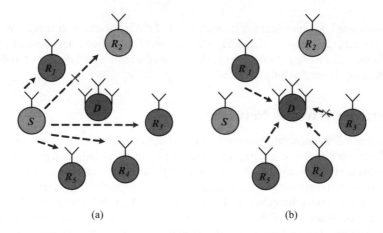

(a)   (b)

Figure 4.22 Two phases in a cooperative ad-hoc network with relay selection: (a) broadcast phase; (b) cooperation phase

After the decoding set, $D(S)$, is determined, the relays that are selected to relay can be different. For the case that all relays in $D(S)$ function with a common destination, they transmit in different frequency bands or separate in time. As the number of relay nodes increases, the system requirement becomes too large to be realistic. For this reason, relay selection is thus an alternative approach to significantly reduce such a large system requirement. In relay selection, only one relay with the best channel condition is selected to relay, which can achieve diversity order m as much as the original one does.

To further increase the diversity order of the cooperative network, we enable the destination to be equipped with multiple antennas. Intuitively, this can further provide diversity gain at only a small increase to cost, including physical antennas and feedback overhead. As MIMO technology becomes mature, a device equipped with multiple antennas no longer automatically equates to high cost, but seems a small and reasonable amount to pay. The feedback overhead required for the destination to select the relays is $\log_2(\min(J,M))$, which does not grow unboundedly as the number of relays increases.

Suppose that an end-to-end data rate R is required for a transmission in one time slot. From the relay perspective and its half duplex nature, it is in total two hops along the relay path. In either hop, at least $2R$ capacity is required. For the first hop, if the capacity is less than $2R$, the relay cannot successfully decode and fails to relay; if the capacity between the relay and the destination is less than $2R$, then the destination cannot decode successfully. That is,

$$\frac{1}{2}\log_2\left(1+|h_{Rj,R}|^2\frac{P}{N_0}\right) = \frac{1}{2}\log_2\left(1+|h_{Rj,R}|^2 SNR\right) \geq R \quad \frac{1}{2}\log_2\left(1+|h_{Rj,D}|^2 SNR\right) \geq R$$

for relay node j

where $SNR = P/N_0$, P is the maximum transmit power of each node and N_0 is the noise power spectral density.

The outage probability, P_{OUT}, is defined as the probability that the mutual information between the source and the destination falls below the required transmission rate, R. In our scenario, an outage event occurs when $I_{sel} < R$, where I_{sel} is the mutual information between the source and the destination for the use of relay selection and is given by

$$I_{sel} = \frac{1}{2}\left(1+SNR|h_{S,D}|^2 + \frac{HKH^*}{N_0}\right) \tag{4.32}$$

where $K = \text{diag}(P_1 \cdots P_m)$ is the power emitted from each of the selected relays, which is assumed equal to P for simplicity, and $H = \begin{bmatrix} h_1^{sel} & h_2^{sel} & \cdots & h_m^{sel} \end{bmatrix}$ indicates the composite channel matrix of selected m vectors, each of which is the channel impinged from a relay to the receiver antenna array at the destination. Note that h_i^{sel} is the ith best channel among $h_{R_1,D}, h_{R_2,D}, \cdots, h_{R_j,D}$, where $h_{R_j,D} = \begin{bmatrix} h_{R_j,1} & \cdots & h_{R_j,M} \end{bmatrix}^T$ denotes the vector of channel impinged from relay j to the destination, and is given by

$$h_i^{sel} = \max_{hj/\left\{h_1^{sel},\ldots,h_{i-1}^{sel}\right\}} h_j h_j^*$$

$$i \in \{1,\ldots,k\}$$

Therefore,

$$\Pr[I_{sel} < R] = \Sigma_{D(S)}\,\Pr[I_{sel} < R|D(S)]\,\Pr[D(S)]$$

■ Probability of the decoding set, $Pr[R_j \in D(S)]$: According to [7], a relay can be in *the decoding set* if it satisfies

$$\Pr[Rj \in D(S)] = \Pr\left[|h_{S,R_j}|^2 > \frac{2^{2R}-1}{SNR}\right] = \frac{1}{2\sigma_{S,R_j}^2}\exp\left(-\frac{1}{2\sigma_{S,Rj}^2}\cdot\frac{2^{2R}-1}{SNR}\right)$$

By independence among relay paths,

$$\Pr[D(S)] = \prod_{R_j \in D(S)}\exp\left(-\frac{1}{2\sigma_{S,R_j}^2}\cdot\frac{2^{2R}-1}{SNR}\right)$$

$$\cdot\prod_{R_j \notin D(S)}\left(1-\exp\left(-\frac{1}{2\sigma_{S,Rj}^2}\cdot\frac{2^{2R}-1}{SNR}\right)\right) \approx \left(\frac{2^{2R}-1}{SNR}\right)^{K-D(S)}$$

$$\cdot\prod_{R_j \notin D(S)}\frac{1}{2\sigma_{S,R_j}^2} \quad \text{as } SNR \to \infty$$

■ Outage probability conditioned on the decoding set, $Pr[I_{sel} < R|D(S)]$: where we have a total number of $\min(J, M)$ relays. For simplicity, we assume that $J > M$, which is realistic in practical wireless networks. In order to find out the conditional probability, we have to explicitly express the mutual information of our selected results. First, we define three random variables, X, Y_i, $Z_{i,j}$, as

$$X = \sum_{i=1}^{M} Y_i = \sum_{i=1}^{M} \sum_{j=1}^{M} Z_{i,j}$$

where Z is the squared value of the channel fading, whose distribution is given by

$$fz\,(z_{i,j}) = |h_{i,j}|^2 \sim \frac{1}{2\sigma_{i,j}^2}\, exp\left(-\frac{1}{2\sigma_{i,j}^2}Z_{i,j}\right),$$

Y_i is the sum of M random variables of $Z_{i,j}$ over js and X is the sum of M largest random variables out of $Y_i's$, for $i \in \{1,\ldots,K\}$. Once the distribution of X can be found, the mutual information of the selection can expressed as

$$I_{sel} = \frac{1}{2}\left(1 + \mathrm{SNR}\left(|h_{S,D}|^2 + X\right)\right) \tag{4.33}$$

■ Outage probability, $Pr\,[I_{set} < R]$, can be easily found based on above.

In addition to the outage behaviours, we also care about how end-to-end capacity, related to I_{sel}, is improved. Similarly, once the distribution of X can be determined, I_{sel} can be obtained. It carries significant meaning in wireless (ad-hoc) networks that we are able to increase aggregated bandwidth in wireless ad-hoc networks via cooperation, instead of increasing diversity in order to combat fading.

4.3.2 Cooperation in Cluster-Based Ad-hoc Networks

In many (mobile) ad-hoc networks (for example, MANET), nodes communicate through a *clustered head* or an *access point* (AP). Furthermore, AP may be connected to certain backbone networks to form a *hybrid infrastructures and ad-hoc* (HIAH) network. In MANET, the AP can collect information about the status of the network nodes, such as channel coefficients, to select an appropriate cooperative mode based on the network performance criterion and to feedback the AP's decision over appropriate control channels (or packets).

Through network architecture such as that shown in Figure 4.23, which is similar to the two-tier network, more reliable and longer range connections for inter-cluster are therefore possible [21]. It invokes a new challenge for AP to match and to activate cooperative links. A typical approach to realise clustered MANET may consist of the following steps.

1. *Partitioning for infrastructure network:* We group nodes (or mobile terminals) into cooperating pairs, which can be considered as the matching problem on graphs. The minimal weighted matching is to find a match of minimal weight, with a cubic complexity of the number of the nodes. A greedy matching algorithm randomly selects a free node and matches with the best remaining partner, which has quadratic complexity. We can also match nodes randomly of linear complexity. Of course, lower complexity algorithms pay the price of outage probability.
2. *Connectivity with cooperation:* Whenever the capacity of aggregated information is sent across the wireless networks, the connectivity (i.e., the pairs of cooperative nodes) to exploit the network capacity should be identified. If we assume that path loss and fading statistics are known, a link is available if the received SNR is above a given threshold. Connectivity has been well-studied for ad-hoc networks in the limit of an infinite number of nodes placed randomly on a two-dimensional region. Assuming that the path-loss is a monotonically increasing function of the distance d_{ij} between two nodes i and j (e.g., $d_{ij}^{-\alpha}$), and denoting by $\mathcal{A}$ the circular area centred at a node i, where all nodes $j \in \mathcal{A}$ have $SNR(d_{ij}) \propto d_{ij}^{-\alpha}$ above the threshold set for connectivity $SNR(d_{ij}) > \eta$, any two nodes that are within a distance from each other that is smaller than the radius of $\mathcal{A}$ are connected. The graph obtained by drawing a line between any two nodes of separation less than the radius of $\mathcal{A}$

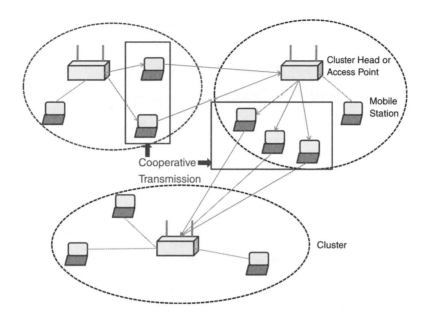

Figure 4.23 Clustering with direct and cooperative transmission

reveals sets of nodes that can communicate with each other directly or through a path consisting of multiple hops, and such a set is called a *cluster*.

In such a setting, there have been two separate definitions of what it means for a network to be connected. In the 'sparse' network setting, the network is defined as connected if a cluster containing an infinite number of nodes (called the 'infinite cluster') is present in the network. In the 'dense' network setting, the network is defined as being connected once all pairs of nodes can communicate with one another. We will refer to the latter definition as the network's being 'fully connected'. Both definitions are intrinsic properties of the network graph, and in clustered multi-hop networks the edges of this graph are used to communicate. In the large sparse network analyses, nodes are generally distributed on the infinite two-dimensional plane with some density λ. The connectivity is amenable to analysis via *percolation theory*. Clearly increasing λ must improve the connectivity. For dense networks, analysis is generally performed for N nodes distributed randomly on a surface of unit area. The work by Gupta and Kumar, considering large N on a unit disk, provides a necessary and sufficient condition to guarantee full connectivity of the network: the area of radio coverage of each node should be at least $A^{non-coop} = N^{-1}[\log N + c(N)]$, where $lim\ inf\ c(N) = +\infty$. The condition is necessary in the sense that a network with nodes communicating with coverage area less than $N^{-1}N[\log N + c(N)]\log$ (where $lim\ sup\ c(N) < +\infty$), is proven to be almost surely not fully connected.

3. *Connectivity in clustered networks with cooperation:* In cooperative networks, clustering can help connectivity by essentially increasing the area in which to search for new neighbours, as shown in Figure 4.24. In fact, if not all nodes are isolated, there will be nodes in a connected cluster that the AP can recruit in finding new neighbours. In dense networks, it can be shown that the cooperative network can be fully connected with a high probability without satisfying the necessity condition for full connectivity in the non-cooperative network. The proof construction relies on sub-dividing the network region into small sections, all of which are likely to have a large cluster of nodes within the area $\mathcal{A}$ with high probability. These clusters can connect not only with all nodes within the section,

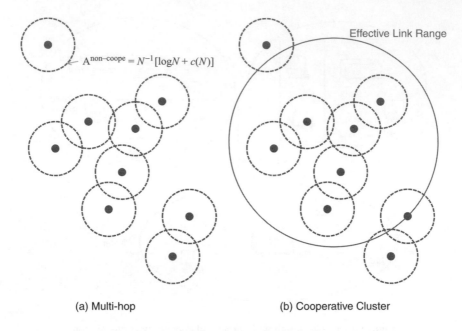

(a) Multi-hop   (b) Cooperative Cluster

Figure 4.24 Connectivity of cooperative cluster

but also with clusters in neighbouring sections, thus fully connecting the network. The required radio coverage area of a given node for such connectivity is given by

$$A^{\mathrm{coop}} \geq N^{-1} 4\pi (4 \log N)^{\frac{\alpha}{\alpha+2}} (\log\log N + \log 2)^{\frac{2}{\alpha+2}}, \qquad (4.34)$$

It suggests a gain of required power for connectivity of $\left(\frac{\log N}{\log\log N}\right)^{\frac{2}{\alpha+2}}$ through cooperative ad-hoc networking.

Actually, we can adopt random cooperative coding and random clustering to facilitate randomised cooperation to improve connectivity in additional ways to the above descriptions. In any case, cooperation is not limited to point-to-point communications, but is an essential network function.

References

[1] T. Cover, J. Thomas, *Information Theory*, John Wiley & Sons Inc., New York, 1991.

[2] D. Tse, *Fundamentals of Wireless Communication*, Cambridge University Press, Cambridge, UK, 2005.

[3] T.M. Cover, A.A.E. Gamal, 'Capacity Theorems for the Relay Channel', *IEEE Trans. Info. Theory*, **25**(5), 1979, 572–584.

[4] J. Laneman, G. Wornell, 'Distributed Space-time-coded Protocols for Exploiting Cooperative Diversity in Wireless Networks', *IEEE Trans. Inf. Theory*, **49**(10), 2003, 2415–2425.

[5] J.N. Laneman, 'Cooperative Diversity in Wireless Networks: Algorithms and Architectures', Ph.D. dissertation, MIT, Cambridge, MA, 2002.

[6] J.N. Laneman, G.W. Wornell, 'Distributed Space-time-coded Protocols for Exploiting Cooperative Diversity in Wireless Networks', *IEEE Transactions on Information Theory*, **49**(10), 2003, 2415–2425.

[7] J.N. Laneman, D.N.C. Tse, G.W. Wornell, 'Cooperative Diversity in Wireless Networks, Efficient Protocols, and Outage Behavior', *IEEE Trans. Inf. Theory*, **50**(12), 2004, 3062–3080.

[8] A. Sendonaris, E. Erkip, B. Aazhang, 'User Cooperation Diversity Part I and Part II', *IEEE Trans. Commun.*, **51**(11), 2003, 1927–1948.

[9] J.N. Laneman, G.W. Wornell, D.N.C. Tse, 'An Efficient Protocol for Realizing Cooperative Diversity in Wireless Networks', *Proc. IEEE ISIT*, Washington, DC, June 2001, p. 294.

[10] J.N. Laneman, G.W. Wornell, 'Exploiting Distributed Spatial Diversity in Wireless Networks', in *Proc. Allerton Conf. Communications, Control, and Computing*, Monticello, IL, Oct. 2000.

[11] T.E. Hunter, A. Nosratinia, 'Cooperative Diversity through Coding', *Proc. IEEE ISIT*, Lausanne, Switzerland, *July* 2002, p. 220.

[12] A. Høst-Madsen, 'Capacity Bounds for Cooperative Diversity', ' *IEEE Trans. Inf. Theory*, **52**(4), 2006, 1522–1544.

[13] E. Beres, R. Adve, 'Selection Cooperation in Multi-Source Cooperative Networks', *IEEE Transactions on Wireless Communications*, **7**(1), 2008, 118–127.

[14] T.E. Hunter, A. Nosratinia, 'Diversity through Coded Cooperation', submitted to *IEEE Trans. Wireless Commun.*, 2004.

[15] J.N. Laneman, G.W. Wornell, 'Distributed space-time-coded protocols for exploiting cooperative diversity in wireless networks', *IEEE Trans. Inform. Theory*, **49**,October 2003, 2415–2525.

[16] A. Wittneben, B. Rankov, 'Distributed Antenna Systems and Linear Relaying for Gigabit MIMO Wireless', *IEEE Vehicular Technology Conference VTC 2004 Fall, Los Angeles, USA*.

[17] A. Wittneben, I. Hammerström, 'Multiuser Zero Forcing Relaying with Noisy Channel State Information', *IEEE WCNC*, 2005.

[18] A. Bletsas, A. Khisti, D.P. Reed, A. Lippman, 'A Simple Cooperative Diversity Method Based on Network Path Selection', *IEEE JSAC*, 2006.

[19] A. Bletsas, H. Shin, M.Z. Win, A. Lippman, 'Cooperative Diversity with Opportunistic Relaying', *IEEE WCNC*, 2006.

[20] A. Nosratinia, T.E. Hunter, A. Hedayat, 'Cooperative Communication in Wireless Networks', *IEEE Communications Magazine*, **42**(10), 2004, 74–80.

[21] A. Scaglione, D.L. Goeckel, J.N. Laneman, 'Cooperative Communications in Mobile Ad Hoc Networks: Rethinking the Link Abstraction', *IEEE Signal Processing Magazine*, September 2006, 18–29.

[22] Raymond W. Yeung, A First Course in Information Theory, Kluwer, 2006.

5

Cognitive Radio Communications

5.1 Cognitive Radios and Dynamic Spectrum Access

Due to the rapid advance of wireless communications, a tremendous number of different communication systems exist in licensed and unlicensed bands, suitable for different demands and applications such as GSM/GPRS, IEEE 802.11, Bluetooth, UWB, Zigbee, 3G (CDMA series), HSPA, 3G LTE, IEEE 802.16, etc. as introduced in Chapter 2. On the other hand, radio propagation favours the use of spectrums under 3 GHz due to non-line-of-sight propagation. Consequently, many more devices, up to 1 trillion wireless devices by 2020, require radio spectrum allocation in order to respond to the challenge for further advances in wireless communications.

In the past, spectrum allocation was based on the specific band assignments designated for a particular service. The spectrum allocation chart used by the Federal Communications Commission (FCC) is shown in Figure 5.1 [3]. This spectrum allocation chart seems to indicate a high degree of utilisation. However, the FCC Spectrum Policy Task Force [4] reported vast temporal and geographic variations in the usage of the allocated spectrum with use ranging from 15% to 85% in the bands below 3 GHz that are favoured in non-line-of-sight radio propagation. The spectrum utilisation is even tighter in the range above 3 GHz. In other words, a large portion of the assigned spectrum is used sporadically, leading to an under utilisation of a significant amount of spectrum.

Although the fixed spectrum assignment policy generally worked well in the past, there has been a dramatic increase in the access to limited spectrum for mobile services and applications in the recent years. This increase is straining the effectiveness of the traditional spectrum polices. The limited available spectrum due to the nature of radio propagation and the need for more efficiency in the spectrum usage necessitates a new communication paradigm to exploit the existing spectrum opportunistically. Inspired by the successful global use of multi-radio co-existing at 2.4 GHz unlicensed ISM bands and others, dynamic spectrum access is proposed as a solution to these problems of current inefficient spectrum usage. The inefficient usage of the existing spectrum can be improved through opportunistic access to the licensed bands by existing users (primary users).

The key enabling technology of dynamic spectrum access is cognitive radio (CR) technology, which provides the capacity to share the wireless channel with the licensed users in an opportunistic way. CRs are envisioned to be able to provide the high bandwidth to mobile users via heterogeneous wireless architectures and dynamic spectrum access techniques. The networked CRs also impose several challenges due to the broad range of available spectrum as well as diverse QoS requirements of applications.

Cognitive Radio Networks   Kwang-Cheng Chen and Ramjee Prasad
© 2009 John Wiley & Sons, Ltd

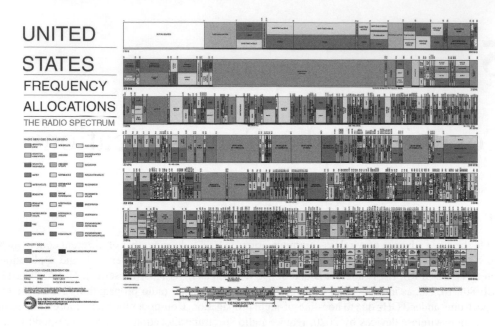

Figure 5.1   FCC spectrum allocation chart

In order to share the spectrum with licensed users without disturbing them, and meet the diverse quality of service requirement of applications, each CR user in a CRN must:

- Determine the portion of spectrum that is available, which is known as *Spectrum sensing*.
- Select the best available channel, which is called *Spectrum decision*.
- Coordinate access to this channel with other users, which is known as *Spectrum sharing*.
- Vacate the channel when a licensed user is detected, which is referred as *Spectrum mobility*.

As mentioned in [5], each CR has the capability of being cognitive, reconfigurable and self-organised to fulfil the functions of spectrum sensing, spectrum decision, spectrum sharing and spectrum mobility that each CR must require.

An overview of different spectrum sharing models, namely open sharing, hierarchical access and dynamic exclusive usage models, was proposed in [7]. Spectrum management is an important functionality in cognitive radio networks, which involves dynamic spectrum access/sharing and pricing, and it aims to satisfy the requirements of both primary and secondary users. Reference 6 also mentions that actual spectrum usage measurements obtained by the FCC's Spectrum Policy Task Force tell a different story: at any given time and location, much of the prized spectrum lies idle. This paradox indicates that spectrum shortage results from the spectrum management policy rather than the physical scarcity of usable frequencies. The underutilisation of spectrum has stimulated a flurry of exciting activities in engineering, economics and regulation communities in searching for better spectrum management policies and techniques.

5.1.1 The Capability of Cognitive Radios

Cognitive radio technology is a key technology that enables the improvement of the spectrum use in a dynamic manner. The term 'Cognitive Radio' is defined by Haykin [8] as follows:

> *Cognitive radio is an intelligent wireless communication system that is aware of its surrounding environment (i.e., outside world), and uses the methodology of understanding-by-building to learn from the environment and adapt its internal states to statistical variations in the incoming RF stimuli by making corresponding changes in certain operating parameters (e.g., transmit-power, carrier-frequency, and modulation strategy) in real-time, with two primary objectives in mind: highly reliable communication whenever and wherever needed; efficient utilization of the radio spectrum.*

The capabilities of cognitive radios as nodes of CRN can be classified according to their functionalities based on the definition of cognitive radio. A cognitive radio shall sense the environment (cognitive capability), analyse and learn sensed information (self-organised capability) and adapt to the environment (reconfigurable capabilities).

We can summarise cognitive capability as follows:

- *Spectrum sensing:* A cognitive radio can sense spectrum and detect 'spectrum holes', which are those frequency bands not used by the licensed users or having limited interference with them. A cognitive radio could incorporate a mechanism that would enable sharing of the spectrum under the terms of an agreement between a licensee and a third party. Parties may eventually be able to negotiate for spectrum use on an ad-hoc or real-time basis, without the need for prior agreements between all parties.
- *Location identification:* Location identification is the ability to determine its location and the location of other transmitters, and then select the appropriate operating parameters such as the power and frequency allowed at its location. In bands such as those used for satellite downlinks that are receive-only and do not transmit a signal, location technology may be an appropriate method of avoiding interference because sensing technology would not be able to identify the locations of nearby receivers. However, such location identification shall be based on relative information for more system flexibility and overall spectrum utilisation, instead of fixed universal location to create system/network control overheads.
- *Network/system discovery:* For a cognitive radio terminal to determine the best way to communicate, it shall first discover available networks around it. These networks are reachable either via directed one hop communication or multi-hop relay nodes. For example, when a cognitive radio terminal has to make a phone call, it shall discover if there is GSM BTSs or WiFi APs nearby. If there is no directed communication link between the terminal and the BTSs/APs but through other cognitive radio terminals some access networks are reachable, it can still make a call in this circumstance. The ability to discovery one hop or multi-hop away access networks is important.
- *Service discovery:* Service discovery usually accompanies network/system discovery. Network or system operators provide their services through their access networks. A cognitive radio terminal shall find appropriate services to fulfil its demands.

5.1.1.1 Reconfigurable Capability

Cognitive radio is obviously generalised from the SDR (described in Chapter 2). However, CR must equip more reconfigurability than SDR as described in the following summary:

- *Frequency agility:* This is the ability of a radio to change its operating frequency. This ability usually combines with a method to select dynamically the appropriate operating frequency based on the sensing of signals from other transmitters or on some other method.
- *Dynamic frequency selection:* This is defined in the rules as a mechanism that dynamically detects signals from other radio frequency systems and avoids co-channel operation with those systems. The method that a device could use to decide when to change frequency or polarisation could include spectrum sensing, geographic location monitoring or an instruction from a network or another

device. Such capability can be generalised as dynamic selection of logic channels and physical channels in wireless communications.

- *Adaptive modulation/coding (AMC):* This has been developed to approach channel capacity in fading channels. AMC can modify transmission characteristics and waveforms to provide opportunities for improved spectrum access and more intensive use of spectrum while 'working around' other signals that are present. A cognitive radio could select the appropriate modulation type for use with a particular transmission system to permit interoperability between systems.
- *Transmit power control (TPC):* This is a feature that enables a device to switch dynamically between several transmission power levels in the data transmission process. It allows transmission at the allowable limits when necessary, but reduces the transmitter power to a lower level to allow greater sharing of spectrum when higher power operation is not necessary.
- *Dynamic system/network access:* For a cognitive radio terminal to access multiple communication systems/networks that run different protocols, the ability to reconfigure itself to be compatible with these systems is necessary. It is therefore useful in co-existing multi-radio environments to exploit heterogeneous wireless networking fully.

5.1.1.2 Self-Organised Capability

So, we know that CRs should have the capability of sensing as well as reconfigurability. With more intelligence to communication terminal devices, CRs should be able to self-organise their communication based on sensing and reconfigurable functions.

- *Spectrum/radio resource management:* A good spectrum management scheme is necessary to manage and organise efficiently spectrum holes information among cognitive radios.
- *Mobility and connection management:* Due to the heterogeneity of CRNs, routing and topology information is more and more complex. Good mobility and connection management can help neighbourhood discovery, detect available Internet access and support vertical handoffs, which help cognitive radios to select route and networks.
- *Trust/security management:* CRNs are heterogeneous networks in essence, and so various heterogeneities (e.g., wireless access technologies, system/network operators) introduce lots of security issues. Trust is therefore a prerequisite for securing operations in CRNs, to support security functions in dynamic environments.

5.1.2 Spectrum Sharing Models of DSA

Representing the opposite direction of the current static spectrum management policy, the term *dynamic spectrum access* (DSA) has broad connotations that encompass various approaches to spectrum reform. The diverse ideas presented at the first IEEE Symposium on New Frontiers in Dynamic Spectrum Access Networks (DySPAN) suggest the extent of this term. As illustrated in Figure 5.2, dynamic spectrum access strategies can be broadly categorised under three models.

5.1.2.1 Dynamic Exclusive Use Model

This model maintains the basic structure of the current spectrum regulation policy: spectrum bands are licensed to services for exclusive use. The main idea is to introduce flexibility to improve spectrum efficiency. Two approaches have been proposed under this model: spectrum property rights and dynamic spectrum allocation. The former approach allows licensees to sell and trade spectrum and to

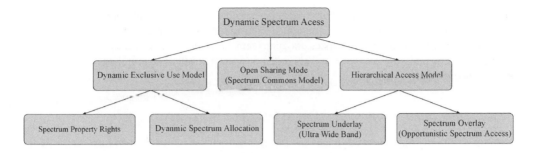

Figure 5.2 A taxonomy of dynamic spectrum access [7]

freely choose technology. Economy and market will thus play a more important role in driving towards the most profitable use of this limited resource. Note that even though licensees have the right to lease or share the spectrum for profit, such sharing is not mandated by the regulation policy.

The second approach, dynamic spectrum allocation, was brought forth by the European DRiVE project. It aims at improving spectrum efficiency through dynamic spectrum assignment by exploiting the spatial and temporal traffic statistics of different services. In other words, in a given region and at a given time, spectrum is allocated to services for exclusive use. This allocation, however, varies at a much faster scale than the current policy. Based on an exclusive-usage model, these approaches cannot eliminate white space in spectrum resulting from the bursty nature of wireless traffic.

5.1.2.2 Open Sharing Model

Also referred to as spectrum commons, this model employs open sharing among peer users as the basis for managing a spectral region. Advocates of this model draw support from the phenomenal success of wireless services operating in the unlicensed industrial, scientific and medical (ISM) radio band (e.g., WiFi). Centralised and distributed spectrum sharing strategies have been initially investigated to address technological challenges under this spectrum management model.

5.1.2.3 Hierarchical Access Model

This model adopts a hierarchical access structure with primary and secondary users. The basic idea is to open licensed spectrum to secondary users while limiting the interference perceived by primary users (licensees). Two approaches to spectrum sharing between primary and secondary users have been considered: spectrum underlay and spectrum overlay.

The underlay approach imposes severe constraints on the transmission power of secondary users so that they operate below the noise floor of primary users. By spreading transmitted signals over a wide frequency band (UWB), secondary users can potentially achieve a short-range high data rate with extremely low transmission power. Based on a worst-case assumption that primary users transmit all the time, this approach does not rely on detection and exploitation of spectrum white space.

Spectrum overlay was first envisioned by Mitola under the term spectrum pooling and then investigated by the DARPA Next Generation (XG) programme under the term opportunistic spectrum access. Differing from spectrum underlay, this approach does not necessarily impose severe restrictions on the transmission power of secondary users, but rather on when and where they may transmit. It directly targets at spatial and temporal spectrum white space by allowing secondary users to identify and exploit local and instantaneous spectrum availability in a nonintrusive manner. Compared to the dynamic exclusive use and open sharing models, this hierarchical model is perhaps the most compatible with current spectrum management policies and legacy wireless systems. Furthermore, the underlay and overlay approaches can be employed simultaneously to further improve spectrum efficiency.

5.1.3 Opportunistic Spectrum Access: Basic Components

We focus on the overlay approach under the hierarchical access model (see Figure 5.2). The term Opportunistic Spectrum Access (OSA) will be adopted throughout. Basic components of OSA include spectrum opportunity identification, spectrum opportunity exploitation and regulatory policy. The opportunity identification module is responsible for accurately identifying and intelligently tracking idle frequency bands that are dynamic in both time and space. The opportunity exploitation module takes input from the opportunity identification module and decides whether and how a transmission should take place. The regulatory policy defines the basic etiquette for secondary users to ensure compatibility with legacy systems.

The overall design objective of OSA is to provide sufficient benefit to secondary users while protecting spectrum licensees from interference. The tension between the secondary users' desire for performance and the primary users' need for protection dictates the interaction across opportunity identification, opportunity exploitation and regulatory policy. The optimal design of OSA thus calls for a cross-layer approach that integrates signal processing and networking with regulatory policy making.

5.1.4 Networking The Cognitive Radios

Recent research (such as [5–7]) suggests that it is not enough to establish a CR link (from CR-Tx to CR-Rx), and we shall develop ways to network the CRs, which is known as CR network (CRN). The CRNs can be deployed in network-centric, distributed, ad-hoc and mesh architectures, and serve the needs of both licensed and unlicensed applications. Also, CRN users can either communicate with each other in a multi-hop manner or access the base station. Thus, CRNs have three different access types as follows:

- *CR network access:* CR users can access their own base station on licensed and unlicensed bands.
- *CR ad-hoc access:* CR users can communicate with each other through ad-hoc communication on licensed and unlicensed bands.
- *Primary network access:* CR users can also access the primary base station through a licensed band.

As mentioned above, CR users can operate in both licensed and unlicensed bands. However, the functionality required for CRN is based on whether the spectrum is licensed or unlicensed. And some properties and constraints of CRN will also differ in spectrums that are licensed or unlicensed. We may categorise the CR application of spectrum into three possible scenarios, which are depicted in Figure 5.3: (i) CRN on a licensed band; (ii) CRN on a unlicensed band; and (iii) CRN on both a licensed band and unlicensed band.

5.2 Analytical Approach and Algorithms for Dynamic Spectrum Access

In order to allow cognitive radio (CR) to take advantage of the opportunity of spectrum hole (or availability) to transmit, which implies the CR system simultaneously operating in the same frequency bands without affecting the primary system (PS), we need to consider possibilities such as *underlay*, *overlay* and *interweave*.

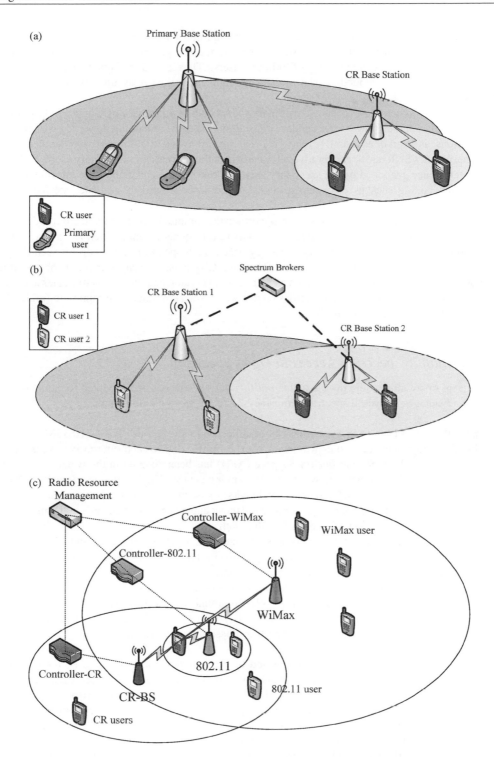

Figure 5.3 (a) CRN on licensed band; (b) CRN on unlicensed band; and (c) CRN on both licensed band and unlicensed band

The underlay approach allows concurrent primary and secondary transmissions by enforcing a spectral mask on the secondary signals so that the interference generated by the secondary users/ nodes is below the acceptable level of primary users. The secondary signal can be typically spread and is usually confined to short-range communications such as ultra wideband (UWB) systems.

The overlay approach allows concurrent primary and secondary transmissions. The secondary nodes/ users can use part of their power for secondary transmissions and the remaining power to assist primary transmissions, as a sort of cooperative relay [13]. Through appropriate splitting power, the increase of the primary user's SNR due to cooperative relay can offset the decrease of the primary user's SNR due to the interference generated by the secondary user(s). Contingent on available side information and sophisticated coding such as dirty paper, codes are commonly used to mitigate interference by secondary nodes/users.

The interweave approach is based on the opportunistic communication, derived from J. Mitola's original idea. There exist temporary *frequency voids* in a frequency band, which are referred to as *spectrum holes*, not used by the licensed/primary users. Such spectrum holes pop up according to changes in time and geographical locations. Therefore, CRs must constantly monitor the spectrum typically via physical layer spectrum sensing and then adopt certain medium access strategies to use the spectrum holes as transmission opportunities for secondary transmissions, with minimum (preferably no) interference to nodes/users in PS.

5.2.1 Dynamic Spectrum Access in Open Spectrum

The concept of cognitive radio in co-existing *multi-radio* systems has been adopted for some time to show its advantages in overall spectrum efficiency. Here are three good examples:

- Bluetooth devices have to co-exist with a lot of unlicensed radio devices at 2.4 GHz ISM (industry, scientific, medical) band, such as wireless LAN (i.e., WiFi) devices and microwave ovens. A key technology called *adaptive frequency hopping* (AFH) has been adopted in the Bluetooth 2.0 and IEEE 802.15.2, which consists of two steps: (i) scanning the ISM band by diving it into several sub-bands (typically three channels) to detect the band(s) in use by other radios (primary users or WiFi devices); (ii) adjust the hopping sequence to avoid the occupied band(s) to ensure good transmission quality to support ACL and SCO links for Bluetooth.
- IEEE 802.11h under development considers a situation at 5 GHz U-NII bands and ISM bands. In certain regions, the primary users may have very high power but low duty-cycle transmission (such as Doppler radar for weather detection). The IEEE 802.11a devices shall detect such primary users' transmissions and utilise the unused band(s) for signal transmissions.
- For global utilisation of *ultrawide band* (UWB) communication with signal transmission over an extremely wide bandwidth, a technique known as *detection and avoidance* (DAA), to detect primary users and then to avoid interference to their transmission, is required in general.

Consequently, communication over unlicensed bands provides good examples of the cognitive radio (CR) concept. Suppose spectrum sensing can be reliably realised (please see Chapter 7). Our next challenge is then how to model the access in open spectrum wireless networks. Figure 5.4 illustrates a representative example having two types of co-existing radio systems in a frequency band [16]. Type A system can use three wide frequency grids and type B system can use nine narrow frequency grids as shown. Suppose both types of radios conduct a listen-before-transmission (LBT, actually a sort of CSMA). The arrival process for each type system is Poisson with rate λ_i, $i = a$, b and service rate is exponential with rate μ_i, $i = a$, b, of course independent between the two types.

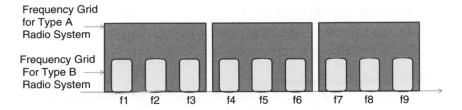

(a) Frequency Channels for Type A and Type B Radios

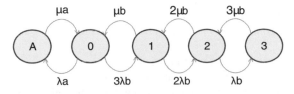

(b) Continuous Markov Chain Model

Figure 5.4 Co-existing multi-radio systems and Markov chain model for access: (a) Frequency channels for type A and type B radios; (b) Continuous Markov chain model

Such unlicensed spectrum access can be intuitively modelled as a continuous-time Markov chain as figure illustrating by defining five states as follows:

- State A: radio system A occupies the spectrum range;
- State 0: all three wide frequency grids are idle;
- State 1: there is only one type B system active;
- State 2: there are two type B systems active;
- State 3: there are three type B systems active.

By defining state-transition matrix Φ for the Markov chain and total probability property for steady-state distribution, we can obtain the steady-state distribution vector $\pi = [\pi_A, \pi_0, \ldots, \pi_3]$ by linear equations:

$$\pi\Phi = 0$$

$$\sum_{i=A}^{3} \pi_i = 1$$

and

$$\Phi = \begin{bmatrix} -\mu_a & \mu_a & 0 & 0 & 0 \\ \lambda_a & -\lambda_a - 3\lambda_b & 3\lambda_b & 0 & 0 \\ 0 & \mu_b & -\mu_b - 2\lambda_a & 2\lambda_a & 0 \\ 0 & 0 & 2\mu_b & -2\mu_b - \lambda_b & \lambda_b \\ 0 & 0 & 0 & 3\mu_b & -3\mu_b \end{bmatrix} \quad (5.1)$$

More networking properties can be analysed based on this model.

5.2.2 *Opportunistic Spectrum Access*

Opportunistic spectrum access (OSA), as a sort of hierarchical DSA, allows the secondary user to access channel(s) when the PS users are not transmitting. We shall optimise spectrum utilisation under the constraint of not affecting PS users.

Consider the hierarchical access class where PS users access the channel according to a scheduled or random access mechanism, and a CR user/node is trying to access the spectrum for opportunistic transmission. Assume there are N parallel channels (indexed from 0 to $N-1$) available within the frequency spectrum. Each channel is independently occupied by a number of PS users, to result in a mathematical model of continuous-time Markov chain described in the later Section , with idle state ($X_i=0$) and busy state ($X_i=1$). The holding times of states are independently and exponentially distributed with parameters λ_i^{-1} for idle states and μ_i^{-1} for busy states. The secondary users (i.e., CRs) execute slotted protocol and conduct actions in each time slot as follows [17]:

1. Sensing the channel at the beginning of each time slot.
2. Determining the channel to transmit among these N channels if there is traffic to send.
3. Receiving acknowledgement for successful transmission (note that this assumption may be worth discussing for CRN operations; practically, it is not possible to assume more actions beyond this acknowledgement as available time window for opportunistic access is very limited).

To simplify the analysis, we assume the sensing is done periodically, which is somewhat reasonable though not precise, as many radio systems are slotted and use beacon signals for timing reference. Each CR senses serially for N channels within N time slots in a protocol period. At the beginning of each time slot, CR senses the channel accordingly (e.g., sensing the channel n at the nth slot) and surely the sensing duration is much shorter than the time slot. Based on current and past sensing results, CR determines to transmit in one of the N channels (not limited to the slot just being sensed), or not to transmit at all.

For the kth time slot, $I_k \equiv [kT_s, (k+1)T_s]$, and the channel $q=k \bmod N$ is sensed, where T_s, denotes the time slot duration. After periodic sensing in I_k, we can define an N-dimensional random process vector $\mathbf{Z}(k) = [Z_0(k), \ldots, Z_{N-1}(k)]^T$ with

$$Z_i(k) = \begin{cases} X_i(kT_s), & \text{if } i = k \bmod N \\ Z_i(k-1), & \text{otherwise} \end{cases} \tag{5.2}$$

where $i=0, 1, \ldots, N-1$ and $k=N, N+1, \ldots$. As a matter of fact, $\mathbf{Z}(k)$ contains the sensing results of the most recent N time slots. If sensing is active at channel n, the nth component of $\mathbf{Z}(k)$ is updated. This Markov chain to model the observations depends on the 'life time' of sensing in terms of time slots. Let $q \equiv k \bmod N$ be the position of the slot in the current N-slot period. If channel i is sensed in slot k, the life time of sensing result is $L(i, k)=1$. In general,

$$L(i, q) \equiv (N+q-i) \bmod N$$

Then, the observed traffic dynamics can be described by a Markov chain model.

Proposition 5.1: Consider the N parallel channels with traffic modelled by independent binary-state continuous-time Markov chains. For channel i, $i=0, 1, \ldots, N-1$, let λ_i^{-1} be the mean holding time for state 0 and μ_i^{-1} for state 1, denote the transition rate matrix as

$$Q_i = \begin{bmatrix} -\lambda_i & \lambda_i \\ \mu_i & -\mu_i \end{bmatrix}$$

Consequently, the vector process $\mathbf{Z}(k)$, $k=N, N+1, \ldots$ defined in Equation (5.2) is a discrete-time Markov chain. We can therefore find the transition probability and thus define the mathematical model deriving engineering solutions.

5.2.3 Opportunistic Power Control

Motivated by the IEEE 802.22 wireless regional area network with TV broadcasting station as the PS, another way to avoid interference from CR to PS users while sharing the same spectrum is to conduct *transmission power control* (TPC) at CR (i.e., secondary user) [18]. We consider a CR system operating in a fading channel, in which CR can collect the instantaneous *channel state information* (CSI) of all links involved. All PS-Tx, PS-Rx, CR-Tx, CR-Rx, are within the communication range with the channel gains as Figure 5.5 shows, while channel fading statistics can be represented by these channel gain coefficients.

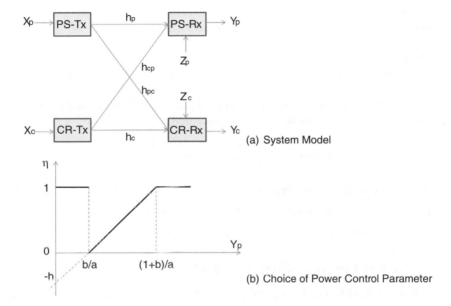

(a) System Model

(b) Choice of Power Control Parameter

Figure 5.5 Opportunistic power control: (a) System model; (b) Choice of power control parameter

At certain time instants, the received signals at both PS-Rx and CR-Rx are

$$Y_p = h_p X_p + h_{cp} X_c + Z_p$$
$$Y_c = h_c X_c + h_{pc} X_p + Z_c$$

where Z_p and Z_c are additive Gaussian noise and interference with zero mean and variance σ_p^2 and σ_c^2. Of course, the PS-Rx has no idea about the existence of the CRs to transmit at power P_p and rate R_p, and thus we can only assume single-user detection without the need to use CSI. With CSI from all involved links, CR-Tx adapts its transmission power P_c within the range of $[0, P_{c, max}]$ to fulfil the following goals:

- try to transmit as much information as possible in CR link;
- maintain the PS's outage probability unaffected by CR transmission.

Now, the SNR at PS-Rx without CR is

$$\gamma_p = \frac{|h_p|^2 P_p}{\sigma_p^2}$$

With CR transmission in the scenario, the SNR (more precisely, SINR) at PS-Rx is now

$$\gamma_p(\eta) = \eta \frac{|h_p|^2 P_p}{|h_{cp}|^2 P_{c,max} + \sigma_p^2}$$

and SINR at CR-Rx is

$$\gamma_c(\eta) = \eta \frac{|h_c|^2 P_{c,max}}{|h_{cp}|^2 P_p + \sigma_c^2}$$

We are using a parameter $0 \le \eta \le 1$ for power control in $[0, P_{c,max}]$. It is obvious that a challenge lies here to obtain all channel gains, which is somewhat an opening task for spectrum sensing.

Opportunistic power control can be considered in two cases:

- Primary link is outage: CR-Tx can use its full power at $P_{c,max}$.
- Primary link is not outage: CR-Tx has to decide an appropriate η without affecting PS, P_p, and R_p.

In order not to introduce outage, the choice of η should guarantee the PS link SINR at PS-Rx higher than $2^{R_p} - 1$ That is, CR-Tx should adjust transmission power in the way of

$$\eta = \min\left\{ \left[\frac{|h_p|^2 P_p}{2^{R_p} - 1} - \sigma_p^2 \right], 1 \right\}$$

Equivalently, for any given h_{cp}, the decision rule to choose η is plotted as in Figure 5.5(b) with $a = \frac{b}{2^{R_p} - 1}$ and $b = \frac{\sigma_p^2}{|h_{cp}|^2 P_{c,max}}$. This power control strategy is optimal to achieve maximal rate of CR users under the constraint of no degradation in PC outage probability.

5.3 Fundamental Limits of Cognitive Radios

As we described earlier, Mitola took the definition of an SDR one step further, and envisioned a radio that could make decisions as to the network, modulation and/or coding parameters based on its surroundings, calling such a 'smart' radio a cognitive radio. Such radios could even make decisions based on the availability of nearby collaborative nodes, or on the regulations dictated by their current location and spectral conditions. Since 2000, the FCC has actively been developing a *Secondary Markets Initiative*, as well as various rulemaking releases regarding the use of cognitive radio technologies. They are interested in removing unnecessary regulatory barriers to new secondary-market-oriented policies such as:

- *Spectrum leasing:* Allowing unlicensed users to lease any part of or all the spectrum of a licensed user.
- *Dynamic spectrum leasing:* Temporary and opportunistic usage of spectrum rather than a longer-term sublease.
- *'Private commons':* A licensee could allow unlicensed users access to his/her spectrum without a contract, optionally with an access fee.
- *Interruptible spectrum leasing:* Suitable for a leaser that wants a high level of assurance that any spectrum temporarily in use, or leased, to an incumbent cognitive radio could be efficiently reclaimed if needed.

A prime example of the latter would be the leasing of the generally unoccupied spectrum allotted to the US government or local enforcement agencies, which in times of emergency could be quickly reclaimed. Interruptible spectrum leasing methods resemble those of *spectrum pooling*. In current

FCC proposals on opportunistic channel usage, the cognitive radio listens to the wireless channel and determines, in either time or frequency, which resources are unused [1]. It then adapts its signal to fill this void in the spectrum domain, by transmitting either at a different time or in a different band, Thus, a device transmits over a certain time or frequency band *only when no other user does.*

Another potentially more flexible, general and spectrally efficient approach would be to allow *two users to transmit simultaneously* over the same time or frequency. Under this scheme, a cognitive radio listens to the channel and, if sensed idle, could proceed as in the current proposals (i.e., transmit during the voids). On the other hand, if another sender is sensed, the radio may decide to proceed with simultaneous transmission. The cognitive radio has not to wait for an idle channel to start transmission. The challenge with this new model would be: (i) is this spectrally more efficient than time sharing the spectrum? (b) what are the achievable rates at which two users could transmit, and how does this compare to when the devices are not cognitive radios, and yet still proceed in the same fashion? N. Devroye, P. Mitran and V. Tarokh [12] provide intelligent exploration towards answering these questions.

Cognitive radios have the ability to listen to the surrounding wireless channel, make decisions on the fly and encode using a variety of schemes. In order to exploit these abilities fully, first consider the simplest example, shown in Figure 5.6(a), of a channel in which a cognitive radio device could be used in order to improve spectral efficiency. As shown on the left, suppose sender $X1$ is transmitting over the wireless channel to receiver $Y1$, and a second incumbent user, $X2$, wishes to transmit to a second receiver, $Y2$. In the current secondary spectrum licensing proposals, the incumbent user $X2$, a cognitive radio that is able to sense the presence of other transmitting users, would either wait until $X1$ has finished transmitting before proceeding, or possibly transmit over a different frequency band. Rather than forcing $X2$ to wait, it has been suggested to allow $X2$ to transmit *simultaneously* with user $X1$ at the same time in the same band of frequencies. The wireless nature of the channel will make interference between simultaneously transmitting users unavoidable. However, by making use of the capabilities of a cognitive radio, it is shown that the cognitive radio is able to potentially mitigate the interference.

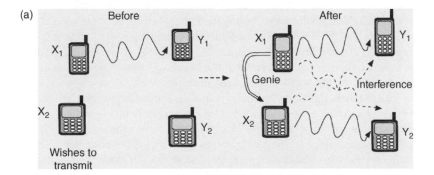

Figure 5.6 (a) The cognitive radio channel is defined as a two-sender ($X1$, $X2$), two-receiver ($Y1$, $Y2$) interference channel in which the cognitive radio transmitter $X2$ is non-causally given by a genie the message $X1$ plans to transmit. $X2$ can then either mitigate the interference it will see, aid $X1$ in transmitting its message or, as we propose, a smooth mixture of both. (b) Rate regions ($R1$, $R2$) for different two-sender, two-receiver wireless channels. Region (1) is the time sharing region of two independent senders. Region (2) is the best known achievable region for the interference channel. Region (3) is the achievable region described for the cognitive radio channel. Region (4) is an outer bound on the cognitive radio channel capacity. All simulations are in AWGN, with sender powers 6 and noise powers 1. The crossover parameters in the interference channel are 0.55 and 0.55 (see Figure 3 in [12])

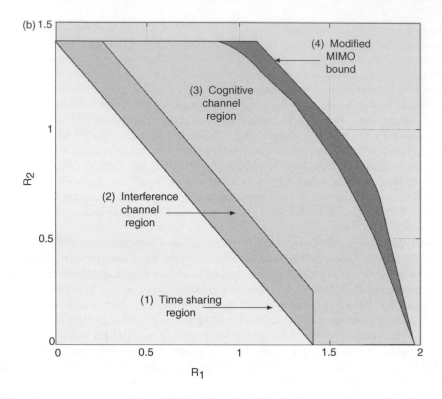

Figure 5.6 *(Continued)*

One of the many contributions of information theory is the notion of *channel capacity*. Qualitatively, it is the maximum rate at which information may be sent *reliably* over a channel. When there are multiple simultaneous information streams being transmitted, we can speak of capacity regions as the maximum set of all reliable rates that can be simultaneously achieved. For example, the capacity region of the channel depicted in Figure 5.6(b) is a two-dimensional region, or set of rates $(R1,R2)$, where $R1$ is the rate between $(X1,Y1)$ and $R2$ is the rate between $(X2,Y2)$. For any point $(R1,R2)$ inside the capacity region, $R1$ on the x-axis corresponds to a rate that can be reliably transmitted simultaneously, over the same channel, with $R2$ on the y-axis. There exist many channels whose capacity regions are still unknown. For such channels, tight inner and outer bounds on this capacity region are research goals. An inner bound is also called an *achievable rate/region*, and consists of suggesting a particular (often random) coding scheme and proving that the claimed rates can be reliably achieved; that is, that the probability of a decoding error vanishes with increasing block size. Notice that this guarantees the existence of schemes that can reliably communicate at these rates. Random coding does not construct explicit practical schemes, and does not guarantee that better schemes do not exist.

What differentiates the cognitive radio channel from a basic two-sender, two-receiver interference channel (in Chapter 4) is the *message knowledge* of one of the transmitters. This message knowledge is possible due to the properties of cognitive radios. If $X2$ is a cognitive radio, and is geographically close to $X1$ (relative to $Y1$), then the wireless channel $(X1, X2)$ could be of much higher capacity than the channel $(X1, Y1)$. Thus, in a fraction of the transmission time, $X2$ could listen to, and obtain, the message transmitted by $X1$. It could then employ this message knowledge, which translates into exact knowledge of the interference it will encounter, to mitigate it intelligently. Although we have used transmitter proximity to motivate the message idealisation assumption, and have proposed a particular transmission scheme for this scenario, different relative distances between transmitting and receiving nodes could dictate different schemes. It is important to note that our introduction is beneficial mostly in the weak interference case, because the strong and very strong interference channels have known capacity regions and known ways of achieving them. The relative node positions determine the type of interference channel results.

We introduce the 'genie' so as to idealise the message knowledge of sender $X2$. That is, we suppose that rather than causally obtaining the message that $X1$ is transmitting, a fictitious genie hands $X2$ this message. Notice that $X1$ is not given $X2$'s message, and so we have an asymmetric problem. This idealisation will provide an upper bound to any real-world scenario, and the solutions to this problem may provide valuable insights into the fundamental techniques that could be employed in such a scenario.

A simple cognitive radio channel is considered as a two-transmitter, two-receiver classical information theoretic interference channel in which sender 2 (a cognitive radio) obtains, or is given by a genie, the message sender 1 plans to transmit. The scenario is illustrated in Figure 5.6(a). The cognitive radio may then simultaneously transmit over the same channel, as opposed to waiting for an idle channel as in a traditional cognitive radio channel protocol. An intuitively achievable region for the rates $(R1, R2)$ at which $X1$ can transmit to $Y1$, and $X2$ to $Y2$, simultaneously has been constructed, which merges ideas used in dirty-paper (or Gel'fand-Pinsker) coding with the Han and Kobayashi achievable region construction for the interference channel, as well as the relay channel. When $X2$ has *a priori* knowledge of what $X1$ will transmit, or the interference it will encounter, one can think of two possible courses of action:

- It can *selfishly* try and mitigate the interference. This can be done using a dirty-paper coding technique. In this case, $X2$ is layering on its own independent information to be transmitted to $Y2$. This strategy yields points of higher $R2$ and lower $R1$ in the cognitive channel region of Figure 5.6(b).

  It can *selflessly* act as a relay to reinforce the signal of user $X1$. Such a scheme, although it does not allow $X2$ to transmit its own independent information, seems intuitively correct from a fairness perspective. That is, since $X2$ infringes on $X1$'s spectrum, it seems only fair that $X1$ should somehow benefit. This strategy yields points of high $R1$ and lower $R2$ in the cognitive channel region of Figure 5.6(b).

The resulting achievable region in the presence of additive white Gaussian noise is plotted as the 'cognitive channel region' in Figure 5.6(b). There, we see four regions:

- The time-sharing region (1) displays the result of pure time sharing of the wireless channel between users $X1$ and $X2$. Points in this region are obtained by letting $X1$ transmit for a fraction of the time, during which $X2$ refrains, and vice versa. These points would be amenable to the current proposals on secondary spectrum licensing.
- The interference channel region (2) corresponds to the best known achievable region of the classical information theoretic interference channel. In this region, both senders encode independently, and there is no message knowledge by either transmitter.
- The cognitive channel region (3) is the achievable region described here. In this case $X2$ received the message of $X1$ non-causally from a genie, and $X2$ uses a coding scheme that combines interference mitigation with relaying the message of $X1$. As expected, the region is convex and smooth. One can think of the convexity as a consequence of time sharing: if any two (or more) schemes achieve certain rates, then by time-sharing these schemes, any convex combination of the rates can be achieved. The region is smooth since our scheme actually involves power sharing at the coding level, which tends to yield rounder edges. We see that both users, not only the incumbent $X2$ that has the extra message knowledge, benefit from using this scheme. This is as expected, because the selfish strategy boosts $R2$ rates, while the selfless one boosts $R1$ rates; thus, gracefully combining the two will yield benefits to both users. The presence of the incumbent cognitive radio $X2$ can be beneficial to $X1$, a point which is of practical significance. This could provide yet another incentive for the introduction of such schemes.

■ The modified MIMO bound region (4) is an outer bound on the capacity of this channel: the 2×2 multiple-input multiple-output (MIMO) Gaussian broadcast channel capacity region, where we have restricted the transmit covariance matrix to be of the form

$$\begin{bmatrix} P_1 & C \\ C & P_2 \end{bmatrix}$$

to more closely resemble our constraints, intersected with the capacity bound on $R2$ for the channel for $(X2, Y2)$ in the absence of interference from $X1$.

5.4 Mathematical Models Toward Networking Cognitive Radios

Cognitive radio research is an emerging research area and there are various approaches to deal with different subjects. Researchers are modelling problems in quite diverse ways. Consequently, we present a few important and useful mathematical models related to cognitive radios (CR) and cognitive radio networks (CRN) in this section. Here, CRN has either a narrow-sense definition as a collection of CRs executing the same set of networking protocols, or a more general-sense definition as a collection of PS nodes and CRs executing the same set of CR networking protocols on top of existing PS networking.

5.4.1 CR Link Model

Assuming a primary communication system (or pair(s) of transmitter and receiver) is functioning, a cognitive radio that is the secondary user for the spectrum explores channel status and seeks the possibility of utilising such spectrum for communication. The channel can be commonly modelled as an Elliot-Gilbert channel [11] with two possible states: (i) existence of primary user(s) (a state not allowing any secondary user to transmit); and (ii) non-existence of primary user (a state allowing secondary user(s) to transmit). Stefan Geirhofer, Lang Tong and Brian M. Sadler [19] have measured WLAN systems to model statistically the 'white space' for cognitive radio transmission by the semi-Markov model shown in Figure 5.7.

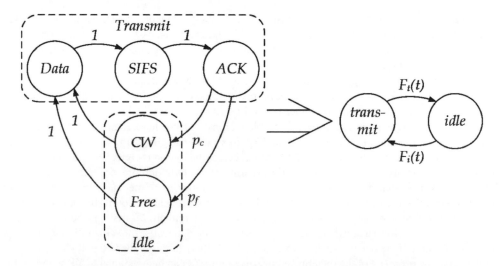

Figure 5.7   White space model from WLAN (Figure 3 in [19])

Assuming that a genie observes operation for both PS and CR, CR must utilise the spectrum hole window to complete transmission of packet(s). Suppose such a spectrum window period is denoted by T_{window}. It is clear that

$$T_{window} \geq T_{sense} + T_{CR-Transmission} + T_{ramp-up} + T_{ramp-down} \qquad (5.3)$$

where T_{sense} stands for minimum sensing duration to ensure CR transmission opportunity and acquisition of related communication parameters; $T_{CR-Transmission}$ is the transmission period for CR packets; and $T_{ramp-up/down}$ means the ramping (up or down) period for transmission. This equation ignores propagation delay and processing delay at the transmitter-receiver pair, which can be considered as a portion of ramp-up/down duration. The maximum duration of spectrum hole (availability) can be considered as the time duration for beacon signals. In the case where the beacon signals have fixed separation, the continuous-time Markov chain in Figure 5.7 can be reduced to a discrete-time Markov chain as Figure 5.8(b).

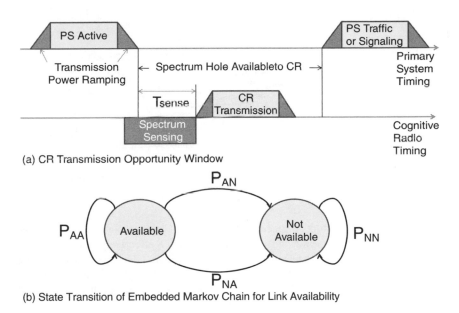

(a) CR Transmission Opportunity Window

(b) State Transition of Embedded Markov Chain for Link Availability

Figure 5.8 CR link availability model: (a) CR transmission opportunity window; (b) State transmission of embedded Markov chain for link availability

5.4.2 Overlay CR Systems

Please recall from Section 5.2 that we may classify CR (or general secondary radio) operation into the *overlay* model of CR based on the assumption of *known interference* and the *underlay* (actually also *interweave*) model based on the concept of *interference avoidance* [13].

5.4.2.1 Overlay Model with Known Interference

Consider CR communications over the frequency band of PS, as shown in the left-hand side of Figure 5.9, while CR-Tx wishes to transmit a message to CR-Rx and has *a priori* knowledge of PS-Tx's transmission message m_1. All channel gains are assumed to be known to both transmitters and receivers. CR-Tx therefore has two strategies:

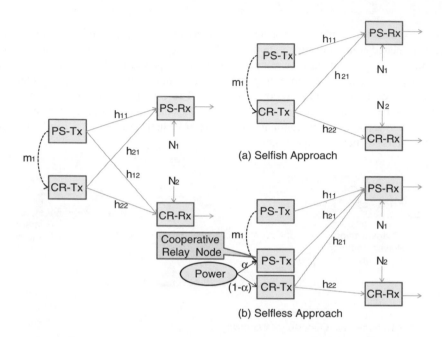

Figure 5.9   Overlay CR model: (a) selfish approach; (b) selfless approach

- *Selfish approach:* CR-Tx adopts a greedy strategy and uses all available power to send its own message to CR-Rx, by using its knowledge of the message in PS effectively to null the interference at CR-Rx by the dirty-paper coding technique [14]. That is why h_{12} does not appear in Figure 5.9(a), while PS-Rx suffers from extra interference created by CR-Tx through h_{12}. To make the CR concept work, we have to limit such interference from CR-Tx to PS-Rx to result in an upper bound of maximal throughput among the CRs (or CR network).
- *Selfless approach:* The CR-Tx allocates part of its power to replay cooperatively the PS-Tx message m_1 to PS-Rx. The rest of the power is used for the transmission of CR-Tx's message.

5.4.2.2 Interweave Model with Interference Avoidance

Under the assumption of knowing all links in overlay models, the *hidden terminal problem* (see Chapter 7 for more) does not exist. The sophisticated coding is useful to characterise the CR performance limit. However, non-causal knowledge of interference information is difficult to achieve. Since CRs have to face interference anyway, we may either build up interference-tolerant receivers or construct mechanisms to avoid interference by CR-Tx using unoccupied PS spectral segments. The *2-switch interweave model* is thus introduced as Figure 5.10. CR-Tx and CR-Rx are composed of a possible CR link, with a number of PS nodes. The dotted line regions are represented as sensing regions for CR-Tx and CR-Rx to identify activities of PS nodes. For example, PS nodes in location A and location B can be detected by CR-Tx and PS nodes in location B and location C can be detected by CR-Rx. To avoid interference in a CR link, we need spectrum sensing at both CR-Tx and CR-Rx to indicate spectrum availability, prior to establishing a CR link as Figure 5.10(a) shows. It also suggests that spectrum availability for CR links and thus for CRN operations should be dynamic and distributed. The 2-switch model is illustrated as Figure 5.10(b). The random variables X and Y represent the input-output relationship for a CR link, with Z as additive interference and noise. Two switching functions S_t and S_r are actually special indicator functions to indicate the activities of primary users based on the sensing by CR-Tx and CR-Rx respectively.

$$S_t = \begin{cases} 1, & \text{switching to CR link transmission as no active PS users being detected} \\ 0, & \text{switch open as CR--Tx detecting active transmission from PS nodes} \end{cases}$$

$$S_r = \begin{cases} 1, & \text{switching to CR link transmission as no active PS users being detected} \\ 0, & \text{switch open as CR--Rx detecting active transmission from PS nodes} \end{cases}$$

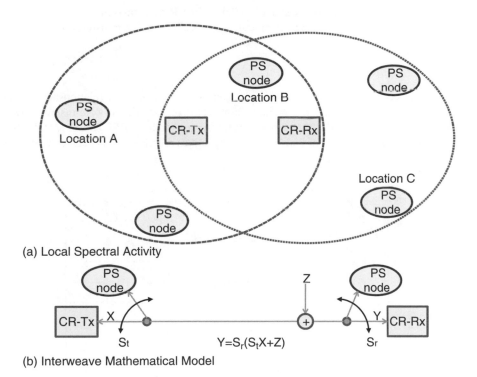

(a) Local Spectral Activity

(b) Interweave Mathematical Model

Figure 5.10   Interweave CR model (a) Local spectral activity; (b) Interweave mathematical model

Consequently, based on the network topology in Figure 5.10, the entire CR link I/O relationship can be represented as

$$Y = S_t(S_r X + Z)$$

Generally speaking, S_t and S_r are operating in an independent way with no information exchange between. The measurement of S_t and S_r might be correlated and the correlation can serve as an indication of the distributed nature. Opportunistic cognitive radios correspond to communication with some *partial side information* from this point of view. It is also easy for us to observe the importance of spectrum sensing to determine the CR link establishment, which will be discussed in more detail in

Chapter 7, especially cooperative sensing. Such a model can be useful in CRN design, with immediate applications to *medium access* of *dynamic (or opportunistic) spectrum access* (DSA). A good example is opportunistic channel selection and therefore network layer functions.

For interference avoidance, we may define a *medium occupancy indicator* (MOI) for the purpose of identifying whether the medium is available to establish a link:

$$1_{\text{MOI}} = \begin{cases} 1, & \text{medium occupied and not available for CR transmission} \\ 0, & \text{medium not occupied and free for CR transmission} \end{cases}$$

MOI can be a function of time, frequency and location, and may represent availability of a physical link or even a logic link. We may also define another indicator, *clear channel indicator* (CCI), by some assessment of the channel. A good example is the *clear channel assessment* (CCA) in the IEEE 802.11 MAC, as the core of carrier-sense multiple access (CSMA) operation:

$$1_{\text{CCI}} = \begin{cases} 1, & \text{medium clear and available for CR transmission} \\ 0, & \text{medium occupied and non available for CR transmission} \end{cases}$$

5.4.3 Rate-Distance Nature

A fundamental aspect of wireless communication systems is the *rate-distance* relationship (or nature), which has not drawn a lot of attention in communication theory but is critical in state-of-the-art wireless communication systems. Let us illustrate this observation from a realistic IEEE 802.11 a/g OFDM PHY and MAC. Figure 5.11 shows a set of realistic operating curves of IEEE 802.11g. Due to the received

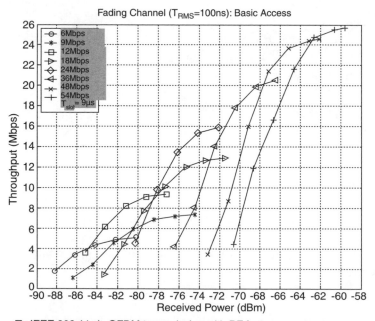

Figure 5.11   Received power versus IEEE 802.11a/g OFDM PHY data rate (and thus MAC throughput)

power level, the system will automatically adjust the PHY transmission rate accordingly, and thus throughput via MAC. It is a common working way in state-of-the-art wireless communication systems. However, we would like to introduce a further concept beyond overlay of CR systems based on it in the following, and it can be shown to further extend overall spectrum utilisation through networking CRs in Chapter 6 and the rate-distance nature that might be considered as a spatial domain extension beyond earlier descriptions about rather ideal overlay CR concepts.

If we consider the propagation distance between transmitter and receiver to have corresponding received power, we may create a new model for such a feature of wireless communications, and we may call it the *rate-distance* feature of wireless communications. We may further consider such a 'distance' as a measure of signal received power, rather than Euclidean distance or propagation distance, to characterise propagation factors for networking operation and cross-layer design. Consequently, distance measure D means any possible location point with received signal power as propagation Euclidean distance D under certain long-term fading. Figure 5.12(a) illustrates the rate-distance feature and the system having two transmission rates as an example. Figure 5.12(b) shows maximum allowable interference caused by the secondary user(s) to the primary users' system at the origin. It may be generally considered that lower rate transmission is more vulnerable to such interference. Consequently, as Figure 5.12(c) shows, a secondary user transmission rate/power can be scheduled without affecting primary user(s). Therefore, in the case where a cognitive radio senses the possible opportunity to transmit, its transmission rate (and thus power) is determined by the following rate-distance conditions:

- channel capacity in fading channel in terms of rate-power allocation;
- interference level by co-existing operating system(s);
- maximal tolerable interference to 'active' primary system user(s);
- effective 'distance' relationship among primary and secondary user devices.

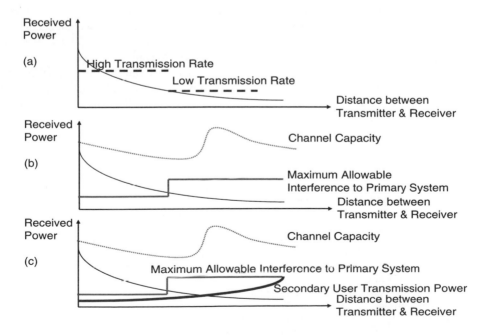

Figure 5.12   Rate-distance feature of cognitive radio

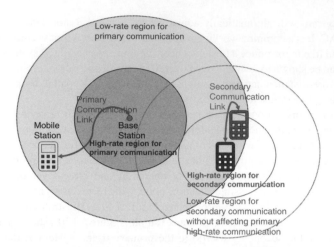

Figure 5.13 Rate-distance nature of co-existing primary/secondary communications

Figure 5.12 actually depicts the worst case scenario of cognitive radio communication, and a more rate-distance nature can be leveraged as shown in Figure 5.13. The base station and mobile station in the primary system are communicating. Due to their effective distance, the low-rate is selected. Near the boundary of the cell (according to the base station), there are two cognitive radio devices wishing to establish communication under the low-level of interference from primary system. As Figure 5.13 shows, high-rate communication might be possible between these two cognitive radios without affecting the primary system, and the interference from active primary system nodes to cognitive radios can be tolerated.

As a matter of fact, multiuser detection (MUD) can be applied here to alleviate co-channel interference for cognitive radios, because cognitive radios know the communication status of the primary system users. From initial synchronisation to user identification, all can be jointly determined. This is not limited to CDMA communications. It is shown that OFDM communications can use MUD to cancel co-channel interference, without precise known primary users. Using the random set theory with MUD, we may further identify and decode signals simultaneously from an unknown number of users. Based on the development of adaptive modulation and coding (AMC), we may summarise a mathematical condition to determine rate-power for secondary cognitive radios, without MUD.

References

[1] I.F. Akyilidiz, W.-Y. Lee, M.C. Vuran, S. Mohanty, 'A Survey of Spectrum Management in Cognitive Radio Networks', Georgia Institute of Technology.

[2] D. Cabric, I.D. O'Donnel, M. S.-W. Chen, R.W. Brodersen, 'Spectrum Sharing Radios', *IEEE Circuits and System Magazine*, 2Q 2006, pp. 30–45.

[3] NTIA, 'U.S. Frequency Allocation', [online] http://www.ntia.doc.gov/osmhome/allochrt.pdf

[4] First Report and Order, Federal Communication Commission Std. FCC 02-48, Feb. 2002.

[5] K.-C. Chen, Y.-J. Peng, N. Prasad, Y.-C. Liang, S. Sun, 'Cognitive Radio Network Architecture: Part I – General Structure', ACM ICUMIC, Seoul, 2008.

[6] K.-C. Chen, Y.-J. Peng, N. Prasad, Y.-C. Liang, S. Sun, 'Cognitive Radio Network Architecture: Part II – Trusted Network Layer Structure', ACM ICUMIC, Seoul, 2008.

[7] Q. Zhao, B.M. Sadler, 'A Survey of Dynamic Spectrum Access: Signal Processing, Networking, and Regulatory Policy', *IEEE Signal Processing Magazine*, **24**(3), 2007, 79–89.

[8] S. Haykin, 'Cognitive Radio: Brain-empowered Wireless Communications', *IEEE Journal on Selected Areas in Communications*, **23**(2), 2005, 201–220.

[9] T. Han, K. Kobayashi, Mathematics of Information and Coding, AMS 2002.

[10] A. Al-Fuqaha, B. Khan, A. Rayes, M.Guizani, O. Awwad, G.B. Brahim, 'Opportunistic Channel Selection Strategy for Better QoS in Cooperative Networks with Cognitive Radio Capabilities', *IEEE Journal on Selected Areas in Communications*, **26**(1), 2008, 156–167.

[11] H.S. Wang, N. Moayeri, 'Finite-State Markov *Channel – A Useful Model for Radio Communication Channels*', *IEEE Tr. on Vehicular Technology*, **44**(1), Feb. 1995.

[12] N. Devroye, P. Mitran, V. Tarokh, 'Limits on Communications in a Cognitive Radio Channel', *IEEE Communications Magazine*, June 2006, 44–49.

[13] S. Srinivasa, S.A. Jafar, 'The Throughput Potential of Cognitive Radio: A Theoretical Perspective', *IEEE Communications Magazine*, May 2007, 73–79.

[14] N. Devroye, P. Mitran, V. Tarokh, 'Achievable Rates in Cognitive Radio Channels', *IEEE Transactions on Information Theory*, **52**, May 2006, 1813–1827.

[15] S.A. Jafar, S. Srinivasa, 'Capacity Limits of Cognitive Radio with Distributed and Dynamic Spectral Activities', *IEEE Journal on Selected Areas in Communications*, **25**, April 2007, 529–537.

[16] Y. Xing, R. Chandramouli, S. Mangold, S. Shankar, 'Dynamic Spectrum Access in Open Spectrum Wireless Networks', *IEEE Journal on Selected Areas in Communications*, **24**(3), 2006, 626–637.

[17] Q. Zhao, S. Geirhofer, L. Tong, B.M. Sadler, 'Opportunistic Spectrum Access via Periodic Channel Sensing', *IEEE Transactions on Signal Processing*, **56**(2), 2008, 785–796.

[18] Y. Chen, G. Yu, H.H. Chen, P. Qiu, 'On Cognitive Radio Networks with Opportunistic Power Control Strategies in Fading Channels', *IEEE Transactions on Wireless Communications*, **7**(7), 2008, 2752–2761.

[19] S. Geirhofer, L. Tong, B.M. Sadler, 'Dynamic Spectrum Access in the Time Domain: Modeling and Exploiting White Space', *IEEE Communications Magazine*, May 2007, 66–72.

[20] E Biglieri, M Lops, 'Multiuser Detection in a Dynamic Environment – Part I: User Identification and Data Detection', *IEEE Transactions on Information Theory*, **53**(9), September 2007, 3158–3170.

6

Cognitive Radio Networks

As we described in Chapter 5, traditional CRs aim at using spectrum holes by dynamic spectrum access to enhance spectrum efficiency. Generalising the concept of cooperative networking with the CR terminal's ability to adapt its communication to connect to various and multiple co-existing radio systems, we can form a cognitive radio network (CRN) in which various systems can be connected and cooperate together, no matter which nodes belong to the primary system (PS) or are CR nodes. An important generalisation of the cognitive radio network is that CR terminals must be able to use the existing PS and/or cooperative/cognitive radio nodes to relay their message, in the form of a cooperative relay (an example is shown in Figure 6.1). We call this form of cooperative relay networking nodes a *cognitive radio relay network* (CRRN). Consequently CRRN can be composed of the network of the PS and/or cooperative/cognitive radio nodes and the CR terminals using the network to transmit data. Please note that cooperative relay nodes here aim at creating a larger aggregated bandwidth for the entire network, instead of just their common purpose of creating diversity gain in cooperative communications.

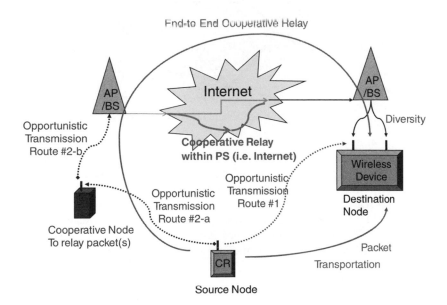

Figure 6.1 Cooperative relay nodes for more aggregated networking bandwidth

Cognitive Radio Networks Kwang-Cheng Chen and Ramjee Prasad
© 2009 John Wiley & Sons, Ltd

In other words, CR is not just a link-level technology to use the spectrum hole. CR technology can work together with cooperative relay (i.e., cooperative networking) to form CRRN by leveraging the PS. Furthermore, messages or packets from the CR-source node can reach the CR-destination node through multi-hop cooperative relay networks, with nodes from co-existing PS or other CR nodes as relay nodes. Such a complete networking scenario represents *cognitive radio network* (CRN) in this book.

6.1 Network Coding for Cognitive Radio Relay Networks

Before going on to the complete networking functions of CRN, we focus on whether CRRN can bring benefits for better network efficiency, which also suggests 'true' spectral efficiency given spectrum bandwidth (rather than spectral efficiency at physical layer). In order to explore this fundamental nature, Huang and Chen [2] adopt network coding in network information theory to study CRRN, while Geng [12] applied network coding to cognitive radio networks composed of just ad-hoc CR nodes.

We use Figure 6.2 to illustrate CRRN. The *PS source* transports packets to the PS sink through the PS network infrastructure consisting of relay nodes (i.e., relay network like any data network). CR source may leverage this (PS) relay network to transport its packets to the CR sink. Therefore, traffic from PS source and CRs co-exist in this relay network to form CRRN with CRs. Since PS's network capacity may decrease due to CR's interference, we analyse the fundamental behaviour of CR's interference to learn how to avoid interference to the PS, so that we can facilitate CRRN by maximising the entire network throughput of CRRN given no interference to PS traffic.

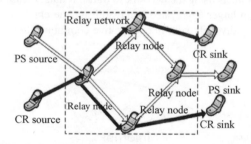

Figure 6.2   Cognitive radio relay networks

In the following, we analyse the interference between CR and PS by examining the *min-cut* capacity variation of the PS in CRRN. Min-cut capacity is the capacity of a network, which is derived in network coding theory. *Superposition coding* is considered in the analysis because CRRN is likely to adopt a *decode-and-forward* (DF) cooperation policy. That is, nodes in the relay network decode the CR's message and forward by superimposing CR's message and PS's message. We analyse under the DF policy whether CR's interference is avoidable or bounded in various CRRN topologies. In spite of diverse scenarios for CRRN, we may consider the following four cases as basic components of CRRN:

- one hop relay network (Figure 6.3(a));
- tandem relay network (Figure 6.3(b));
- cooperative relay network (Figure 6.4(a));
- parallel cooperative relay network (Figure 6.4(b)).

These network topologies are used in cooperative communication with network coding. Their analysis of bounded and avoidable interference is then generalised to arbitrary CRRN topologies with a DF

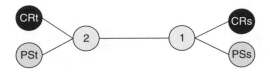

(A) One-Hop Relay:
Interference to PS from CR is unavoidable and unbounded.

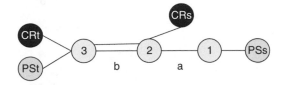

(B)Tandem Relay:
The network capacity of such network is constrained by the smallest link capacity among all links. Hence CR can connect to the network and transmit through those links other than the smallest capacity.

Figure 6.3   (a) One-hop relay and (b) tandem relay

cooperation policy. We target at tolerable interference analysis due to CR's nature of avoiding interference to PS. First, we derive the conditions to enable CR to avoid interference to the PS in CRRN. Second, we restrict the CR and PS to be uni-cast in CRRN to simplify the maximum capacity analysis. Under the constraint of avoiding interference to the PS, we derive the CR's maximum network capacity, and formulate the link allocation problem to achieve the maximum network capacity as a *multi-commodity flow* problem. Third, we release the constraint on uni-cast, assume the PS is multicast and the CR is uni-cast, and we show that the problem of link capacity allocation to maximise the CR's network capacity in this situation becomes another linear programming problem. Therefore, we can decide whether an arbitrary network can be used by the CR without interference and maximise the CR's network capacity in the CRRN. Finally, unavoidable interference and interference bounded conditions are also analysed. Based on these steps, we can develop operating principles to examine the 'opportunity' to enhance network throughput without interfering with the PS.

6.1.1 System Model

We therefore develop the following system model as shown in Figure 6.2 to study the network capacity of CRRN.

6.1.1.1 Assumptions for Our CRRN

We make the following assumptions to focus our analysis on interference in CRRN and simplify the analysis procedures:

- There is one PS source, one CR source, one or a few PS sinks and one or a few CR sinks in the CRRN we analyse.
- The CR source does not transmit directly to its sinks. That is, we consider only the CR traffic relayed by relay network.
- The links in CRRN are generally uni-directional since CR links are likely to exist by chance for a short period of time. Therefore, the networks are able to be modelled as directed graphs.

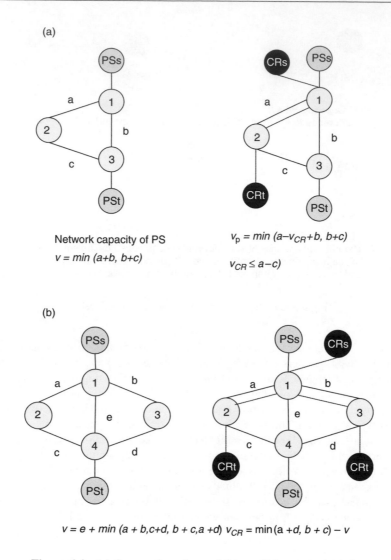

Figure 6.4 (a) Cooperative relay and (b) parallel cooperative relay

- Building new links between the CR nodes and the nodes of relay network does not alter the link capacity of other links.
- Every edge in the relay network is contained in at least a path from the PS source to PS sink. In other words, every link in the relay network that should relay the traffic comes from the PS source.

We denote the directed graph of the network without CR terminals that use the relay network as $G = (V, E)$. V is the set of nodes in G, E is the set of links in G and capacity matrix $\boldsymbol{R} = [R_{ij}]$ whose entries correspond to link $(i,j) \in E$. Similarly, we denote the directed graph of CRRN $G'' = (V'', E'')$, capacity matrix $\boldsymbol{R'} = [R'_{ij}]$, whose entries correspond to link $(i,j) \in E'$. Consequently, the two graphs have following relationships:

$$1. \ V' = V \cup s_{CR} \cup t_{CR}, \ E' = E \cup E_{CR-V},$$

$$2. \ R'_{ij} = R_{ij} \ if \ (i,j) \in E.$$

s_{CR}, t_{CR} are the set of the CR source and CR sinks respectively, and E_{CR-V} is the link between the CR nodes and relay network nodes. We call $G = (V, E)$ the original network and $G'' = (V'', E'')$ the CRRN in the following sections.

6.1.1.2 Decode-and-Forward Cooperation Policy: Superposition coding

Based on network coding theory model in [10, 11], we define the following components for the network code we consider here:

1. Message set:

$$\Omega_{PS} = \{1, \ldots \lceil 2^{nh_{PS}} \rceil\}, \quad \text{and} \quad \Omega_{CR} = \{1, \ldots \lceil 2^{nh_{CR}} \rceil\}$$

2. Encoding functions on each link:
   For the links that do not relay message from CR:

$$f_{ij} : \prod_{(i',i) \in E'} A_{i'i} \rightarrow A_{ij} \tag{6.1}$$

For the links that relay both the CR's and PS's messages:

$$f_{ij,PS} : \prod_{(i',i) \in E'} A_{i'i,PS} \rightarrow A_{ij,PS} \tag{6.2}$$

$$f_{ij,CR} : \prod_{(i',i) \in E'} A_{i'i,CR} \rightarrow A_{ij,CR} \tag{6.3}$$

For the links that connect node i to sink:

$$g_i : \prod_{(i',i) \in E'} A_{i'i} \rightarrow A_i \tag{6.4}$$

3. Decoding in sinks:
   For PS

$$g_{PS} : \prod_i A_i \rightarrow \Omega_{PS} \tag{6.5}$$

and for CR

$$g_{CR(i)} : \prod_i A_i \rightarrow \Omega_{CR} \tag{6.6}$$

Message sets for the PS and CR are Ω_{PS} and Ω_{CR}. Sources uniformly choose an index from their sets and then transmit it on the network: n is the block code length, h_{PS} and h_{CR} are the code rates and f_{ij} is the encoding function on link (i,j). In our CRRN, relay nodes decode messages from the PS and CR separately and superimpose them to transmit on the links. So the encoding functions of the links that relay both the CR's and PS's messages are $f_{ij,PS}$ and $f_{ij,CR}$ and they separately code the CR's and PS's messages. Then $A_{ij} = A_{ij,PS} \times A_{ij,CR}$ would be sent on link (i,j). According to these settings, the code rate on each link (i,j), which relays both data from the CR and PS, is

$$\begin{aligned} R_{ij} &= n^{-1} \log_2 |A_{ij}| \\ &= n^{-1} \log_2 |A_{ij,PS}| + n^{-1} \log_2 |A_{ij,CR}| \end{aligned} \tag{6.7}$$

This is known as *superposition coding*. This coding scheme superimposes the two network codes together. Although superposition coding sometimes cannot achieve multi-source network capacity bounds derived in [11], we use it because it is probably not feasible to compress and code messages jointly from the PS and CR. And we do not need to alter the PS's network code in CRRN if the PS's network capacity is maintained. In other words, in CRRN, the PS can use the same network code as if the CR terminals do not connect to the network under superposition coding. Superposition coding is a DF cooperation policy.

6.1.1.3 Network capacity under DF cooperation policy

Under the DF cooperation policy, for those links that contain both traffic from the CR and PS, we can denote the capacity of the link $R_{ij} = R_{ij,CR} + R_{ij,PS}$, while $R_{ij,CR} > 0$ and $R_{ij,PS} > 0$. By network coding theory, under our settings, we can derive network capacity by calculating the min-cut capacity respective to the PS and CR. We analyse interference by the follow procedures. First, we want to derive the network capacity v of original network, $G = (V, E)$ and $\boldsymbol{R} = [R_{ij}]$. Then we can derive the network capacity of the PS v_p, the network capacity of the CR v_{CR} and the total network capacity $v_t = v_{PS} + v_{CR}$ in the CRRN, $G'' = (V'', E'')$ and $\boldsymbol{R}' = [R'_{ij}]$. Thus we are able to know the conditions enabling us to achieve $v_{PS} = v$ and $v_{CR} > 0$, which means the CR's interference is avoidable. Moreover, we want to maximise v_{CR} and maintain $v_{PS} = v$ by appropriately allocating link capacity. We can also derive whether v_{PS} is always larger than zero under any link capacity allocation. If yes, we call the interference bounded.

6.1.2 Network Capacity Analysis on Fundamental CRRN Topologies

In different CRRN topologies, the CR under the DF cooperative policy may cause interference to the PS to different extents. Depending on topology and link capacity, CR's interference can be avoidable and unavoidable, bounded and unbounded. We analyse these interference properties in the following fundamental topologies, and generalise the analysis to arbitrary topology. We set the capacity of links between relay nodes and sources or sinks to be infinite. Hence we can focus on the interference in the relay network.

6.1.2.1 One-hop Relay Network

We start from the simplest topology analysis: one hop relay network. There is only one link in this network. Intuitively, as long as the CR is transmitting, the PS would be interfered with by the CR's transmission, and not be able to transmit if the CR occupies full link capacity. Therefore, interference is unavoidable and unbounded in this network.

6.1.2.2 Tandem Relay Network

A tandem relay network is formed by connecting relay nodes in series. The network capacity of such a network is constrained by the smallest link capacity among all links. Hence the CR can connect to the network and transmit through those links other than the smallest capacity one, i.e., if $b > a$ in Figure 6.3(b). Unless we allocate too much capacity to the CR in a link so that the link becomes the smallest capacity link for the PS, the CR will not interfere with the transmission of PS, the hence interference is avoidable. However, if the CR occupies full capacity in any link, the PS cannot transmit anything by this relay network. Therefore, the CR's interference is unbounded if we do not put any constraint on it.

6.1.2.3 Cooperative Relay Network

Like cooperative communications, we form a cooperative relay network by adding a cooperative relay node (node 2 in Figure 6.4), thus there are three links and two cuts in the network. Let us first consider the network capacity of the original network (i.e., without CRs as in the scenario at the left hand side of Figure 6.4(a)). Let the capacity of link $(1,2),(1,3),(2,3)$ be a,b,c respectively. The cuts of the PS are $\{(1,2),(1,3)\}, \{(2,3), (1,3)\}$. Network capacity in this PS network is

$$v = \min(a+b, b+c) \tag{6.8}$$

Now, we calculate the network capacity of cooperative relay as depicted in the scenario at the right hand side of Figure 6.4(a)). We connect the CR sink to node 2, the cooperative relay node. Denote the network capacity of the CR v_{CR}, capacity of the PS v_{PS}, and the total network capacity of CRRN (CR plus PS) v_t. The cut of CR is $(1,2)$. Network capacities of the PS and CR are

$$v_{PS} = \min(a - v_{CR} + b, b + c), \quad v_{CR} \leq a \tag{6.9}$$

If $a + b > v$, v_{CR} can be larger than zero while $v_P = v$. Under this condition

$$v_{PS} = \begin{cases} a - v_{CR} + b & \text{if} \quad v_{CR} \geq a - c \\ v & \text{if} \quad v_{CR} < a - c \end{cases}, \quad v_{PS}, v_{CR} > 0 \tag{6.10}$$

$$v_t = \begin{cases} a + b & \text{if} \quad v_{CR} \geq a - c \\ v + v_{CR} & \text{if} \quad v_{CR} < a - c \end{cases} \tag{6.11}$$

If we set $v_p = v$, we know that

$$v_{CR} \leq a - c, \tag{6.12}$$

Hence $a - c$ is the maximum network capacity of the CR when $v_{PS} = v$. Note that $v_{CR} \leq a$, therefore, we always have $v_{PS} \geq b$. This shows that interference of the CR to the PS is bounded in this case. Under this setting, one CR source and one CR sink, situations are the same when we add them in $(2,3)$, just change a to c in the previous analysis.

From the above analysis, if the CR's message is relayed by the links that do not belong to min-cut of PS, then the CR's interference to the PS is avoidable. Moreover, the interference to the PS is bounded in this case. This observation results from the fact that adding the cooperative relay node creates a new route and more cuts, and hence the CR's interference to the PS is avoidable and bounded. In this case, there are two scenarios of the relationship between the CR's and PS's network capacity. When the CR's network capacity has not yet reached its maximum value, we can increase the CR's network capacity without altering the PS's network capacity. After reaching the maximum value of the CR's network capacity, network capacities of the CR and PS can be allocated in a certain range, but the capacity we add to the CR is equal to the capacity we subtract from the PS, and vice versa.

6.1.2.4 Parallel Cooperative Relay Network

Up to this moment, we have considered only one CR sink. Now we extend our analysis to multiple CR sinks to investigate the interference. In the cooperative relay network, we add an extra cooperative relay node, and only one CR sink can be added in under the constraint of avoiding interference to the PS. Now we add another node (node 3 in Figure 6.4(b)) in to form a cooperative relay structure parallel to the original one. Then we add another CR sink into the CRRN. In this topology, we have four cuts, each with two links.

Let us again consider the network capacity of the original PS network as in left hand side of Figure 6.4(b). Denote the capacity of link (1,2),(1,3),(2,4),(3,4),(1,4) to be a,b,c,d,e respectively. The cuts of the PS are $\{(1,2),(1,3),(1,4)\}$, $\{(2,4),(3,4),(1,4)\}$, $\{(1,2),(3,4),(1,4)\}$, $\{(1,3),(2,4),(1,4)\}$. Network capacity of this network is:

$$v = e + \min(a+b, c+d, b+c, a+d) \tag{6.13}$$

We then analyse the network capacity of the parallel cooperative relay with the CRs. The cuts of the CR are $\{1,2\},\{1,3\}$ to give the network capacities of the PS and CR as

$$v_{PS} = \min(a+b-2v_{CR}, a-v_{CR}+d, b-v_{CR}+c, c+d) + e, \quad v_{CR} \leq \min(a,b) \tag{6.14}$$

Therefore, if $a+b, a+d, b+c > v$, that is, $a > c, b > d$, then v_{CR} can be larger than zero while $v_{PS} = v$. Under this condition and assuming $a+d > b+c$, we have

$$v_{PS} = e + \begin{cases} a+b-2v_{CR} & \text{if } v_{CR} > b-d \text{ and } v_{CR} > a-c \\ b-v_{CR}+c & \text{if } v_{CR} > b-d \text{ and } v_{CR} > a-c \\ c+d & \text{if } v_{CR} < b-d \text{ and } v_{CR} > a-c \end{cases} \tag{6.15}$$

$$v_t = e + \begin{cases} a+b-v_{CR} & \text{if } v_{CR} > b-d \text{ and } v_{CR} > a-c \\ b+c & \text{if } a-c > v_{CR} > b-d \\ c+d+v_{CR} & \text{if } v_{CR} < b-d \text{ and } v_{CR} < a-c \end{cases} \tag{6.16}$$

These cases are symmetrical to $a+d < b+c$. If we set $v_{PS} = v$, we have

$$v_{CR} = \min(a+d, b+c) - v \tag{6.17}$$

This topology can accommodate two CR sinks in the two parallel relay nodes, and the CR's interference is still avoidable and bounded. We can infer that more CR sinks can exist in the parallel cooperative relay CRRN due to the parallel structure created by the cooperative relay nodes. However, in the case of two CR sinks and one PS sink, increasing network capacity of the CR may possibly decrease total network capacity. This is due to the two links connecting to the two CR sinks belonging to the same min-cut of PS in this case. Therefore if we increase the CR's network capacity by x, we decrease the PS's network capacity by $2x$, and thus total network capacity is decreased by x. But if we appropriately choose the network capacity of the CR, we still can improve total network capacity and avoid interference to PS.

From the above analysis, we can observe some crucial relationships between the interference properties and network topologies as follows

- *Avoidable interference:*
  - In one hop relay network, the interference is always unavoidable because the min-cut includes all the edges in the network.
  - In tandem relay network, if the cut of CR is not included in the min-cut of PS, interference is avoidable.
- *Bounded interference*:
  - Interference in the cooperative relay network is bounded because the CR cannot occupy all edges in any cut of PS.
- *Multiple CR sinks:*
  - In the parallel cooperative relay structure with multiple CR sinks, the above properties still hold, but total network capacity in the CRRN may decrease when the constraint on avoiding interference to the PS is released.

Now we generalise the above observations to arbitrary network topologies under the DF cooperation policy. In the following analysis, we denote the network capacity of the CR as ν_{CR}, the network capacity of the PS as ν_{PS}, the network capacity of the original network as ν, and total network capacity in the CRRN as $\nu_t = \nu_{PS} + \nu_{CR}$. We first elaborate the *avoidable interference*, then release the constraint on avoiding interference and analyse *unavoidable interference*, and finally we derive the condition(s) for *bounded interference*.

Regarding *avoidable interference*, in most cases the CR tries to use the communication resource under the constraint of avoiding interference to the PS. Hence we emphasise this aspect and below provide more details about how the CR can avoid interference when using the PS to relay. Lemma 1 derives the condition for the relay network topologies, and Theorem 1 further derives the condition for the CRRN topologies based on Lemma 1. In these network topologies, we derive the CR's maximum network capacity in the uni-cast PS network in Theorem 2, and then provide algorithms to achieve the CR's maximum network capacity in both uni-cast PS and multicast PS. Finally, we briefly analyse the total network capacity variation when the CR begins interfering with the PS.

Lemma 6.1: Given that every link belongs to at least one min-cut in the CRRN, if $\nu_{PS} = \nu$, then $\nu_{CR} = 0$.

Proof: Denote the set of PS cuts φ, φ contains all the PS cuts whose capacities are equal to ν. And the capacity of a cut c is denoted $R(c)$ in the original network and $R'(c)$ in the CRRN. Suppose $\cup_{c_i \in \varphi} c_i = E$. If $\nu_{CR} > 0$, there is at least one link (i,j) such that $R_{ij,CR} > 0$. The link capacity of PS on edge (i,j) becomes

$$R'_{ij,PS} = R'_{ij} - R_{ij,CR} = R_{ij} - R_{ij,CR} < R_{ij} \qquad (6.18)$$

This edge (i,j) must belong to some cut c_i in φ. Because $R'_{ij,PS} < R_{ij}$, we have $R'(c_i) < R(c_i)$. Therefore, we can infer $\nu_{PS} = R'(c_i) < R(c_i) = \nu$. Hence there is a contradiction. Thus our lemma is proved.

Theorem 6.1: Set $\nu_p = \nu$. Then $\nu_{CR} > 0$ *if and only if* there is no cut contained in the edge set that contains all min-cuts of PS.

Proof: We first prove the 'if' part. Let link set $\varphi_p = \{e_i | e_i$ belong to any cut whose capacity $= \nu_{PS}\}$, let set of cuts ψ_{CR} include all cuts of the CR and let h_{PS} be the smallest PS cut capacity among the PS cuts not belonging to φ_{PS}. If $c_i \not\subset \varphi_{PS}$ for all $c_i \in \psi_{CR}$, allocate small capacity $h_{PS} - \nu_{PS}/m > 0$ to the CR in every link $e_i \not\in \varphi_p$. Let m be the number of edges in the relay network. No more than m edges can belong to the same cut, hence the capacity of every element in ψ_{CR} is larger than or equal to $h_{PS} - m \times h_{PS} - \nu_{PS}/m$. Hence $\nu_{PS} = \nu$, $\nu_{CR} \geq h_{PS} - \nu_{PS}/m > 0$.

Now we prove the 'only if' part. Assume $\nu_{CR} = \Delta > 0$, then the capacity of every element in ψ_{CR} is larger than or equal to Δ. But $\nu_{PS} = \nu$, so the link capacity for every element in φ_p allocates no capacity to the CR, hence $c_i \not\subset \varphi_{PS}$ for all $c_i \in \psi_{CR}$.

Note that the 'only if' part may not be necessarily true if we consider the compress-and-relay cooperative policy instead of the DF cooperative relay policy.

Next we derive the CR's maximum network capacity under the constraint of avoiding interference to the PS, as well as link capacity allocation algorithms in the PS uni-cast network. To simplify the problem, we first consider only one PS sink and one CR sink when we derive the CR's maximum network capacity. We define the cut of the CRRN to mean the edge set that divides the CRRN into two disconnected networks, one containing the CR source and PS source, the other containing the CR sink and PS sink. Consequently, a cut of the CRRN must contain the CR cut and PS cut. We assume that every edge in the relay network is contained in at least one path from the PS source to PS sink. According to this assumption, every cut of the CRRN is a cut of the PS.

Theorem 6.2: Let $\nu_{PS} = \nu$. The CR's maximum network capacity in the CRRN is the CRRN's minimum cut capacity minus original network capacity.

Proof: Let the CRRN's minimum cut capacity be ν_{CRRN}. We can add a supersource connected to the CR source and PS source and a supersink connected to the CR sink and PS sink, hence the network capacity of the supersource is ν_{CRRN}, and $\nu_{PS} + \nu_{CR} = \nu_{CRRN}$. Therefore,

$$\nu_{CR} = \nu_{CRRN} - \nu_{PS} = \nu_{CRRN} - \nu \tag{6.19}$$

Thus our lemma is proved.

6.1.3 Link Allocation

After we derive the CR's maximum network capacity, we should find a way to allocate link capacity to achieve the CR's maximum network capacity without interfering with the PS. The *link allocation* problem in the CRRN can be formulated as a *multi-commodity flow* problem. The messages from the CR and PS are the commodities, and the CR's maximum network capacity $\nu_{CRRN} - \nu$ and PS's network capacity ν are the commodities' demands (i.e., flow value). Therefore, we can allocate the link capacity to achieve the CR's maximum network capacity by solving the multi-commodity flow problem through linear programming.

Then we release the assumption of uni-cast, let the PS be multicast and the CR still uni-cast. For a multicast network, the network capacity is limited by its min-cut capacity. Hence we can formulate the problem of maximum network capacity of the CR by extending the multi-commodity flow problem.

Proposition 6.1: In the CRRN of multicast PS and uni-cast CR, the problem of link allocation to achieve maximum network capacity can be formulated as a linear programming problem. We show Proposition 6.1 is true in the following. We consider the flow from the PS source to each sink separately. Every flow should exceed or at least equal ν to guarantee $\nu = \nu_{PS}$. Then we choose the maximum flow among the flows to each PS sink on each edge to be the link capacity allocated to the PS, and maximise the CR's network capacity.

Now we can show that this problem can also be formulated as a linear programming problem. We change the notation of original network capacity from ν to α in order to distinguish from the notation of vertices ν. Hence our link allocation algorithm in the PS multicast network can be formulated as follows. s_{PS} and s_{CR} denote the sources of the PS and CR, t_{PSl} and t_{CR} denote the sinks of the PS and CR (PS has multiple sinks), f_{PSi} and f_{CR} denote the link capacity (flow) allocated to the PS and CR and ν denotes the relay nodes. The linear programming problem of link allocation in the PS multicast CRRN is summarised as follows:

Maximise

$$\sum_{\nu \in V_{CRs}} f_{CR}(s, \nu), \quad V_{CRs} = V - \{t_{PSl} | l = 1 \ldots k\} \tag{6.20}$$

subject to

$$(i) \quad \sum_{\nu \in V_{is}} f_{PSi}(s_{PS}, \nu) \geq a$$

$$V_{is} = V - \{s_{CR}, t_{CR}, t_{PSl} | l = 1 \ldots k\} \tag{6.21}$$

(ii) $\sum_{v \in V_i} f_{PSi}(u, v) = 0$ where

$$V_i = V - \{t_{CR}, t_{PSl} | l = 1 \ldots k \text{ except } i\},$$

$$u \in V - \{s_{CR}, s_p, t_{CR}, t_{PSl} | l = 1 \ldots k\}$$

(6.22)

(iii) $\sum_{v \in V_{CR}} f^z_{CR}(u, v) = 0$ where

$$V_{CR} = V - \{t_{PSl} | l = 1 \ldots k\},$$

$$u \in V - \{s_{CR}, s_{PS}, t_{CR}, t_{PSl} | l = 1 \ldots k\}$$

(6.23)

(iv) $f_{PSi}(u, v) = -f_{PSi}(u, v)$ (6.24)

(v) $f_{CR}(u, v) = -f_{CR}(v, u)$ (6.25)

(vi) $f_{PSi}(u, v) \le c(u, v) - f_{CR}(u, v)$ $for\ i = 1 \ldots k$ (6.26)

We can solve the above linear programming by many well-known algorithms such as the *simplex algorithm*.

The above results provide us with a framework to form the CRRN in which the CR can avoid interference to the PS. Lemma 6.1 shows that the original network should not have links on the bottleneck of the network. Min-cut is the bottleneck of a network. Theorem 6.1 derives the condition that the CR should satisfy: min-cuts of the PS should not include any cut of the CR. That is, we should be able to find at least one route from the CR's source to sink that does not go through the bottlenecks of PS. Lemma 6.1 and Theorem 6.1 give us a complete characterisation of the CRRN topology. Given the CRRN's topology and assuming the PS and CR are uni-cast, Theorem 6.2 gives us the CR's maximum network capacity under the constraint of avoiding interference to PS. This is the upper bound of the resource in terms of network capacity that the CR can get by using relay network. Then we formulated the link allocation problem to achieve the CR's maximum network capacity as a multi-commodity flow problem, which can be solved by linear programming algorithms. Finally, we extended our work to the multicast PS and uni-cast CR, showing that the CR maximum capacity link allocation problem is still a linear programming problem and hence can be solved accordingly.

We then investigated the variation of the total network capacity corresponding to the variation of the CR's network capacity when the CR begins to interfere with PS. By Theorem 6.2, the CR's min-cut must be contained in PS's min-cut, which is also the CRRN's min-cut when achieving the maximum network capacity. Hence a small variation of link capacity allocation without changing the positions of the CR's min-cut and the PS's min-cut would not change the total network capacity. Total network capacity is still the CRRN's min-cut capacity. But when we consider that the CR is multicast, variation of the link capacity allocation may decrease the total network capacity, which is stated in the following lemma.

Lemma 6.2: If the CR is multicast and PS is uni-cast in the CRRN, increasing allocation of network capacity to the CR decreases total network capacity if the links of different min-cuts of different sinks of the CR occupy more than one link belonging to the same min-cut of PS.

Proof: We only prove the case for two CR sinks and occupying two links belonging to the same min-cut of PS because other cases are trivial extensions of this proof. Denote the two links belonging to the different CR's min-cuts e_1, e_2, and these two links belong to the same PS min-cut φ_1. Let the capacity allocation of link e_i be $c_{ei} = c_{ei,PS} + c_{ei,CR}$. Now, if we increase the link capacity allocated to the CR on e_1 and e_2 by $\delta > 0$, we have

$$c_{e1} = (c_{e1,PS} - \delta) + (c_{e1,CR} + \delta) \qquad for\ I = 1, 2$$ (6.27)

Because e_1 and e_2 are on the min-cut of the different sinks of the CR and belong to the same min-cut of PS, we have

$$\nu'_{CR} \leq \nu_{CR} + \delta. \tag{6.28}$$

$$\nu'_{PS} = \nu_{PS} - 2\delta \tag{6.29}$$

$$\nu'_t = \nu'_{CR} + \nu'_{PS} \leq \nu_{CR} + \nu_{PS} - \delta < \nu_t \tag{6.30}$$

The inequality in Equation (6.28) is due to the fact that there may be another CR's min-cut other than e_1 and e_2 belonging. By Equation (6.30), the total network capacity of the CRRN is decreased by at least δ.

Next we investigate the CR's interference to the PS when the interference is unavoidable. We restrict our discussion on small capacity allocation to the CR, and estimation to the bound of capacity interference to the PS. The idea and the proof of the following lemma come from the proof of Theorem 6.1 with some modifications.

Lemma 6.3: Assume the CR's interference is unavoidable. Let link set $\varphi_{PS} = \{e_i | e_i$ belongs to the cut whose capacity $= \nu_{PS}\}$. There are m cuts of the CR contained in φ_{PS}. Then the CR's interference to the PS's network capacity is bounded by $\nu_{CR} \times m$, under the constraint that φ_{PS} is not altered by allocating link capacity to CR.

Proof: Assume the CR's cuts in φ_{PS} are not overlapped. We allocate link capacity to the CR to achieve some small network capacity ν_{CR} under the constraint that φ_{PS} is not altered. Therefore, the PS min-cuts are still contained in the original φ_{PS}, and there are m CR's cuts to get ν_{CR} link capacity in φ_p. Assume these CR's cuts are contained in only one PS min-cut; then this cut's capacity will be decreased by $\nu_{CR} \times m$, and the PS's network capacity decreases proportionally. If we release the assumption of non-overlapping CR's cuts or interference on only the PS min-cut, interference will be distributed and the effects on network capacity will be decreased. Consequently, the interference will be bounded by $\nu_{CR} \times m$ under our assumptions.

Now we investigate the condition for bounded interference generalised from the observation list.

Lemma 6.4: The CR's interference to the PS is bounded if the edge set that is the union of all the CR's paths does not contain any of the PS cuts.

Proof: We state that the CR occupying all the links on all its paths between its source and sinks is the worst case for the PS. If the edge set that is union of all the CR's paths does not contain any of PS cut, the min-cut capacity of the PS is larger than zero even in the worst case. Hence the PS's network capacity is always larger than zero, in other words, the CR's interference is bounded.

In the networks satisfying the above condition, the PS can always maintain its transmission even when the CR transgresses the access etiquette.

6.1.4 Numerical Results

To verify that our algorithm applies in randomly generated CRRN topologies, we conduct simulations on randomly generated CRRN topologies in a 7 by 7 grid graph, as shown in Figure 6.5(a). The relay nodes appear on each grid point on the graph with probability 0.5. We consider the multicast PS with one source and two sinks, and uni-cast CR, one source and one sink, as shown in Figure 6.5(a).

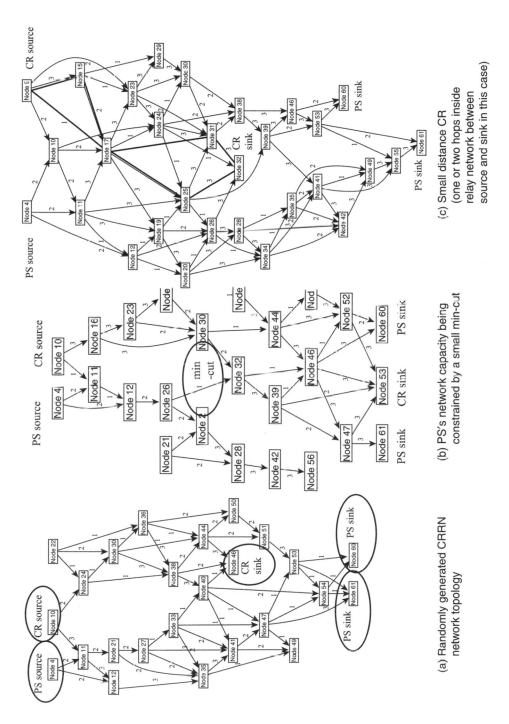

(a) Randomly generated CRRN network topology

(b) PS's network capacity being constrained by a small min-cut

(c) Small distance CR (one or two hops inside relay network between source and sink in this case)

Figure 6.5 Simulation scenarios: (a) Randomly-generated CRRN network topology; (b) The PS's network capacity being constrained by a small mini-cut; (c) Small distance CR (one of two hops inside the relay network between source and sink in this case)

Simulation parameters are set up as follows. The capacity on each link is a randomly selected integer from an interval (1,3). Link exists between two nodes if their distance is smaller than 2 in terms of grid in the graph. Our algorithm executes on 1000 randomly generated CRRN topologies. The histogram of the numerical result ratio of the CR's maximum network capacity to the PS's network capacity is shown in Figure 6.6. We use the ratio to present our result to show a comparison between the PS's communication resource and CR's communication resource. Ratio -0.5 is defined for the CR's maximum network capacity being zero. Simulation results show that the CR can use the PS to relay packets with a probability of 92% among randomly generated CRRN topologies. We also observe that the cases in which the CR can get network capacity equal to, or with only a small difference from the PS's network capacity, occur most frequently. And on average, the CR's network capacity is 1.3 times that of the PS to enhance spectrum efficiency at networking throughput. By this result, we are able to know approximately the probability distribution of the CR's maximum network capacity when it uses the PS network to relay.

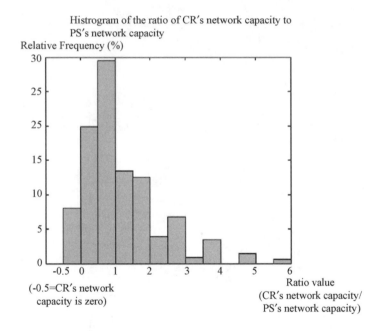

Figure 6.6   Histogram of capacity gain

By investigating the CRRN topologies in which the CR acquires higher network capacity, we observe that either the PS's network capacity is constrained by one or a few small min-cuts as shown in Figure 6.5(b) or the distance between the CR's source and sink happen to be small as in the example in Figure 6.5(c) (the CR only needs to go through one or two hops inside the relay network to reach its sink as marked in the network). We can intuitively infer these characteristics from our theoretical results. If the number of PS min-cuts is small and their capacities are small, the PS can use only small amounts of resource in the network. Consequently, CR has more opportunities to utilise such resources. On the other hand, if CR uses fewer hops in the CRRN, it is able to utilise more resource in those hops if such hops are not on the bottlenecks of the PS. In these situations, we can allocate more capacity to the CR without interfering with the PS. The CR may have a maximum network capacity of zero because of the condition stated in Theorem 6.1 or there is no route existing between the CR source and sink.

In this section, we analysed the interference characteristics corresponding to network topologies when the CR uses the PS to relay in the CRRN. When the CR uses the PS to relay, we adapt the DF cooperative policy. We started from fundamental network topologies to see whether the CR's interference is avoidable and bounded. We derived that interference is unavoidable in the one-hop relay, avoidable but unbounded in the tandem network and bounded in the cooperative relay structure. Then we extended our work to multiple CR sinks in the parallel cooperative relay network. When generalising the theoretical results, we put emphasis on avoidable analysis due to the CR's nature of avoiding interference to PS. First we derived the generalised condition on the CRRN topologies for the CR's interference to be avoidable: min-cuts of PS should not include any cut of the CR. Hence we should have at least one route of the CR that bypasses the bottlenecks of PS when designing the CRRN. In this kind of topology, we derive the CR's maximum network capacity in the PS uni-cast CRRN under the constraint of avoiding interference to the PS. We also show that link capacity allocation to achieve maximum network capacity can be formulated as a linear programming problem in both uni-cast and multicast PS. When the CR transcends the interference constraint, we described total network capacity variation corresponding to the CR's network capacity variation. We then analysed unavoidable interference and estimated the bound of interference in a small CR network capacity range. Finally, we derived the generalised condition on the CRRN topologies for the CR's interference to be bounded. We simulated the randomly generated CRRN topologies and executed our link allocation algorithm to derive the CR's maximum network capacity. On average, the CR's maximum network capacity is 1.3 times that of the PS's, which is a significant gain. By such analysis on interference characteristics in the CRRN, routing and scheduling can be developed to enhance the network efficiency of applying CRRN. Therefore, CRs using the PS to relay packets cooperatively are shown to achieve spectrum efficiency at networking throughput and at high availability, and the foundation of cognitive radio networking can be established.

6.2 Cognitive Radio Networks Architecture

It has been widely recognised that CR can efficiently improve spectrum utilisation at link level. We also demonstrate that cooperative relay among CRs and nodes in the PS can greatly enhance the network capacity by constructing a general sense CRN. This suggests that a cognitive radio shall sense available networks and communication systems around it, to complete networking functions beyond utilising the spectrum hole at link level. Thus, CRN are not just another network with interconnecting cognitive radios. They are composed of various kinds of co-existing multi-radio communication systems, including cognitive radio systems. CRNs can be viewed as some sort of heterogeneous networks composed of various communication systems. The heterogeneity exists in wireless access technologies, networks, user terminals, applications, service providers and so on. The design of the cognitive radio network architecture aims towards the objective of improving network utilisation. From the users' perspective, the network utilisation means that they can always fulfil their demands anytime and anywhere through accessing CRNs. From the operators' perspective, they can not only provide better services to mobile users, but also allocate radio and network resources in a more efficient way.

6.2.1 Network Architecture

The CRN can be deployed in network-centric, distributed, ad-hoc and mesh architectures, and serve the needs of both licensed and unlicensed applications. The basic components of CRNs are the *mobile station* (MS), *base station/access point* (BSs/APs) and *backbone/core networks*. These three basic components compose three kinds of network architectures in CRNs: Infrastructure, Ad-hoc and Mesh architectures, which are introduced in the next three sections.

6.2.1.1 Infrastructure Architecture

In the Infrastructure architecture (Figure 6.7), a MS can only access a BS/AP in the *one-hop* manner. MSs under the transmission range of the same BS/AP shall communicate with each other through the BS/AP. Communications between different cells are routed through backbone/core networks. The BS/AP may be able to run one or multiple communication standards/protocols to fulfil different demands from MSs. A cognitive radio terminal can also access various kinds of communication systems through their BS/AP.

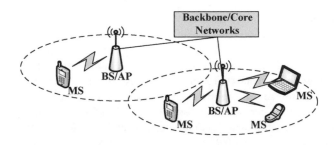

Figure 6.7   Infrastructure architecture of a CRN

6.2.1.2 Ad-hoc Architecture

There is no infrastructure support in ad-hoc architecture (Figure 6.8). The network is set up on the fly. If a MS recognises that there are some other MSs nearby and they are connectable through certain communication standards/protocols, they can set up a link and thus form an ad-hoc network. Note that these links between nodes may be set up by different communication technologies. In addition, two cognitive radio terminals can either communicate with each other by using existing communication protocols (e.g., WiFi, Bluetooth) or dynamically using spectrum holes.

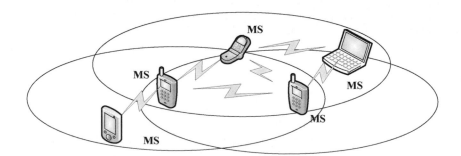

Figure 6.8   Ad-hoc architecture of a CRN

6.2.1.3 Mesh Architecture

This architecture is a combination of the infrastructure and ad-hoc architectures plus enabling the wireless connections between the BSs/APs (Figure 6.9). This network architecture is similar to the *Hybrid Wireless Mesh Networks*. In this architecture, the BSs/APs work as wireless routers and form *wireless backbones*. MSs can either access the BSs/APs directly or use other MSs as multi-hop relay nodes. Some BSs/APs may connect to the wired backbone/core networks and function as gateways. Since BSs/APs can be deployed without necessarily connecting to the wired backbone/core networks, there is more flexible and less cost in planning the locations of BSs/APs. If the BSs/APs have cognitive radio capabilities, they may use spectrum holes to communicate with each other. Due to the inefficiency of current spectrum utilisation, there may be lots of spectrum holes detected. So the capacity of wireless communication links between cognitive radio BSs/APs may be large and it makes the wireless backbone feasible to serve more traffic.

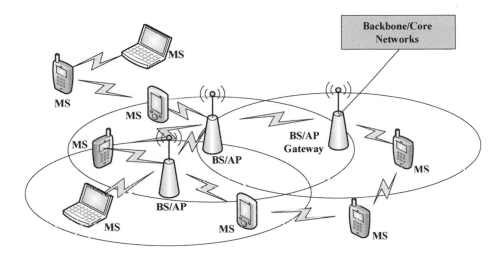

Figure 6.9   Mesh architecture of a CRN

6.2.2 Links in CRN

We can recall two kinds of wireless communication systems in CRNs: *Primary System* (PS) and *Cognitive Radio* (CR) systems, which are classified by their priorities on frequency bands. A primary system is referred to as an existing system which operates in one or many *fixed* frequency bands. Various kinds of primary systems work either in licensed or unlicensed bands, either in the same geographical location or in the same frequency band (or the same set of frequency bands) and are described as follows:

- **Primary System in Licensed Bands:**
  A primary system operated in the licensed band has the highest priority to use that frequency band (e.g., 2G/3G cellular, digital TV broadcast). Other unlicensed users/systems can neither interfere with the primary system in an intolerable way nor occupy the licensed band.
- **Primary System in Unlicensed Bands:**
  A primary system operating in the unlicensed band (e.g., ISM band) is called an unlicensed band primary system. Various primary systems should use the band compatibly. Specifically, primary systems operating in the same unlicensed band shall *co-exist* with each other while considering interference to each other. These primary systems may have different levels of priorities which may depend on some regulations.

A cognitive radio system does not have the privilege to access certain frequency band. Entities of the CR system must communicate with each other by dynamically using spectrum holes and opportunistic access. There are two components in CR systems: Cognitive Radio Base Station (CR-BS) and Cognitive Radio Mobile Station (CR-MS):

- **Cognitive Radio Base Station (CR-BS):**
  A CR-BS is a fixed component in the cognitive radio system and has cognitive radio capabilities. It represents the infrastructure side of the CR system and provides supports (e.g., spectrum holes management, mobility management, security management) to CR-MSs. It provides a gateway for CR-MSs to access the backbone networks (e.g., Internet). CR-BSs can also form a mesh wireless backbone network by enabling wireless communications between them, and some of them act as gateway routers if they are connected with wired backbone networks. If a CR-BS can run PR system protocols, it can provide access network services to PR-MSs.
- **Cognitive Radio Mobile Station (CR-MS):**
  A CR-MS is a portable device with cognitive radio capabilities. It can reconfigure itself in order to connect to different communication systems. It can sense spectrum holes and dynamically use them to communicate with CR-MS or CR-BS.

Since the CR system can provide interoperability among different communication systems, some inter-system connections should be enabled. We list the possibilities in Table 6.1 and illustrate them in Figure 6.10.

Table 6.1 Summary of links in the CRN

Rx\Tx	CR-MS	CR-BS	PR-MS	PR-BS
CR-MS	•	•	•	•
CR-BS	•	•	•	
PR-MS	•	•	•	•
PR-BS	•		•	•

• = possible link

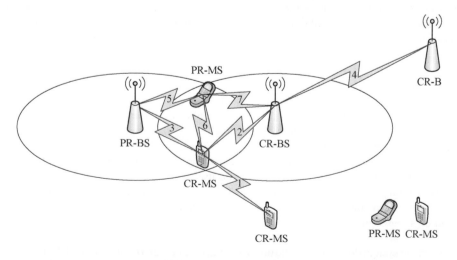

Figure 6.10   Links in CRNs

- **CR-MS←→CR-MS:**
  A CR-MS can communicate with other CR-MSs in direct links. They may cooperatively sense spectrum holes in different frequency bands, which may be licensed or unlicensed, and utilise them as their operating frequency band. A common control channel may be necessary for them to exchange spectrum hole information.
- **CR-MS←→CR-BS:**
  A CR-BS can dynamically sense an available frequency band around it and gather other MSs' sensing results and provide one-hop access to CR-MSs under its coverage area. This may need the cooperative sensing technique. Under the coordination of CR-BS, the CR-MS can either access the backbone networks or communicate with other communication systems.
- **CR-MS←→PR-BS:**
  If there is a need for a CR-MS to connect to a PR-BS, it will reconfigure itself and become one part of the primary system (i.e., PR-MS). In this case, it will become a primary user on that band.
- **CR-BS←→CR-BS:**
  While enabling direct wireless links between CR-BSs, they can form a mesh wireless backbone network. Because of their cognitive radio capability, they can dynamically choose an operating frequency band and communicate with each other. Since CR-BSs may have much more air interfaces, the link capacity between CR-BSs may be large. Another benefit of this kind of link is the reduced cost in placing the CR-BSs. This is because setting up a CR-BS in some environment with a physical wired link is not feasible.
- **PS-MS←→PS-BS:**
  It is the typical one-hop connection between mobile stations and base stations. The PR-BS is responsible for coordinating communications in its coverage and providing backbone network access to the PR-MS. This link is bi-directional all the time, which is fundamentally different from other links.
- **PS-MS←→CR-MS:**
  In order to provide interoperability between different communication systems, this kind of link may be necessary. In this case, the CR-MS shall reconfigure itself to be one part of the primary system.
- **PS-MS←→CR-BS:**
  In order to provide interoperability between different communication systems, this kind of link may be necessary. If the CR-BS can run the protocol of primary system, it can provide access service to the PR-MS.
- **PS-MS←→PS-MS:**
  This type of communication may exist in the PS as a sort of ad-hoc network in wireless networking systems. However, it may also be prohibited under infrastructure mode in some systems, which holds in our definition. However, if both nodes are transformed into CRs this case folds back to CR-MS ←→ CR-MS.

Please note a special feature of CR Links in the above list. Except the link between PS-MS and PS-BS that warrantees bi-directional, each of the other seven types of links is available in only one direction, during an opportunity of spectrum access. This is not hard to understand, as the opportunity window in time might be too short to warrantee bi-directional exchange of packets and the next opportunity of available time is not warranted either. This *unidirectional link* property plays a more critical role if when consider other network operations such as network security, which will be discussed in detail in Chapter 9.

6.2.3 IP Mobility Management in CRN

As we have mentioned, CRNs are *heterogeneous networks* in many aspects, such as wireless communication technologies and protocols, backbone network types, user terminal types, network operators, and so on. A CR-MS selects the best communication system(s) to fulfil user demands. Since different wireless systems have different medium access controls (MAC) and physical layer (PHY), how to integrate these systems to provide upper layers with better services is an important task. Since the network layer is the interface between available communications interfaces (or access technologies) that operate in a point-to-point fashion, and the end-to-end (transport and application) layers, it has a fundamental role in this integration process.

The Internet protocol (IP) and its extensions have been recognised as a technology that allows integration of heterogeneous networks into a single, all-IP based, integrated network platform, and mobile IP is considered as the most relevant extension. There are two important entities in mobile IP: *home agent* (HA) and *foreign agent* (FA), and they are Internet routers on the home network or foreign network, respectively. A *mobile node* (MN) accesses the Internet via an HA or FA. The node that has a connection to the MN is called the correspondent node (CN). We expect mobile IP to keep its important role in CRNs as mobile ad-hoc networks.

In the conventional infrastructure architecture, there is only one hop between a BS and MS, and no links between MSs and no multi-hop routes are allowed. Since all MSs only directly connect to BSs, a centralised mobility management scheme is feasible. For example, for a mobile IP network, all BSs can advertise their Care-of-Address (CoA) directly to MSs via *agent advertisement*. It is easy for an MS to acquire a CoA from the foreign network and register it to its home agent (HA). Then the HA can tunnel all packets to the latest CoA of the MS. However, mobility management under the mesh network architecture of CRNs is a much more challenging task open to research, especially when enabling multi-hop relay function between MSs. Related issues include location management and handover.

6.2.3.1 Location Management

Location management is a two-stage process that enables the network to discover the current attachment point of the mobile user for call delivery. The two stages are *location registration* and *call delivery*. When an MS visits to a foreign network (FN) and wants to get the Internet access service, it will first discover the mobile agents of the FN by detection agent advertisements. After getting the agent advertisement, the MS is able to form a CoA and inform the HA of the association between the current CoA and MN's home address. However, in CRNs, a CR-MS can simultaneously connect with many different wireless systems, which may belong to different FNs, and it should acquire a CoA from each of them in order to route packets to/from them. So there is a need to develop new schemes to deal with multiple CoAs. Specifically, a CR-MS can acquire many CoAs from each of those connectable FNs so that a CR-MS may no longer use a single CoA to represent its current position and to route packets. Moreover, multiple CR-MSs can form an ad-hoc network and some of them can connect to BSs/APs and access backbone/core networks. We call these nodes 'gateway nodes'. Due to the limited coverage of BSs/APs, some of the MSs can only get the BSs' service through multi-hop relay MSs, as shown in Fig. 6.11. The cooperation and integration of mobile IP and ad-hoc network [15] is a challenging task.

6.2.3.2 Handover Management

Handover management enables the network to maintain a user's connection as the mobile terminal continues to move and change its access point to the network [14]. Three stages are included: initiation, new connection generation and data flow control. Due to the multi-hop characteristic of the CRN, handover management is no longer an issue between a single MS and FNs. It is about multiple MSs and FNs. For example, if some of the gateway nodes move away from the BSs' coverage area, they shall inform those nodes in the ad-hoc network about its loss of connection. So nodes in the ad-hoc network can prepare to perform handover if they have active connections through that gateway node. Moreover,

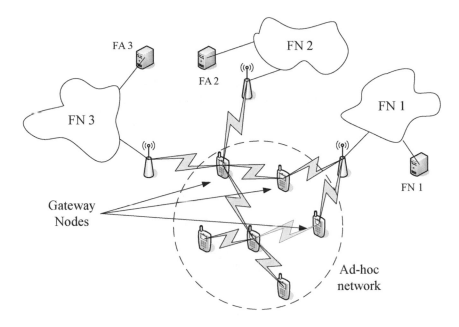

Figure 6.11 Integration of ad-hoc network and IP mobility management

some FNs can never be connectable due to the lost of gateway nodes. This can induce those CoAs issued from those FNs to be invalid. New CoA registration mechanisms may also be necessary.

6.3 Terminal Architecture of CRN

Realisation of CRs and thus CRNs is based on intelligent/smart terminal devices with powerful capability to handle distributed networking functions. In contrast to conventional cellular communication and network structure, CRN is therefore composed of a lot of localised and self-organised optimisation to reach global spectrum efficiency at link level and network level [21].

6.3.1 Cognitive Radio Device Architecture

Figure 6.12 depicts the device architecture of our proposed self-organised cognitive radio, which consists of several major functional blocks:

- *Cognitive Radio:* CR recognises wireless communication environments and co-existing systems/ networks.
- *Software-Defined Radio (SDR):* Based on the decision of the self-coordinator, SDR configures to the appropriate transceiver parameters for communication of the mobile device. Chapter 2 supplied several examples for fully programming SDR.
- *Re-configurable MAC:* The self-coordinator also determines the best possible routing among available systems/networks, and the re-configurable MAC adjusts to proper subroutines in a universal access protocol machine.
- *Network-layer procedures:* The self-organised coordinator instructs the right network layer functions such as radio resource allocation, mobility management, etc. to complete the wireless network operation.

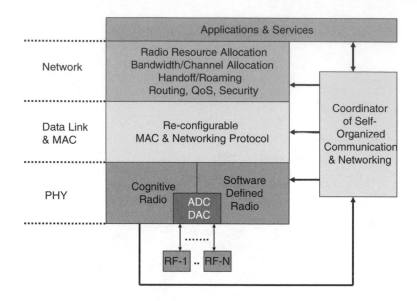

Figure 6.12   Terminal architecture of a CRN

- *Self-organised communication/networking coordinator:* The brain of terminal device determines: (i) decent access network route based on cognitive radio information; (ii) configuration of proper hardware and software of cognitive radio; and (iii) maintenance of a user's communication need for the terminal device.
- *RF:* RF might consist of several sub-band RFs to cover the right frequency range, with the capability of adjustable RF filtering to fit selected system parameters.

The cognitive radio along with interaction of the self-organised coordinator and SDR is shown in Figure 6.13. We temporarily do not consider the circuit reuse situation under this scheme.

It is well known that cognitive radio is centred around spectrum sensing. However, as Figure 6.13 shows, there is a lot more information needed to make good sensing in practice, not only in spectrum, but also in some networking functions, as a generalised sensing (or cognition). We categorise such spectrum/network sensing features into the following:

- *RF Signal Processing* includes (carrier) frequency, signal bandwidth, signal strength (RSSI), SINR estimation;
- *BB Pre-detection Signal Processing* includes symbol rate, carrier and timing, pilot signal, channel fading;
- *BB Post-detection Processing* (some may be done before the detection stage) includes system/user identification, modulation parameters, FEC type and rate, MIMO parameters, transmission power control.

Networking processing information includes multiple access protocol or MAC, radio resource allocation (such as time slot, subcarrier, code, allocation), automatic repeated request (ARQ) and traffic pattern (arbitary bit rate (ABR), constant bit rate (CBR) or variable bit rate (VBR)), and routing or mobility information. The purpose of the above list is to execute spectrum sensing, identification of co-existing systems/networks, and then operation of such co-existing systems/networks. The research literature supplies a good number of examples to facilitate partial functions in the list. The cognitive radio cycle to show working flow is therefore summarised in Figure 6.14.

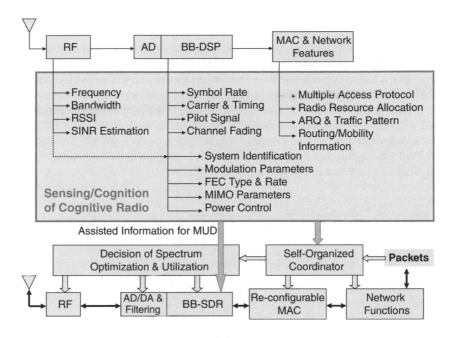

Figure 6.13   CRN terminal hardware structure

J. Mitola [1] and Haykins [40] have developed different but similar cognitive cycle concepts. Since we generalise to the cognitive radio 'network' along with the rate-distance concept, novel features are primarily distinguished here. Cognitive radio functions not only sense the spectrum and fitting spectrum resource but also the networking environment, and adapt into cognitive routing in the network level.

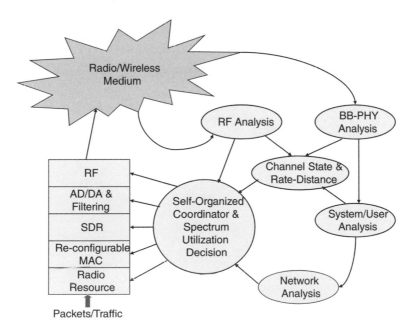

Figure 6.14   Cognitive cycle for the CRN terminal

6.3.2 Re-configurable MAC

Medium access control (MAC) of wireless networking for mobile/ubiquitous computing is a fundamental element in addition to the radio transceiver. Chen [4] described some fundamental challenges for wireless networks in fading channels and suggested principles to resolve them. After a series of efforts, a unified MAC algorithm was presented in [5] to execute most well-known access protocols. It leveraged the intellectual concept of R. Gallager that multiple access conducts either carrier sensing (generalised as collision avoidance) or collision detection, to form CATE and CRTE (collision avoidance/resolution tree structures to generally represent all protocols) [6]. The following algorithm and Table 6.2 tailor the appropriate multiple access protocol by adjusting the corresponding parameters; more discussions can be found in Chapter 8.

```
   Re-configurable MAC Algorithm

RP_1
if(access method = blocked) {
  allow new arrivals during previous cycle -> DN; }
if (memoryless_after_lost is set) {
  have all noted in DN call CATE (type_CATE;) }
else{
  unmarked nodes in DN call CATE (type_CATE);
  associate marked nodes in DN to group
    number # (original group number -g);}
  unmarked nodes in CN call CRTE (type_CRTE);
  associate marked nodes in CN to group
    number # (original group numbers -g);
  if (report grouping result is set) {
    all nodes report the grouping result back;}
set g=1; // start to process each group
RP_2
if(access method=free){
  nodes with new arrival packets during the
  processing of group # (g-1) -> TX(g);}
nodes in group #g -> TX(g);
process group #g with GP(gp_scheme);
if (there is no transmission) {
  g++;
if (G is set ) { // G is the maximum TE size
    if (g>G) { goto RP_1; }
    else{goto RP_2;}}
  else{goto RP_2;}}
elseif (there is transmission) {
  if (access method=freee) {
    nodes in group # (g+1)
        to group # (g+t) -> DN; }
      //t is the duration of the transmission
      If(the transmission is a success) {
        the successful node removes the
          transmitted packet from buffer;
      if (completeness is set) {
        g++;
```

```
            if(g>G){goto RP_1;}
            else{goto RP_2;}}
        else{
            if memoryless_after_lost is set){
            mark the loser in CN;}
        else{mark the loser in CN and DN;}
            current cycle ends and goto RP_1;}}
        else{
                collided nodes -> CN;
                  if(completeness is set){
                    g++;
                    if(g>G){goto RP_1;}
                  else{goto RP_2;}}
        else{
            if(memoryless_after_lost is set){
                mark the loser in CN;}
            else{mark the loser in CN and DN;}
                current cycle ends and goto RP_1;}}}
```

6.3.3 Radio Access Network Selection

Following the above architecture, the self-organised coordinator schedules right networking functions in routing to control QoS, and decides appropriate configuration of MAC, software radio communication parameters and RF parameters. The typical approach towards self-organised wireless communications looks into topology control of the entire possible networks/systems, and optimises based on different criteria. Towards a practical realisation, we consider the problem from a different angle, that is, a terminal determines its routing based solely on information available and to select the right radio access networks, including CR. The radio access network can be part of digital cellular network such as UTRAN, an access point of Wireless LANs connected to Internet or a base station (or a subscriber station in mesh network) in WiMAX, etc. We therefore generally assume there are users from K systems that are operating within a certain geographical area, and devices can access all operating frequency bands. Traditionally one mobile device capable of operating in one system cannot operate in another system, and resources within these K systems may not be evenly distributed in that some of the systems may be crowded whereas others may have no or little traffic. Through cognitive radio, we shall leverage possible cooperation among these systems to improve the individual and overall performance. The primary challenge is to determine proper cooperation among various combinations of systems to enhance performance or QoS. Please note that the users may want 'cost' as a performance measure in practical applications. Without loss of generality, we consider a circuit switching (CS) network (such as 2G/3G cellular) with n_1 users and a packet switching (PS) network (such as WiFi) with n_2 users. For these $N = n_1 + n_2$ users to operate the cognitive mode between such systems, we want to demonstrate effective routing to enhance overall cognitive radio network performance as in Figure 6.15. The packet loss is due to collision with retransmission, and the number of users in the network is also assumed to be relatively steady.

- *Access Network 1:* CS network, modelled as a multi-server queue with N_1 servers. The service rates of N_1 servers are all deterministic and equal to μ_1. A certain user is served by only one server and no more than N_1 users can be admitted into this network.
- *Access Network 2:* PS network, modelled as a single-server queue with a multiple access device in front of the server to decide which user has the right to access media. The service rate of this server is deterministic with rate μ_2, and the multiple access scheme is assumed to be slotted ALOHA with retransmission probability q.

Table 6.2 Re-configuration of MAC parameters

	ALOHA with random backoff	Basic Q-ary CRA	p-persistent CSMA	CSMA/CA	GRAP
Slot time	One transmission + One feedback	One transmission + One feedback	One propagation eelay	Defined in Spec.	One propagation delay
Access method	Free	Free	Free	Free	Blocked
Completeness	No	No	No	No	Yes
Memory-less after loss	No	Yes	No	Yes	No
Report grouping result	No	No	No	No	Yes
Group process scheme	Two way handshaking	Two way handshaking	Two way handshaking	Four way handshaking	Polling
Type of CATE	None	None	Geometric.CATE	BEB.CATE	Uniform.CATE
Type of CRTE	Geometric.CRTE or BEB.CRTE	Q-ary CRA.CRTE	Geometric.CRTE	BEB.CRTE	Uniform.CRTE

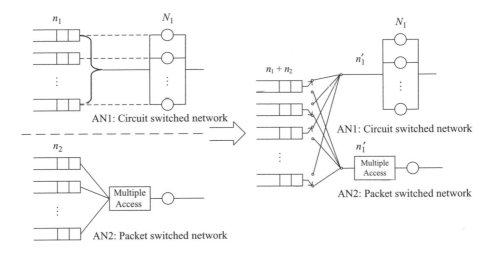

Figure 6.15   Queuing model of cooperative access networks

Based on the queuing analysis, the terminal can make a right selection of access network to execute the best possible routing for its traffic. Of course, this is the simplest case for illustration. More general results for CRN are still open for research.

6.4 QoS Provisional Diversity Radio Access Networks

To extend the network coverage, provide the reliability and increase the data rate, cooperation/collaboration between multiple communication nodes has been considered as an emerging design paradigm for future wireless networks such as CRN. We define cooperation and collaboration among nodes as follows:

> *Cooperation among multiple nodes means that multiple nodes work together to achieve individual goals, and collaboration between multiple nodes means that multiple nodes work together to achieve a common goal.*

We can further divide the cooperative/collaborative scenarios into *symmetry* and *asymmetry* cases. A symmetry cooperation/collaboration means that a node (network) helps other nodes (networks) while other nodes (networks) also help the node. However, in the asymmetry case, a node (network) helps other nodes (networks) but other nodes (networks) do not help the node (network). A general network topology of diversity networks is shown in Figure 6.16. We clarify the figure as follows:

- N1 has traffic transmitted to N3 and N2 has traffic transmitted to N4. N1 and N2 relay traffic for each other with different destinations. Thus N1, N2, N3 and N4 form the symmetry cooperation diversity network.
- N5 has traffic transmitted to N6 and the traffic is relayed via N2. Since N2 transmits the traffic to N4 without the relay of N5, N2, N5 and N6 form the asymmetry cooperation diversity network.
- N7 and N8 work together to transmit one traffic flow to N1, thus N1, N7 and N8 form the symmetry collaboration diversity network.
- N6 transmits the traffic to N10 and N9 helps the relay, while N9 does not have its own traffic to transmit. Therefore, N6, N9 and N10 form the asymmetry collaboration diversity network.

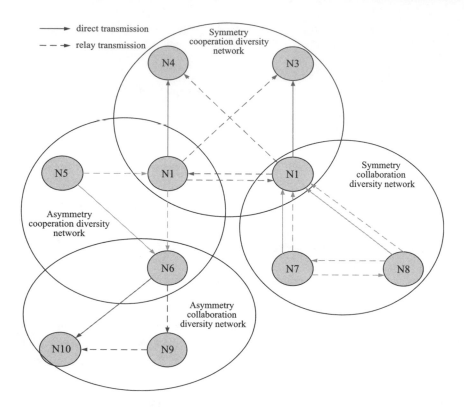

Figure 6.16   General diversity network topology

For these definitions, various diversity network contexts have been proposed but only for the asymmetry collaboration diversity network scenario.

To guarantee the QoS over wireless communications and networks is a permanent challenge [34–37]. The major reason is that the wireless capacity is time-variant and bursty, which makes the data rate sensitive to the channel variations. Once the source rate exceeds the data rate provided by the system, the packet delivery time may exceed its maximum tolerable value and this packet is considered as invalidated for the delay-sensitive traffics. As original concept of user cooperation, Sendonaris, Erkip and Aazhang indicated that the cooperative/collaborative relay decreases the sensitivity of the data rate to the channel variations, and that is the most significant part for the QoS guarantee, even if the cooperation/collaboration brings no other benefits such as data rate increase. This is because of the minimum data-rate requirements of some real-time applications, such as audio or video, and the resulting lower probability of outage, and thus better QoS, due to cooperation/collaboration. However, providing QoS on the cooperative/collaborative relay only receive a few attentions. In this section, we review some existing work on providing QoS over diversity networks. Based on this previous work, we point out various substantial open issues in providing QoS.

6.4.1 Cooperative/Collaborative Diversity and Efficient Protocols

A basic cooperative/collaborative network topology is composed of a source node, a relay node and a destination node as shown in Figure 6.17. There are operation schemes of the relay node as follows:

- *Facilitation:* the relay node does not actively help the source, but rather facilitates the source transmission by inducing as little interference as possible.

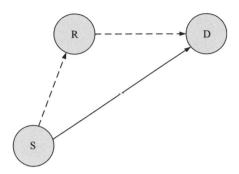

Figure 6.17   Basic cooperative/collaborative network topology

- *Cooperation/collaborative:* (i) the relay node fully decodes the source message and retransmits (known as the decode and forward, DF), or (ii) the relay node only amplifies the source message and retransmits (known as the amplify and forward, AF).
- *Observation:* the relay node encodes a quantised version of its received signal and retransmits (known as the compress and forward, CF).

Since a full-duplex receiver design is complicated, removing the interference between the nodes at the destination radio substantially simplifies the receiver design. To ensure half-duplex operation, the channel should be divided into orthogonal subchannels, as shown in Figure 6.18. For facilitation operation, the time-division channel allocation can be as in Figure 6.18(a) and (b); otherwise, the time-division channel allocation as Figure 6.18(c) is chosen.

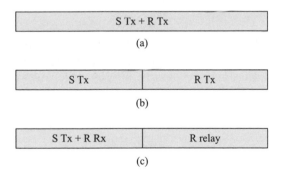

Figure 6.18   (a) The source node and the relay node simultaneously transmit. (b) The source node transmits first, and then the relay node transmits. (c) The source node transmits packets to the relay node, and the relay node relays the packets to the destination node.

As described in Chapter 4, existing relay protocols can be divided into three categories:

- *Fixed relay:* The relays are allowed to either amplify their received signals subject to their power constraint, or to decode, re-encode and retransmit the messages.
  - *AF:* Under the assumption that the additive noise is modelled as independent zero-mean, circularly symmetric complex white Gaussian with unit variance at each receiver, the maximum mutual

information and thus the maximum achievable rate is given as

$$R_{AF} = \left(\frac{T_f B}{2}\right) \log_2 \left(1 + 2\gamma_1 P_s + \frac{4\gamma_2 P_s \gamma_3 P_c}{1 + 2\gamma_2 P_s + 2\gamma_3 P_c}\right) \qquad (6.31)$$

where B is the system spectrum, P_s and P_c are average transmission powers of the source node and the cooperative node, respectively, T_f is the frame duration, and $\gamma_1 = |h_{s,r}|^2$, $\gamma_2 = |h_{s,d}|^2$ and $\gamma_3 = |h_{r,d}|^2$ are the channel gain of links between source and cooperative node, source and destination node, and cooperative and destination node.

- *DF:* In DF mode, the relay forwards the message to the destination if it decodes successfully; the maximum mutual information and thus the maximum achievable rate is given by [11] as

$$R_{DF} = \left(\frac{T_f B}{2}\right) \min(\log_2(1 + 2\gamma_2 P_s), \log_2(1 + 2\gamma_1 P_s + 2\gamma_3 P_r)) \qquad (6.32)$$

- *Selection relay:* If the measured $|h_{s,r}|^2$ falls below a certain threshold, the source simply continues its transmissions to the destination, in the form of repetition or more power codes. If the measured $|h_{s,r}|^2$ lies above the threshold, the relay forwards what it received from the source, using either AF or DF, in an attempt to achieve diversity gain.
- *Incremental relay:* The fixed and selection relaying make inefficient use of the degree of freedom of the channel, especially for high rates, because the relay repeats all the time. The incremental relaying protocol can be viewed as extensions of incremental redundancy, or hybrid automatic-repeat-request (HARQ), to the relay context. In ARQ, the source retransmits if the destination provides a negative acknowledgement via feedback. In the incremental relay, the relay retransmits in an attempt to exploit spatial diversity.

6.4.2 Statistical QoS Guarantees over Wireless Asymmetry Collaborative Relay Networks

Real-time multimedia services such as video and audio require the bounded delay, or equivalently, the guaranteed bandwidth. Once a received real-time packet violates its delay bound, it is considered as useless and will be discarded. However, over the mobile wireless networks, a hard delay bound guarantee is practically infeasible due to the impact of the time-varying fading channels. For example, over the Rayleigh fading channel, the only lower-bound of the system bandwidth that can be *deterministically* guaranteed is a bandwidth of zero. Thus, we should consider an alternative solution by providing the *statistical* QoS guarantees that guarantee the delay-bound with a small violation probability. For this purpose, Wu and Negi [33] developed a powerful concept termed *effective capacity*, which evolved from the theorem of *effective bandwidth* for statistical QoS guarantee. In addition to providing the statistical QoS guarantee in general wireless communications, effective capacity also provides a new design paradigm for the QoS provision in relay networks. In this section, we review the theories of effective bandwidth and effective capacity and the application of effective capacity on the radio resource for the asymmetry collaborative relay networks with the target of meeting the delay QoS guarantee.

6.4.2.1 Effective Bandwidth and Effective Capacity

During the 1990s, a framework of statistical QoS guarantees was developed in the context of the effective bandwidth theory. In this literature, researchers use a parameter ϑ to represent the degree of QoS guarantee of the system. Specifically, C.-S. Chang [36] showed that for a queuing system with a stationary ergodic arrival and service process, the queue length process $Q(t)$ converges to a random variable $Q(\infty)$ such that

$$- \lim_{x \to \infty} \frac{\log(\Pr\{Q(\infty) > x\})}{x} = \theta \qquad (6.33)$$

The above theorem states that the probability of the queue length exceeding a certain value x decays exponentially fast as x increases. The parameter ϑ ($\vartheta > 0$) in Reference (1) indicates the exponential decade rate of the QoS violation probabilities. A small ϑ corresponds to a slow decay rate, which implies that the system provides a looser QoS guarantee. On the other hand, a larger ϑ corresponds to a fast decay rate, and thus the system provides a strict QoS guarantee.

From the effective bandwidth theory, Wu and Negi [33] proposed a concept termed *effective capacity*, which is defined as

the maximum constant arrival rate that a given service rate can support in order to guarantee a QoS requirement specified by ϑ

and is expressed as

$$E_c(\theta) = -\frac{1}{\theta}\log(E[e^{-\theta R}]) \qquad (6.34)$$

where R is the independent identical distributed (i.i.d.) service process and $E[y]$ is taking expectation over y. Specifically, if $\theta \geq -\log\varepsilon/D_{max}$, then

$$\sup_t \Pr\{D(t) \geq D_{max}\} \leq \varepsilon \qquad (6.35)$$

where D_{max} is the maximum tolerable delay of the traffic and $D(t)$ is the delay at time t. Reference (3) state that the probability that the delay of the traffic exceeds the maximum tolerable delay is below ε.

6.4.2.2 Dynamically Resource Allocation for AF and DF Relay under a Given QoS Constraint

The cross-layer radio resource allocation scheme can be built on effective capacity for asymmetry collaborative relay networks with the target of meeting the delay QoS guarantee for wireless multimedia communications. To maximise the effective capacity under a given QoS constraint represented by ϑ, the problem can be modelled as a power allocation for the source power and the relay power. Thus the maximisation problem of the AF relaying can be expressed as

$$\arg\max_{P(\gamma)}\left\{ -\frac{1}{\theta}\log(E_\gamma[\exp(-\theta R_{AF}(\gamma))])\right\} \qquad (6.36)$$

subject to the power constraints

$$\begin{cases} E_\gamma[P_s(\gamma) + P_r(\gamma)] \leq \bar{P} \\ P_s(\gamma) \geq 0 \text{ and } P_r(\gamma) \geq 0 \end{cases} \qquad (6.37)$$

where $P(\gamma) \underline{\underline{\Delta}} (P_s(\gamma), P_r(\gamma))$ and $\bar{P}$ is the average power constraint.

The optimal power allocation policy is determined as

$$
\begin{cases}
P_s(\gamma) = u P_r(\gamma) \\
P_r(\gamma) = \dfrac{1}{v}\left(\left[\left(\dfrac{\gamma_0}{\gamma_3}\right)\left(\dfrac{\gamma_3 + c}{\gamma_1 + c}\right)^2 \right]^{-\frac{2}{\beta+2}} - 1 \right)
\end{cases}
\tag{6.38}
$$

where $c = \sqrt{\gamma_1\gamma_3 + \gamma_2\gamma_3 - \gamma_1\gamma_2}$, γ_0 is a cutoff threshold determined by the mean total network power constraint, $\beta \underset{=}{\Delta} \theta T_f B/\log 2$, and if both $P_s(\gamma) > 0$ and $P_r(\gamma) > 0$

$$
\begin{cases}
u = \dfrac{\gamma_3(\gamma_1 + c)}{(\gamma_3 - \gamma_1)\gamma_2} \\
v = \dfrac{2c\gamma_3(\gamma_1 + c)^2}{(\gamma_3 - \gamma_1)\gamma_2(\gamma_3 + c)}
\end{cases}
\tag{6.39}
$$

Otherwise, the policy reduces to direct transmission and $P(\gamma)$ is determined by

$$
\begin{cases}
P_s(\gamma) = \dfrac{1}{2}\left[\left(\gamma_0^{\frac{2}{\beta+2}} \gamma_1^{\frac{\beta}{\beta+2}} \right)^{-1} - \gamma_1^{-1} \right]^{+} \\
P_c(\gamma) = 0
\end{cases}
\tag{6.40}
$$

The maximisation problem of the AF relaying can be expressed as

$$
\arg \max_{P(\gamma)} \left\{ -\frac{1}{\theta} \log(E_\gamma[\exp(-\theta R_{DF}(\gamma))]) \right\}
\tag{6.41}
$$

and the optimal power allocation is

$$
\begin{cases}
P_s(\gamma) = \dfrac{1}{2}\left[\left(\gamma_0^{\frac{2}{\beta+2}} \gamma_1^{\frac{\beta}{\beta+2}} \right)^{-1} - \gamma_2^{-1} \right]^{+} \\
P_c(\gamma) = 0
\end{cases}
\tag{6.42}
$$

As above, Tang and Zhang [38] maximise the system supporting arrival rate while statistically guaranteeing the delay by modelling this optimisation problem as a power allocation problem. This solution is mainly for the asymmetry collaborative relay case where the relay node does not have traffic to transmit. Although the effective capacity is promising for the statistical QoS guarantee in direct transmissions without relay and asymmetry collaborative relaying, it is complicate to apply to the symmetry collaborative relay case where both relay node and the source have packets to transmit and they help each other to guarantee the diverse QoS of each other.

6.4.2.3 Asymmetry Collaborative Relay Over OFDMA

In addition to general physical transmission, asymmetry collaborative relays over orthogonal frequency division multiple access (OFDMA) has started to receive research interest due to the adoption of IEEE 802.16j for the multiple-hop relay network. However, since inter-carrier interference (ICI) cancellation, synchronisation and channel estimation in OFDMA remains challenging, especially for the uplink, the collaborative relaying is mainly for the downlink context. Pischella and Belfiore [26] proposed an OFDMA downlink cooperation context that two base station (BS) using the non-orthogonal AF cooperative protocol to serve and maintain the QoS for a mobile station (MS). In this downlink collaboration framework, two BSs use the same FFT size, N_{FFT}, and total bandwidth is B. Each MS, k, is served by its original BS, $BS_{s,k}$, and may be relayed by its closest BS (in term of path loss), $BS_{r,k}$. The QoS driven resource allocation problem is that each BS aims at providing the target data rate to the guaranteed performance (GP) users, while maximising the sum rate of the Best Effort (BE) users. Therefore, the radio resource allocation problem can be modelled as per the following optimisation problem:

$$\min \sum_{k=1}^{K_1} \sum_{m=1}^{n_{SC}} c_{k,m} p_{BS,k,m}$$

$$s.t.\ D_k = D_{\text{target},k} \text{ for } k \in [1, K_1]$$

$$\max \sum_{k=K_1+1}^{K_1+K_2} D_k \tag{6.43}$$

$$s.t. \sum_{k=1}^{K_1+K_2} \sum_{m=1}^{n_{SC}} c_{k,m} p_{BS,k,m} \le p_{\max}$$

where m is the subcarrier index, $p_{\max}$ is the maximum power constraint and $p_{s,k,m}$ (resp. $p_{r,k,m}$) is the transmission power from the source (resp. the relay) to user k in the mth user. D_k is the data rate of the user k, $c_{k,m}$ is a indicator of the power allocation for user k on the mth subcarrier and $c_{k,m} \in [0,1]$. The first K_1 users are GP users and the next K_2 users are BE users. The GP users can be either direct or relayed users.

The aforementioned diversity network contexts are mainly for asymmetry collaboration. How to apply the cooperative relay to guarantee QoS in the CRN remains open for more research.

6.5 Scaling Laws of Ad-hoc and Cognitive Radio Networks

As the CRN allows possible multiple transmissions, the *sum rate* of the CRN might better represent its performance, as suggested by M. Vu, N. Devroye, V. Tarokh [22]. The scaling law of a (wireless) network is usually used to establish the upper bound and lower bound of the sum rate [23]. Lower bounds may be obtained by certain transmission strategies, which provide an achievable rate. Upper bounds can be obtained by theory. Upon scaling the law of the lower bound to meet that of the upper bound, we obtain the precise sum-rate scaling law of this network.

6.5.1 Network and Channel Models

Suppose there are n pairs of wireless devices (or users or nodes) located on a plane to communicate with each other. Each pair of nodes consists of a transmitter and a receiver. We may have two common network models: *dense network* and *extended network*. In dense networks, the network area stays a constant as the number of nodes grows. In extended networks, the node density remains the same, which means the network area grows linearly with the number of nodes.

The figure of merit is the total network capacity, that is, the sum rate or the throughput. The sum rate is defined as

$$C(n) = \sum_{i=1}^{n} R_i$$

where R_i is the information rate of pair i. Alternatively, the capacity of each pair is

$$R(n) = \frac{C(n)}{n}$$

We would like to know how the capacity scales as n goes to infinity. We also consider propagation distance, shadowing and fading into the channel model. Given the distance between transmitter and receiver, the channel gain is

$$h(d) = \frac{e^{-\gamma^d}}{d^{\alpha/2}} h_s$$

where d is the transmitter-receiver distance, α is the power of path loss exponent, γ is the absorption/delay constant and h_s is the shadowing and fading component. To simplify the study, we can set $h_s = 1$ and $\gamma = 0$, and rely on path loss exponent to represent propagation effects, which is reasonable by choosing $\alpha \geq 2$ [22].

6.5.2 Ad-hoc Networks

Let us consider the ad-hoc network with randomly and uniformly distributed nodes, with random pairing of transmitters and receivers. For $\alpha > 2$, each pair of transmission requires the minimum received SINR as

$$\frac{P/d_i^\alpha}{\sigma_n^2 + \sum_{i \neq k} P/d_{ik}^\alpha} \geq \beta$$

where α_n^2 is the noise power, d_k is the distance for the kth transmitter-receiver pair, and d_{ik} is the distance between the receiving node k and interference node i. The transmission rate per node is lower bounded:

$$R(n) \geq \frac{c(\beta)}{\sqrt{n\log n}}$$

$c(\beta) \propto \beta^{-1/2\alpha}$ is a constant, which is achieved by the *nearest-neighbour forwarding* in multi-hop routing. The upper bound can be obtained without interference constraint as

$$R(n) < \frac{c_1}{\sqrt{n}}$$

where c_1 is a constant. The throughput per node of an extended network can be shown by $1/\sqrt{n}$ [24].

When nodes are distributed according to a Poisson point process, the throughput per node can be upper bounded by $1/\sqrt{n}$, by a new routing based on percolation theory. If we further allow cooperation among nodes by physical layer processing, the throughput per node can stay asymptotically constant by hierarchical clustering with ad-hoc communication intra-clusters and MIMO communication inter-clusters. For extended networks, $2 < \alpha \leq 3$,

$$C(n) \sim n^{2-\alpha/2}$$

For $\alpha > 3$, the nearest-neighbouring multihop scheme is optimal and

$$C(n) \sim \sqrt{n}$$

6.5.3 Cognitive Radio Networks

We now investigate the CRN with two types of nodes, in the PS and CR (i.e., secondary users of the spectrum), by considering Figure 6.19. We consider again the extended network with the PS's transmitter (PS-Tx) at the centre with radius R_0 as the *primary exclusive region* which the PS receiver (PS-Rx) locates inside. The CRs are randomly and uniformly distributed in the cognitive region outside the primary exclusive region with density λ. Cognitive communication occurs in single-hop with maximum propagation distance D_{max}. Any interfering transmitter has non-zero distance $\in$ in this scenario.

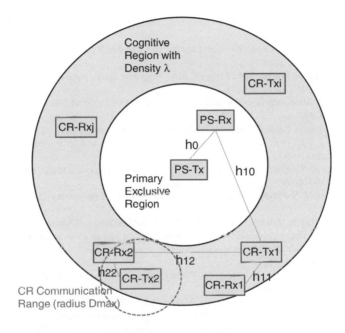

Figure 6.19   Cognitive radio network setup

We introduce a concept of *outage probability* that the received signal of the PS user is below a certain level. Assume that PS-Tx transmits with power P_0 and each CR-Tx transmits with power P_c. Without MUD, each receiver treats other transmissions (actually, interference) as noise. With $\alpha > 2$, the average sum rate of the CRN scales linearly with the number of nodes n. With high probability, the rate per node stays constant to satisfy

$$R(n) \geq \log\left(1 + \frac{P_{min}}{\sigma_{max}^2 + I_{worst}}\right)$$

where $P_{min} = P_c/D_{max}^\alpha$, $\sigma_{max}^2 = \sigma_n^2 + P_0/R_0^\alpha$; and σ_n^2 is noise power. The worst case interference, I_{worst}, is

$$I_{worst} = \frac{2\pi\lambda P_c}{(\alpha - 2)\varepsilon^{\alpha - 2}}$$

$\varepsilon > 0$ plays a critical role in achieving constant throughput per node, which is equivalent to an exclusive region around each CR-Rx where no other CR can operate or transmit. In the worst case, the distance between transmitter and receiver can grow as $\sqrt{n}$ leading to $1/\sqrt{n}$ throughput per node.

With C_0 as the PS-Tx's transmission rate and η as the given threshold, the outage constraint in the worst case is

$$P[C_0 \leq \eta] \leq b$$

The outage comes from the random placement of the CRs and fading. By upper-bounding the average interference to PS-Rx, we can bound the radius R_0 of the primary exclusive region.

References

[1] J. Mitola, G.Q. Maguire, 'Cognitive Radio: Making Software Radios More Personal', *IEEE Personal Communications*, **6**(4), 1999, 13–18.

[2] C.-H. Huang, Y.-C. Lai, K.-C. Chen, 'Network Capacity of Cognitive Radio Relay Networks', *Physical Communications (PHYCOM)*, **1**(2), 2008.

[3] K.-C. Chen, Y.-J. Peng, N. Prasad, Y.-C. Liang, S. Sun, 'Cognitive Radio Network Architecture: Part I – General Structure', ACM ICUIMC, Seoul, 2008.

[4] K.C. Chen, 'Medium Access Control of Wireless Local Area Networks for Mobile Computing', *IEEE Networks*, September 1994, 50–64.

[5] C.M. Teng, K.C. Chen, 'A Unified Algorithm for Wireless MAC Protocols', *Proc. IEEE VTC*, 2002.

[6] Y.K. Sun, K.C. Chen, D.C. Twu, 'Generalized Tree Multiple Access Protocol for Wireless Communications', *Proc. IEEE PIMRC*, Helsinki, 1997.

[7] D. Raychaudhuri *et al.* 'CogNet – An Architecture for Experimental Cognitive Radio Networks within the Future Internet', *Proc. of MobiArch 2006*, December 2006, pp. 11–16.

[8] Z. Qing, B.M. Sadler, 'A Survey of Dynamic Spectrum Access', *Signal Processing Magazine, IEEE*, **24**(3), 2007, 79–89.

[9] Y. Xing, C.N. Mathur, M.A. Haleem, R. Chandramouli, K.P. Subbalakshmi, 'Dynamic Spectrum Access with QoS and Interference Temperature Constraints', *IEEE Transactions on Mobile Computing*, **6**(4), 2007, 423–433.

[10] R. Ahlswede, C. Ning, S.-Y.R. Li, R.W. Yeung, 'Network information flow', *IEEE Transactions on Information Theory*, **46**(4), 2000, 1204–1216.

[11] L. Song; R.W. Yeung, N. Cai, 'Zero-error Network Coding for Acyclic Networks', *IEEE Transactions on Information Theory*, **49**(12), 2003, 3129–3139.

[12] L. Geng, Y.-C. Liang, F. Chin, 'Network Coding for Wireless Ad Hoc Cognitive Radio Networks', *Personal, Indoor and Mobile Radio Communications, 2007 (PIMRC 2007), IEEE 18th International Symposium*, September 2007, pp. 1–7.

[13] L. Xiao; T. Fuja, J. Kliewer, D. Costello, 'A Network Coding Approach to Cooperative Diversity', *IEEE Transactions on Information Theory*, **53**(10), 2007, 3714–3722.

[14] S. Fu, K. Lu, Y. Qian, M. Varanasi, 'Cooperative Network Coding for Wireless Ad-Hoc Networks', *Global Telecommunications Conference, 2007 (GLOBECOM'07), IEEE*, 26–30 November 2007, pp. 812–816.

[15] J. Mitola, 'Cognitive Radio for Flexible Mobile Multimedia Communications', *1999 IEEE International Workshop on Mobile Multimedia Communications (MoMuC '99)*, 1999, pp. 3–10.

[16] L.R. Ford, Jr., D.R. Fulkerson, *Flows in Networks*, Princeton University Press, New Jersey, 1962.

[17] T.C. Hu, 'Multicommodity Network Flows', *Operations Research*, **11**,1963, 344–360.

[18] R.E. Gomory, T.C. Hu, 'Synthesis of a Communication Network', *J. SIAM*, **12**,1964, 348–369.

[19] R. Hassin, 'On Multicommodity Flows in Planar Graphs', *Networks*, **14**(2), 1984, 225–235.

[20] T.H. Cormen, C.E. Leiserson, R.L. Rivest, C. Stein, *Introduction to Algorithms*, The MIT Press, Cambridge, MA, 2001, 788–789.

[21] K.C. Chen, *et al.* 'Terminal Architecture for Cognitive Radio Networking Devices', *Proceeding Wireless Personal Multimedia Communication Conference*, 2007.

[22] M. Vu, N. Devroye, V. Tarokh, 'An Overview of Scaling Laws in Ad-hoc and Cognitive Radio Networks', *Wireless Personal Communications*, **45**,2008, 343–354.

[23] P. Gupta, P.R. Kumar, 'The Capacity of Wireless Networks', *IEEE Transactions on Information Theory*, **2**(46), 2000, 388–404.

[24] P. Tarasak, Y.H. Lee, 'Joint Cooperative Diversity and Scheduling in OFDMA Relay Systems', *IEEE Wireless Communications and Networking Conference (WCNC)*, 2008.

[25] M. Pischella, J.C. Belfiore, 'QoS-Based Resource Allocation with Cooperative Diversity in OFDMA', *IEEE Vehicular Technology Conference (VTC)*, 2008.

[26] M. Pischella, J.C. Belfiore, 'Power Control in Distributed Cooperative OFDMA Cellular Networks', *IEEE Trans. Wireless Commun.*, **7**(5), 2008, 1900–1906.

[27] T.Q.S. Quek, H. Shin, M.Z. Win, 'Robust Wireless Relay Networks: Slow Power Allocation with Guarantee QoS', *IEEE Journal of Selected Topics in Signal Processing*, **1**(4), 2007, 700–713.

[28] S.-J. Kim, X. Wang, M. Madihian, 'Optimal Resource Allocation in Multi-hop OFDMA Wireless Networks with Cooperative Relay', *IEEE Trans. Wireless Commun,*, **7**(5), 2008, 1833–1838.

[29] A. Fallahi, E. Hossain, A.S. Alfa, 'QoS and Energy Trade Off in Distributed Energy-limited Mash/Relay Networks: A Queuing Analysis', *IEEE Trans. Parallel and Distributed Systems*, **17**(6), 2006, 576–592.

[30] S. Ren, K.B. Letaief, 'Optimal Effective Capacity for Cooperative Relay Networks with QoS Guarantees,' *IEEE International Conference on Communications*, 2008.

[31] C. Sun, K.B. Letaief, 'User Cooperative in Heterogeneous Cognitive Radio Networks with Interference Reduction', *IEEE International Conference on Communications*, 2008.

[32] J.N. Laneman, D.N.C. Tse, G.W. Wornell, 'Cooperative Diversity in Wireless Networks: Efficient Protocols and Outage Behavior', *IEEE Trans. Information Theory*, **50**(12), 2004, 3062–3079.

[33] D. Wu, R. Negi, 'Effective Capacity: A Wireless Link Model for Support of Quality of Service', *IEEE Trans. Wireless Commun.*, **2**(4), 2003, 630–643.

[34] D. Wu, 'Providing Quality of Service Guarantees in Wireless Networks,' Ph.D Dissertation, Carnegie Mellon University, PA, 2003.

[35] D. Wu, R. Negi, 'Downlink Scheduling in a Cellular Network for Quality of Service Assurance', *IEEE Trans. Veh. Technol.*, **53**(5), 2004, 1547–1557.

[36] C.-S. Chang, 'Stability, Queue Length, and Delay of Deterministic and Stochastic Queuing Networks', *IEEE Trans. Automat. Contr.*, **39**(5), 1994, 913–931.

[37] A.I. Elwald, D. Mitra, 'Effective Bandwidth of General Markovian Traffic Sources and Admission Control of High Speed Networks', *IEEE/ACM Trans. Networking*, **1**(3), 1993, 329–341.

[38] J. Tang, X. Zhang, 'Cross-layer Resource Allocation Over Wireless Relay Networks for Quality of Service Provisioning', *IEEE Journal of Selected Areas in Communications*, **25**(4), 2007, 645–656.

[39] R.U. Nabar, H. Bölcskei, F.W. Kneubühler, 'Fading Relay Channel: Performance Limits and Space-time Signal Design', *IEEE Journal of Selected Areas in Communications*, **22**,2004, 1099–1109.

[40] S. Haykin, 'Cognitive Radio: Brain-empowered Wireless Communications', *IEEE Journal on Selected Areas in Communications*, **23**(2), 2005, 201–220.

7

Spectrum Sensing

To allow reliable operation of cognitive radios, we must be able to detect precisely the spectrum holes at link level (that is, certain frequency bands are not used for transmission at certain times), which gives spectrum sensing a critical role and results in plenty of research in the literature. For overlay or co-existing multi-radio systems, we may want to sense more spectrum information to control the proper amount of interference. When networking cognitive radios are involved, we need more information beyond link establishment to achieve better spectrum utilisation at the network level.

7.1 Spectrum Sensing to Detect Specific Primary System

We start spectrum sensing with the simplest case, one primary system (PS) transmitter-receiver pair and one secondary cognitive radio (CR) transmitter-receiver pair. In addition to effective spectrum sensing, there are more challenges. The secondary CR nodes, even at the edge of the system or guard band, should be able to detect the primary signal even if decoding the signal is impossible. Furthermore, the secondary CR nodes are in general not aware of the exact transmission scheme used by the PS nodes, and the secondary CR nodes/users may not have access to training and synchronisation signals for the PS transmissions. This suggests that the secondary users would be constrained to use noncoherent energy-based detectors with much poorer performances than coherent receivers under low signal-to-noise ratio (SNR).

7.1.1 Conventional Spectrum Sensing

Since CR is usually considered as an intelligent wireless communication radio device, to be aware of the environment and to adapt itself to optimise the transmission and spectral utilisation, spectrum sensing is an important function to enable CRs to detect the underutilised spectrum of primary systems and improve the overall spectrum efficiency. Conventional spectrum sensing at link level targets at a single primary system is used to decide between the two hypotheses namely

$$
y[n] = \begin{cases} w[n] & H_0 \\ hs[n] + w[n] & H_1 \end{cases} \quad n = 1, \ldots, N \tag{7.1}
$$

where $y[n]$ is the complex signal received by the cognitive radio, $s[n]$ is the transmitted signal of the primary user, $w[n]$ is the additive white Gaussian noise (AWGN), h is the complex gain of an ideal channel and N is the observation interval. If the channel is not ideal, h and $s[n]$ are convolved instead of multiplied. H_0 represents the null hypothesis that no primary user is present and H_1 represents the alternate hypothesis that a primary user signal exists, respectively. Wang *et al.* summarised spectrum sensing techniques into energy-based and feature-based ones [4]. In the following, we give an overview of some well-known spectrum sensing techniques for a single system: *energy detection, matched filter, cyclostationary detection* and *wavelet detection* (energy-based).

7.1.1.1 Energy Detection

When the primary user signal is unknown, the energy detection method is optimal for detecting any unknown zero-mean constellation signals and can be applied to CRs. In the energy detection approach, the radio frequency energy or the received signal strength indicator (RSSI) is measured over an observation time to determine whether the spectrum is occupied or not. Energy detection is usually realised in time domain and sometimes in frequency domain; we focus on the time-domain approach hereafter. The received signal is squared and integrated over the observation interval. Then the output of the integrator is compared to a threshold to decide whether the primary user exists or not. That is, the following binary decision is made:

$$\begin{cases} H_0, & \text{if } \sum_{n=1}^{N} |y[n]|^2 \leq \lambda \\ H_1, & \text{otherwise} \end{cases} \tag{7.2}$$

where λ is the threshold which depends on the receiver noise.

In terms of implementation, there are a number of options for energy detection based sensors. Analogue implementations require an analogue pre-filter with fixed bandwidth which becomes quite inflexible for simultaneous sensing of narrowband and wideband signals. Digital implementations offer more flexibility by using FFT-based spectral estimates. This architecture inherently supports various bandwidth types and allows sensing of multiple signals simultaneously. Figure 7.1 shows the architecture for digital implementation of an energy detector.

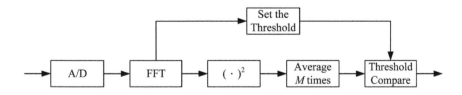

Figure 7.1   Digital implementation of an energy detector

Although the energy detection approach can be implemented without any *a priori* knowledge of the primary user signal, it has some difficulties. First of all, it can only detect the signal of the primary user if the detected energy is above the threshold. The threshold selection for energy detection can also be problematic since it is highly susceptible to the changing background noise and especially interference level. Another challenging issue is that the energy approach cannot distinguish the primary user from the other secondary users sharing the same channel. This is a critical challenge when the primary users of multiple systems co-exist in cognitive radio networks, which we discuss later in this chapter.

7.1.1.2 Matched Filter

A matched filter is an optimal detection method as it maximises the SNR of the received signal in the presence of additive Gaussian noise. However, a matched filter requires *a priori* knowledge of the primary user signal at both physical and medium access control (MAC) layers, e.g., pulse shaping, modulation type and the packet format. Consequently, the matched filter mechanism is a sort of feature-based spectrum sensing. A matched filter is facilitated by correlating a known signal with an unknown signal to detect the presence. This is equivalent to convolving the unknown signal with a time-reversed version of the assumed signal. The output of a matched filter is then compared to a threshold to decide whether the primary user signal exists or not. Therefore the following binary decision is made:

$$\begin{cases} H_0, & \text{if } \sum_{n=1}^{N} y[n]x[n]^* \leq \lambda \\ H_1, & \text{otherwise} \end{cases} \tag{7.3}$$

where λ is the threshold.

Matched filters are commonly used in radio communications and radar transmission. In the cognitive radio scenario, however, the use of the matched filter can be severely limited if the information of the primary user signal is hardly available at the cognitive radios, which falls into the *signal interception* problem. Fortunately, most licensed systems include pilots, preambles, synchronisation words or spreading codes to achieve the coherent detection. If partial information of the primary user signal is known, such as pilots or preambles, the use of a matched filter is still possible for coherent detection. Here we take the coherent pilot detection as an example. In this case, $x[n]$ in Equations (7.1) and (7.3) is replaced by $x_p[n]$ which is the pilot signal of the primary user. Figure 7.2 presents the architecture for digital implementation of a coherent pilot detector. The major advantage is that the matched filter needs less time to achieve high processing gain due to coherent detection. For demodulation, however, CR has to perform timing, carrier synchronisation and even channel equalisation. Therefore, cognitive radios need a dedicated receiver for every kind of primary system, which increases complexity as a significant challenge for the use of the matched filter when the target system is out of multiple possible systems, even for programmable realisation.

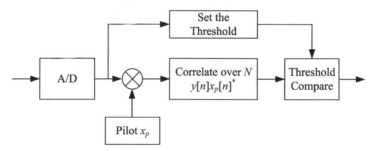

Figure 7.2 Digital implementation of a coherent pilot detector

7.1.1.3 Cyclostationary Detection

The idea of the cyclostationary detection is to exploit the built-in periodicity of the modulated signal such as sine wave carriers, pulse trains, repeating spreading, hoping sequences or cyclic prefixes. A signal is said to be cyclostationary (in the wide sense) if its autocorrelation is a periodic function of time t with some period. The cyclostationary detection can be performed as follows. First, one can calculate the cyclic autocorrelation function (CAF) of the observed signal $x(t)$, $R_x^\alpha(\tau)$, as

$$R_x^\alpha(\tau) = \lim_{T \to \infty} \frac{1}{T} \int_{-T/2}^{T/2} x\left(t + \frac{\tau}{2}\right)x\left(t - \frac{\tau}{2}\right)e^{-j2\pi\alpha t}\,dt \tag{7.4}$$

where α is called cyclic frequency. Second, the discrete Fourier transformation of the CAF can be computed to obtain the spectral correlation function (SCF):

$$S_x^\alpha(f) = \int_{-\infty}^{\infty} R_x^\alpha(\tau)e^{-j2\pi f\tau}d\tau \tag{7.5}$$

Specifically, it is shown that

$$S_x^\alpha(f) = \lim_{T\to\infty}\lim_{Z\to\infty}\frac{1}{TZ}\int_{-Z/2}^{Z/2} X_T\left(t,f+\frac{\alpha}{2}\tau\right)X_T^*\left(t,f-\frac{\alpha}{2}\tau\right)dt \tag{7.6}$$

where

$$X_T(t,f) = \int_{t-T/2}^{t+T/2} x(u)e^{-j2\pi fu}du \tag{7.7}$$

Spectral correlation function $S_x^\alpha(f)$ is also called cyclic spectrum, which is a two-dimensional function in terms of frequency and cyclic frequency. We also note that power spectrum density is a special case of a spectral correlation function for $\alpha = 0$. Finally, the detection is completed by searching for the unique cyclic frequency corresponding to the peak in the SCF plane.

Cyclostationary detectors are usually implemented in digital domain. Direct algorithms first compute the spectral components of the data through FFT, and then perform the spectral correlation directly on the spectral components. The digital implementation of a cyclostationary detector for spectrum sensing is shown in Figure 7.3. The main advantage of the cyclostationary detection is that it can distinguish the noise energy from the signal energy. This is because noise has no spectral correlation whereas the modulated signals are usually cyclostationary with spectral correlation due to the embedded redundancy of signal periodicities. Cyclostationary detection is more robust to noise uncertainty than an energy detector. Furthermore, it can work with lower SNR than energy detectors since the latter do not exploit the information embedded in the received signal whereas feature detectors do. It is known that cyclostationary detection can achieve a larger processing gain than energy detection. However, the implementation of cyclostationary detection is more complicated. Cyclostationary detectors also require longer observation time than energy detectors. Therefore, the spectrum holes with short time duration may not be exploited efficiently. When we consider the multiple systems co-existing case, the system identification of cyclostationary detector becomes more complex and we need a efficiency methodology to deal with it.

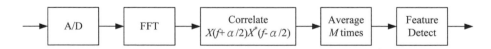

Figure 7.3   Digital implementation of a cyclostationary detector

7.1.1.4 Wavelet Detection

For the detection of wideband signals, the wavelet approach offers advantages in terms of both implementation cost and flexibility, in adapting to the dynamic spectrum as opposed to the conventional use of multiple narrowband bandpass filters (BPF). In order to identify the locations of vacant frequency bands, the entire wideband is modelled as a train of consecutive frequency sub-bands where the power spectral characteristic is smooth within each sub-band but changes abruptly on the border of two neighbouring sub-bands. By employing a wavelet transform of the power spectral density (PSD) of the observed signal $x[n]$, the singularities of the PSD $S(f)$ can be located and thus the vacant frequency bands can be found. One critical challenge of implementing the wavelet approach in practice is the high sampling rates for characterising the large bandwidth. The digital implementation of a wavelet detector for spectrum sensing is shown in Figure 7.4. When multiple systems co-exist in spectrum, it is difficult for the wavelet detector to identify the multiple systems because the wavelet detector is used to identify the locations of the vacant frequency bands and it is hard to cope with the inter-system interference environment.

Figure 7.4 Digital implementation of a wavelet detector

The pros and cons of the aforementioned spectrum sensing techniques are summarised in Table 7.1.

Table 7.1 Advantages and disadvantages of spectrum sensing techniques

Spectrum sensing technique	Advantage	Disadvantage
Energy detection	1. Does not need any *a priori* information 2. Low computational cost	1. Cannot work with low SNR 2. Cannot distinguish users sharing the same channel
Matched filter	1. Optimal detection performance 2. Low computational cost	1. Requires *a priori* knowledge of the primary user 2. Design for each kind of primary system signal
Cyclostationary detection	1. Robust in low SNR 2. Robust to interference	1. Requires partial information of the primary user 2. High computational cost
Wavelet detection	1. Effective for wideband signal	1. Does not work for spread spectrum signals 2. High computational cost

7.1.2 Power Control

When a cognitive radio operates as a secondary system in CRNs, it has a lower priority of spectrum access than the pre-existing primary system. As a consequence, the cognitive radio should be designed to prevent the primary system from harmful interference during transmission. However, it is not an easy task to reliably guarantee service of the primary system because of the *hidden terminal problem*, which is well-studied in *carrier sense multiple access* networks. When two terminals are within the range of the station but out of the range of each other, or separated by some physical obstacles transparent to radio signals, the two such terminals are said to be hidden from each other. In wireless LAN or similar wireless data networks, the hidden terminal problem induces possible collisions at the receiver when the transmitters are hidden (i.e., could not sense the transmission of each other) from each other and then transmit. It is improved by the four-way handshaking methodology in that a short request signal is sent first and the data packets are transmitted only when the receiver permits. Therefore the collision only happens during the transmission of the request signal not in the transmission of the traffic signal. Through its testing and observation process, the hidden terminal problem is alleviated, as in IEEE 802.11 MAC.

Similarly, the hidden terminal problem occurs in CRNs. When a cognitive radio transmits, it needs to know the impact of its transmission on the primary system. More specifically, the interference at the primary receiver (PS-Rx) should be guaranteed to below a given threshold. We regard PS-Rx as a hidden terminal when its existence cannot be detected by cognitive radios. Figure 7.5 shows the hidden terminal problem in CRNs because of the path-loss nature and the shadowing nature, respectively. The CRNs may suffer from the hidden terminal problem seriously because unpredictable interference is induced to the primary system.

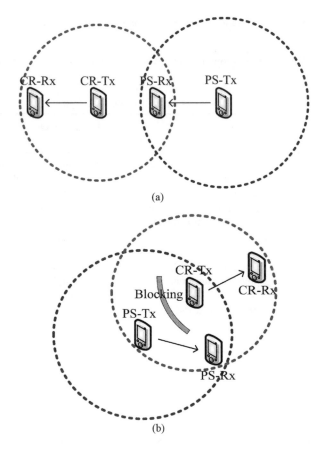

(a)

(b)

Figure 7.5   The hidden terminal problem due to (a) path-loss and (b) shadowing

7.1.3 Power-Scaling Power Control

The conventional spectrum sensing techniques primarily detect the signals from primary transmitters (PS-Tx) passively. Since it is usually assumed that the cognitive radio transmitter (CR-Tx) does not know its channel to PS-Rx, the sensitivity of cognitive radios should be much larger than PS-Rx. In general, it is required to budget for the worst case, in which PS-Rx is located at the closest point to the CR-Tx in the primary system service coverage as shown in Figure 7.6. The corresponding transmission power control of CR-Tx, called power scaling power control, can be accomplished through sensing the signals transmitted by PS-Tx.

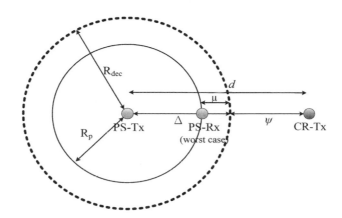

Figure 7.6   Conventional power scaling power control of the cognitive radio

To be specific, we assume that the primary system has a minimum required SINR γ_{dec} to decode successfully at its target rate. In the absence of interference, γ_{dec} occurs at a radius R_{dec} from PS-Tx. The idea is to guarantee service to PS-Rx within some protected radius R_p so that PS-Rx has a guaranteed successful reception rate even in the presence of the secondary system transmission. Define Δ (dB) as the signal attenuation between PS-Tx and the decidability radius R_{dec}, μ (dB) as the margin of protection that represents how much interference above the noise floor the primary system can tolerate, d as the distance between PS-Tx and CR-Tx, and $\psi \equiv \lambda - \Delta$ (dB) where λ (dB) is the path loss due to the distance d. Assume that the path-loss and the propagation power attenuation of the channel is characterised by $g(r) = r^{-\alpha}$ where α represents the power loss exponent. With the presence of cognitive radio transmission, the quality of service for PS-Rx should be guaranteed:

$$\frac{Q_{PS}}{Q_{CR}+\sigma^2} \geq 10^{\frac{\gamma_{dec}}{10}} \tag{7.8}$$

where σ^2 is the noise power of PS-Rx and Q_{PS} and Q_{CR} are the received signal power from PS-Tx and CR-Tx, respectively.

Considering the worst case that PS-Rx is located at the closest point to CR-Tx in the boarder of protection region, we can obtain

$$10\log\left(\frac{P_{CR}}{\sigma^2}\right) \leq \Delta + 10\log\left(10^{\frac{\mu}{10}} - 1\right) + 10\alpha\,log\left(\left(10^{\frac{\psi}{10}}\right)^{\frac{\mu}{10}} - \left(10^{\frac{\mu}{10}}\right)^{\frac{1}{d}}\right) \equiv f(\psi) \tag{7.9}$$

for the constants α and μ. Thus the maximum transmission power of CR-Tx can be determined. Please note that the *a priori* information of the CRs includes σ^2, Δ, μ and α. Therefore, ψ is the target of spectrum sensing that can be inferred from the estimation of false alarm probability. When considering the case in which CR-Tx may lie in the shadow (signal loss of β dB) with respect to the primary signal, CR-Tx must measure a margin of $\mu + \beta$. Adjusting the equation to account for shadowing, we have

$$10\log\left(\frac{P_{CR}}{\sigma^2}\right) \leq \Delta + 10\log\left(10^{\frac{\mu}{10}} - 1\right) + 10\alpha \log\left(\left(10^{\frac{\psi-\beta}{10}}\right)^{\frac{1}{\alpha}} - \left(10^{\frac{\mu}{10}}\right)^{\frac{1}{\alpha}}\right) \tag{7.10}$$

7.1.4 Cooperative Spectrum Sensing

The major challenge for the CR's spectrum sensing to the primary system (PS) is the hidden terminal problem, just as for any transmission sensing mechanism, such as the widely used CSMA in wireless networks. Figure 7.7 illustrates the hidden terminal problem for spectrum sensing. The PS operating transmission power range is as at the left big circle and the right small circle represents the CR

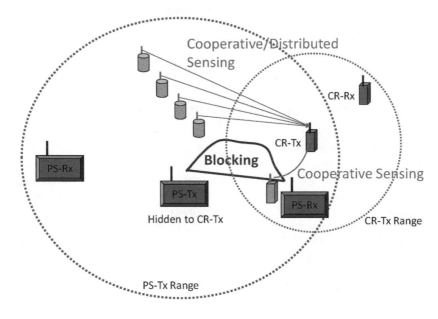

Figure 7.7   Hidden terminal problem of spectrum sensing and cooperative strategy

transmission range. The CR transmitter (CR-Tx) wishes to sense the spectrum hole and to access dynamically the channel for transmission under a constrained probability γ to interfere with the PS (ideally $\gamma \rightarrow 0$). However, certain blocking resulting in shadow fading prohibits CR-Tx's effective spectrum sensing, which is exactly like the hidden terminal problem in CSMA. The immediate solution is to adopt the cooperative communication strategy into sensing by placing a set of sensors scattering in different locations to detect PS-Tx's possible transmission and by relaying such detected information from the distributed cooperative sensors to CR-Tx. The pioneer work in cooperative sensing was carried out by researchers at Georgia Tech [8,9]. However, in the following, we shall model the problem as a sensor detection problem.

Regarding user-cooperation for cognitive radio systems, researchers have considered two kinds of schemes: (i) some kind of joint detection is employed among all the cooperating users; and (ii) the final decision is made based on hard decisions made by each of the cooperating users. Following [3], we focus on the more feasible system in which the individual secondary users make independent decisions about the presence of the primary signal in the frequency band that they are monitoring. They communicate their decisions to a *fusion centre* that makes the final decision about the occupancy of the band by fusing the decisions made by all cooperating nodes/radios in that area that are monitoring the same frequency band. In practice, the fusion centre would act as the centralised controller to manage the channel assignment and scheduling for the secondary CR nodes/users. Such a system scenario could also help the secondary nodes/users to exchange their decisions while each secondary node/user performs its own fusion of all the decisions.

We assume that the fusion centre knows the geographic locations of all cooperating secondary users and hence can construct the correlation between their observations. However, the primary's location is not known. The sensing problem now becomes a binary hypothesis testing problem to decide whether the mean received power at the receiver location is higher than the power expected at the edge of the transmission range or not. When the primary is actively transmitting and the secondary users are within the range, the power they receive is the sum of the primary signal power and the noise power. In this case, we model the received power levels as common shadow fading being log-normally distributed.

The observations of power level from total N cooperative sensors form an observation vector Y. The binary hypothesis problem is

$$H_0 : Y \sim G(0, \sigma_0^2 I)$$
$$H_1 : Y \sim G(\vartheta u, C)$$

where $G(m, V)$ represents Gaussian vector distribution with mean vector m and covariance matrix V; u is unit vector and I is the identity matrix; ϑ is a variable parameter representing the mean of observed distributions under H_1 and μ is the mean total power with $\mu = E[10 \log_{10}(1 + SNR)]$; σ_0^2 presents noise power; C is the covariance matrix with $C_{ij} = \sigma_1^2 \rho^{d(i,j)}$ and $d(i,j)$ is the distance between node i and node j. ρ is a measure of correlation coefficient between nodes separated by unitary distance and is related to correlation distance D_c by $\rho = e^{-1/D_c}$.

Our sensing system should guarantee that the probability of making an erroneous decision under hypothesis H_1 should be lower than the constraint on the probability of interference γ. Furthermore, this constraint shall hold for all values ϑ greater than μ. This is thus a composite binary Neyman-Pearson testing since we do not have any *a priori* information about ϑ. A *robust* or *uniformly most powerful* (UMP) detection is needed.

We reduce our original detection into a simple Neyman-Person hypothesis testing problem between the following modified hypotheses:

$$H_0 : Y \sim G(0, \sigma_0^2 I)$$
$$H_1 : Y \sim G(\mu u, C)$$

In this system model, the final decision about the hypothesis is made at the fusion centre that has obtained only the binary-valued decisions made by the sensors based on individual observations $\{Y_i\}_{i=1}^N$. We use $\{D_i\}_{i=1}^N$ to represent the decisions made at the individual sensors and D to represent the vector of decisions made by all sensors. Hence, we have the *decentralised Neyman-Pearson* hypothesis testing problem illustrated in Figure 7.8.

For the sensor node i, the optimal test to determine its decision D_i is the likelihood test (L) based on its observation Y_i of the form as

$$D_i = 1_{\log[L(Y_i)] > \eta} \tag{7.11}$$

For the typical sensor approach, each cooperative sensor quantises (decides) the observation and then proceeds to the final decision δ at the fusion centre. One of the simplest sub-optimal solutions to

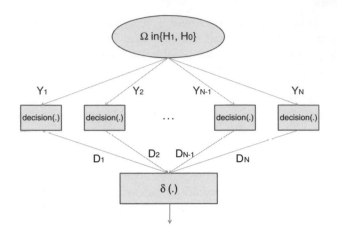

Figure 7.8 Cooperative sensing as sensor detection

such a data fusion problem is the *Counting Rule* (also referred to as the *Voting Rule*), which just counts the number of sensor nodes that vote in favour of H_1 and compares them with a threshold. Such a methodology is more like DF in cooperation. More sub-optimal schemes can be found in Reference 5.

Another cooperative sensing scenario can be depicted as in Figure 7.9 to result in an adaptive filtering of cooperative observations to reach a decision, more like AF in cooperation.

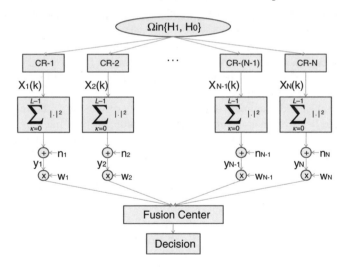

Figure 7.9 Cooperative sensing as data fusion

For these secondary cooperative nodes in a CRN, the binary hypothesis test for spectrum sensing at the kth observation instant is

$$H_0 : x_i(k) = v_i(k) \, i = 1, 2, \ldots, N$$
$$H_1 : x_i(k) = h_i s(k) i + v_i(k) \, i = 1, 2, \ldots, N$$

where $s(k)$ denotes the transmitted signal from the primary user with channel gain h_i, and $x_i(k)$ is the received signal by the ith secondary cooperative user. The channel embedded noise is AWGN and thus $v_i(k) \sim G(0, \sigma_i^2)$ is independent of $s(k)$, while $\sigma = [\sigma_1^2, \sigma_2^2, \ldots, \sigma_N^2]^T$.

Each cooperative secondary node/user conducts non-coherent detection over the interval of L samples (L is determined by time-bandwidth product):

$$u_i = \sum_{k=0}^{L-1} |x_i(k)|^2 \quad i = 1, 2, \ldots, N$$

The above equation suggests local sensing at the ith cooperative secondary node/user based on energy detection, which is the sum of squared Gaussian random variables. Please recall the central chi-square distribution and we can have

$$\frac{u_i}{\sigma_i^2} \sim \begin{cases} \chi_L^2, & H_0 \\ \chi_L^2(\eta_i), & H_1 \end{cases}$$

and the local SNR is defined as

$$\eta_i = \frac{E_s|h_i|^2}{\sigma_i^2}$$

As long as we define the decision threshold for each node, the decision rule for single-cooperative-CR can be easily obtained. However, we wish to adjust the weighting coefficients $w = [w_1, w_2, \ldots w_N]^T$ so that we can reach optimal spectrum sensing based on the observation vector $y = u + n$ at the fusion centre, and then make a final decision about the appearance of the primary system. That is, $y_{fusion\ centre} = w^T y$. Such a weighting can adjust according to individual fading from cooperative secondary paths. Let us define

$$\begin{aligned} h &= [|h_1|^2, |h_2|^2, \ldots, |h_N|^2] \\ v^2 &- [Var(n_1), Var(n_1), \ldots, Var(n_1)]^T \end{aligned}$$

Since the linear combination of Gaussian variables is Gaussian, $y_{fusion\ centre}$ is Gaussian with mean and variance as

$$\begin{aligned} \mu_{fc} &= \begin{cases} L\sigma^T w & H_0 \\ (L\sigma + E_s g)^T w & H_1 \end{cases} \\ \Sigma_0 &= (2L)\text{diag}^2(\sigma) + \text{diag}(v^2) \\ \Sigma_1 &= (2L)\text{diag}^2(\sigma) + \text{diag}(v^2) + 4E_s\text{diag}(g)\text{diag}(\sigma) \end{aligned}$$

At the fusion centre, with a test threshold γ_{fc}, we can easily construct the test as

$$y_{fusion\ center} \underset{H_0}{\overset{H_1}{\underset{<}{\gtrless}}} \gamma_{fc}$$

And, it is straightforward to calculate the probability of detection and probability of a false alarm, as the performance index of the test, so that we may proceed with performance optimisation. However, please note that some information here might be hard to obtain in practice, such as channel gains, although we already require very minimum information to determine mean and variance at fusion centre. More details can be found in Reference 6. As a matter of fact, in more practical situations in co-existing multi-radio systems, we have to consider interference aggregated into spectrum sensing and other CR operations. Ghasemi and Sousa [7] have summarised a mathematical model of aggregated interference for such a concern and demonstrated its applications to cooperative spectrum sensing.

7.2 Spectrum Sensing for Cognitive OFDMA Systems

After introducing the fundamental spectrum sensing techniques, we would like to design the spectrum sensing for a practical OFDMA system using the overlay rate-distance that was introduced in Chapter 5 [10]. We consider a cell-structured OFDMA system. Base station (BS) lies at the centre of a cell with coverage radius R as in Figure 7.10. Let W be total bandwidth, which contains N_F sub-bands. One sub-band is allocated to one cell and is further divided into N channels. Similarly, timing axis is segmented by OFDMA symbol period. Then, radio resource, composed of channels and OFDMA symbols, is allocated to active users in the cell. This information is usually specified in the frame header. Without loss of generality, we adopt the frame structure in IEEE 802.16 as an example. The frame starts with preamble, which is mainly used for synchronisation and channel estimation, and then followed by the frame header, including DL_MAP and UL_MAP. In addition, the frame header also defines the transmission parameters, such as forward error control code (FEC) rate, and modulation in DL burst and UL burst respectively. To simplify the model without loss of generality, assume there are two data transmission rates, which follows specific adaptive modulation and coding (AMC). Although the data transmission rate can be adaptively adjusted, the fundamental symbol rate is rather fixed in most systems.

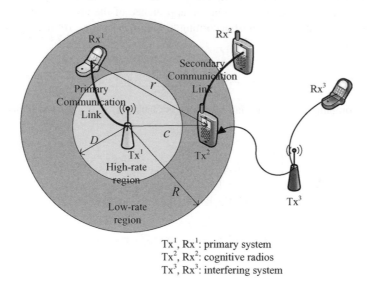

Tx^1, Rx^1: primary system
Tx^2, Rx^2: cognitive radios
Tx^3, Rx^3: interfering system

Figure 7.10   OFDMA CR system model

We categorise received signals at CRs within a sub-band of the PS into six classes:

- C_0: $w_{LP}(t)$ background noise only;
- C_1: $s_t(t)$ primary system traffic signal;
- C_2: $s_c(t)$ primary system control signal;
- C_3: $i_s(t)$ interfering signal of the same fundamental symbol rate as PS;
- C_4: $i(t)$ interfering signal of different fundamental symbol rate as PS;
- C_5: $i_c(t)$ interfering control signal.

In the above, $w_{LP}(t)$ denotes equivalent baseband (BB) additive white Gaussian noise (AWGN) with zero mean and two-sided power spectrum density N_0, and $s_t(t)$ denotes traffic signal of the primary system. In addition, $i_s(t)/i(t)$ denotes traffic signal from the interfering system with/without the same fundamental symbol rate as the primary system and is assumed white within a sub-band. By the *central limit theorem*, traffic signals are modelled as independent circular symmetric complex Gaussian processes. In the mean time, $s_c(t)$ and $i_c(t)$ denote control and management signals from the primary and interfering systems respectively and are periodically transmitted. In the following, we just denote 'control and management signal' by 'control signal' for simplicity. By analysing received signals, CRs determine the operation state of the primary system where there are five possible states:

- S_0 (non-existent): neither C_1 nor C_2 exist;
- S_1 (existent, inactive): C_2 exists but C_1 does not exist;
- S_2 (existent, active, high rate): C_1 exists and traffic is transmitted at high rate in a channel;
- S_3 (existent, active, low rate): C_1 exists and traffic is transmitted at low rate in a channel;
- S_4 (existent, active, idle): C_1 exists and no traffic is transmitted in a channel.

Here, we will further consider the state S_1 (existent, inactive), in which users of the primary network are in the system but not transporting traffic, and thus cognitive radios could aggressively adjust system parameters and connect to the primary BS through cooperative (multi-radio) communications. In addition, to find out the potential radio resource of secondary system(s) within occupied sub-bands, we require information about data transmission rate and sub-band utilisation, which is governed by the MAC protocol.

7.2.1 Cognitive Cycle

Operation of cognitive radios can be viewed as a finite-state machine and thus be considered by cognitive cycle, as in J. Mitola III and S. Haykins. We facilitate our cognitive cycle, as shown in Figure 7.11, with two sensing paths. Path A is responsible for detecting existence and activity of the primary system while Path B is used to decode the frame header and determine frequency-time utilisation. For each OFDMA symbol, we establish two $N_F \times N$ tables, channel state table and radio resource table, where the (n, m)th element records operation state of the primary system and maximum transmission power at cognitive radios of the mth channel within the nth sub-band respectively. The networking (cognitive) terminal device collects sensing information and makes optimal decisions of operating mode, including configuration of software and hardware in MAC and SDR, network routing and hand-off, and maintenance of requirements from upper-layer applications. Based on this system architecture, CRNs can be practically implemented in future mobile communications.

7.2.1.1 Cognitive Cycle – Path A

Received signal strength indicator (RSSI) is a simple and widely applied method in spectrum sensing; however, it is not enough to distinguish the signal of the primary system (C_1) from interfering signals (C_3 and C_4). For example, in the 2.4 GHz ISM band, large RSSI may result from microwave inside the band, rather than the IEEE 802.11b network. In addition, noise uncertainty degrades energy detector performance significantly. Therefore, we have to extract features of the traffic signal by transformation, which suppresses noise and interference signals. By the fact that the fundamental symbol rate is invariant or belongs to a finite set (e.g., scalable WiMAX), we model traffic signals as cyclostationary processes, which are induced by pulse trains and detect such features by the spectral correlation function. Though preliminary spectrum sensing exists, we adopt spectral-line generator to meet the need of quickness in cognitive radios. However, in our system model, C_1 and C_3 have the same

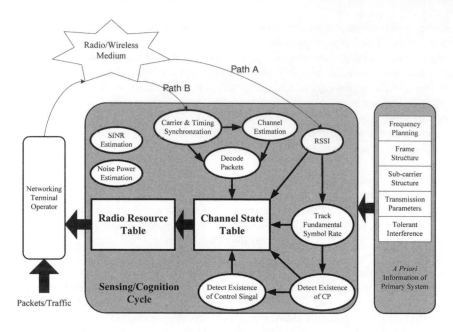

Figure 7.11   Cognitive cycle

fundamental symbol rate. By means of cyclic prefix (CP) in OFDMA systems, we can precisely detect the existence of the active primary system. Finally, since base stations broadcast control signals (e.g., beacons in IEEE 802.11) periodically to maintain mobile network operation, we can identify the existence of the inactive primary system by this property.

7.2.1.2 Cognitive Cycle – Path B

In Path A, we determined the state of the primary system: non-existent (S_0), existent and inactive (S_1) and existent and active (S_2, S_3, S_4). We further considered the channels that are occupied by transmissions and applied rate-distance nature so that simultaneous communication is possible. The information about the channel occupation and data transmission rate can be obtained by analysing the frame header after the cognitive radios achieve synchronisation with the primary BS.

We summarise a set of cognitive information in spectrum sensing under OFDMA systems as follows:

- *RF signal processing:* (carrier) frequency, bandwidth, RSSI, signal-to-noise-and-interference ratio (SINR), noise power;
- *BB pre-detection signal processing:* fundamental symbol rate, carrier and timing, pilot signal, channel fading;
- *BB post-detection signal processing:* system identification, modulation parameters, FEC type and rate;
- *Network processing information:* multiple access protocol, radio resource allocation.

Let us suppose that the cognitive radios know *a priori* information about potential primary systems including frequency planning, frame structure, subcarrier structure (FFT size, pilot positions), transmission parameters (fundamental symbol rate, CP length, FEC type) and tolerant interference of primary systems, since spectrum usage is regulated and system specifications are well-defined.

7.2.2 Discrimination of States of the Primary System

The next step of design is to identify the existence and activity of the primary (OFDMA) system within one sub-band by RSSI, fundamental symbol rate, cyclic property and control signal of the primary system, where RSSI is simply an energy detector. In general, we do not know *a priori* probability of existence of the primary system. Hence, we adopt the *Neymann-Pearson* criterion to design the spectrum sensing procedure.

7.2.2.1 Fundamental Symbol Rate

If there exist traffic signals from the primary system, the received signal at cognitive radios may be written as

$$z(t) = \Re\left\{ e^{j2\pi f_c t}\left(\sum_{n=-\infty}^{\infty} x(n)h(t - nT_s) + w_{LP}(t) \right) \right\} = \Re\{e^{j2\pi f_c t}(s_t(t) + w_{LP}(t))\} = \Re\{y(t)\} \quad (7.12)$$

Here, f_c denotes carrier frequency, T_s denotes reciprocal of fundamental symbol rate, $h(t)$ denotes pulse shaping filter and $x(n)$ denotes transmitted data in time domain, or

$$x[n + (N_{FFT} + N_{CP})m]$$

$$= \begin{cases} \dfrac{1}{\sqrt{N_{FFT}}} \displaystyle\sum_{k=0}^{N_{FFT}-1} X_m(k)e^{j2\pi\frac{kn}{N_{FFT}}} & n \in \{0, 1, \dots, N_{FFT} - 1\} \\[2mm] x[n + (N_{FFT} + N_{CP})m + N_{FFT}] & n \in \{-N_{CP}, -N_{CP}+1, \dots, -1\} \end{cases} \quad (m \in \mathbb{Z}) \quad (7.13)$$

In Equation (7.13), $X_m(k)$ denotes modulated data at the kth subcarrier in the mth OFDMA symbol, N_{FFT} is the FFT size and N_{CP} is the CP length. Without loss of generality, $h(t)$ is assumed to be a square-root raised cosine filter with roll-off factor less than 100% and we can have

$$E[z^2(t)] \approx \frac{1}{2}\left(\frac{\sigma_X^2}{T_s} + \sigma_w^2 \right) + \frac{\sigma_X^2}{T_s}\Re\left\{ Z_1 e^{j2\pi\frac{t}{T_s}} \right\} \quad (7.14)$$

where

$$Z_m = \frac{1}{2\pi}\int_{-\infty}^{\infty} H(j\omega)H^*(j(\omega - 2\pi m/T_s))d\omega \quad (7.15)$$

noise power is σ_w^2, $Z_{-m} = Z_m^*$ and $Z_0 = 1$. Note that the component at the fundamental symbol rate is independent of noise power and thus the challenge of noise uncertainty in an energy detector is released. However, received signal power in the second term of Equation (7.14) is reduced by factor Z_1. Therefore, performance of the detector depends on Z_1. From Equation (7.15), we conclude that if $h(t)$ has larger bandwidth (i.e., higher roll-off factor), the detector has better performance. A similar phenomenon also happens in timing tracking systems. Furthermore, since this algorithm can be implemented at the radio frequency (RF) part in front of the analogue-to-digital converter (ADC), it can be done quickly.

7.2.2.2 Cyclic Prefix of OFDMA Signal

The initial setup to detect the existence of CP is similar to Reference 38. Collect $2N_{FFT} + N_{CP}$ samples with sampling rate $1/T_s$ and assume this region contains one complete OFDMA symbol. The detection problem becomes

$$H_0 : r(n) = w_0(n) \qquad n = 0, 1, \ldots, 2N_{FFT} + N_{CP} - 1$$
$$H_1 : r(n) = s(n) + w_1(n) \quad n = 0, 1, \ldots, 2N_{FFT} + N_{CP} - 1$$

Under H_1, there exists timing offset ϑ and frequency offset ε due to lack of synchronisation. Then, the traffic signal of the primary system becomes

$$s(n) = x(n - N_{CP} - \theta)e^{j2\pi\varepsilon n/N_{FFT}}$$

Let I and $\tilde{I}$ be two sampling intervals, containing CP and its replica respectively, that is

$$I = \{\theta, \theta + 1, \ldots, \theta + N_{CP} - 1\}$$
$$\tilde{I} = \{\theta + N_{FFT}, \theta + N_{FFT} + 1, \ldots, \theta + N_B - 1\}$$

If the primary system exists, the samples in CP and their copies are correlated, or

$$E[r(n)r^*(n+m)|H_1] = \begin{cases} \sigma_s^2 + \sigma_w^2 & m = 0 \\ \sigma_s^2 e^{-j2\pi\varepsilon} & m = N_{FFT}, \quad n \in I \\ 0 & \text{otherwise} \end{cases}$$

We model $w_1(n)$ as white Gaussian process with zero mean and variance σ_w^2 and $E[|s(n)|^2] = E[|x(n)|^2] = \sigma_s^2$. On the other hand, $w_0(n)$ denotes superposition of interfering signals and background noise and is modelled as white Gaussian process with zero mean and variance $\sigma^2 = \sigma_s^2 + \sigma_w^2$, i.e.,

$$E[r(n)r^*(n+m)|H_0] = \begin{cases} \sigma^2 & m = 0 \\ 0 & \text{otherwise} \end{cases}$$

We assume that total powers under both hypotheses are the same, which leads to the worst case since the energy detector does not work. Then, likelihood ratio test (LRT) becomes

$$M(\mathbf{r}) = |S(\theta)|\cos(2\pi\varepsilon + \angle S(\theta)) - \frac{\rho}{2}P(\theta) \underset{H_0}{\overset{H_1}{\gtrless}} \tau_{CP} \tag{7.16}$$

where

$$S(\theta) = \sum_{n \in I} r(n)r^*(n + N_{FFT})$$

$$P(\theta) = \sum_{n \in I} [|r(n)|^2 + |r(n + N_{FFT})|^2]$$

$$\rho = SINR/(SINR + 1),$$

and $SINR = \sigma_s^2/\sigma_w^2$. The decision metric $M(\mathbf{r})$ only depends on the samples in the CP and their copies whose energy $P(\vartheta)$ and correlation $S(\vartheta)$ are considered. With unknown parameters (i.e., ϑ, ε, σ^2, $SINR$), it becomes a *composite detection* problem, in which it is desirable to find the *uniformly most powerful* (UMP) test. However, the UMP test does not exist because the decision region depends on θ. In absence of the UMP test, *generalised LRT* (GLRT) can be applied. We assume $SINR$ is known and have

$$M(\mathbf{r}) = \frac{\sum_n |r(n)|^2}{\sum_n |r(n)|^2 - \frac{2\rho}{1 - \rho^2}\max_\theta \left\{|S(\theta)| - \frac{\rho}{2}P(\theta)\right\}} \underset{H_0}{\overset{H_1}{\gtrless}} \tau_{CP} \tag{7.17}$$

Note that this detector is just the ratio of estimations of total energy under two hypotheses and frequency offset does not affect Equation (7.19). In addition, we evaluate the probability of false alarm to reach

$$P_F = \int_{\tau_{CP}}^{\infty} f(M(\mathbf{r})|H_0)dM \approx Q\left(\frac{-\mu_1 - \mu_2}{\sqrt{\sigma_1^2 + \sigma_2^2}}\right) \tag{7.18}$$

where $f(\cdot)$ is probability density function, $Q(x)$ denotes right-tail probability of a Gaussian random variable with zero mean and unit variance and

$$\mu_1 = (2N_{FFT} - N_{CP})(1 - \tau_{CP})$$

$$\sigma_1^2 = (2N_{FFT} - N_{CP})(1 - \tau_{CP})^2$$

$$\mu_2 = 2N_{CP} - \frac{2N_{CP} - \rho\sqrt{\pi N_{CP}}}{1 - \rho^2}\tau_{CP}$$

$$\sigma_2^2 = 2N_{CP} - \frac{4N_{CP} - 4N_{CP}\rho\sqrt{\pi}(\sqrt{N_{CP}+1} - \sqrt{N_{CP}})}{1 - \rho^2}\tau_{CP} \tag{7.19}$$

$$+ \frac{2N_{CP} + 4N_{CP}\rho^2(1 - \frac{\pi}{4}) - 4N_{CP}\rho\sqrt{\pi}(\sqrt{N_{CP}+1} - \sqrt{N_{CP}})}{(1 - \rho^2)^2}\tau_{CP}^2$$

By taking the inverse of $Q(x)$, we can obtain the optimal threshold under the NP criterion. Next, we investigate a special case, $\rho = 0$, which results in detection of white random signal under AWGN with unknown noise power and the decision metric becomes $M(\mathbf{r}) = 1$. This explains why the simple energy detector does not work with unknown noise power. Finally, to deal with unknown SINR, note that Equations (7.18) and (7.19) depend on *SINR* via ρ; therefore, we design a robust CP detector at the target SINR, $\overline{SINR}$. The parameters, ρ and τ_{CP}, can be calculated in advance by with $SINR = \overline{SINR}$. Furthermore, since this detector explores the CP property of the target OFDMA system, it can be used to distinguish OFDM-based systems with different system parameters, N_{FFT} and N_{CP} (e.g., IEEE 802.11a and IEEE 802.16 in the 5 GHz unlicensed band.)

7.2.2.3 Control and Management Signal

We do not specify control signals of potential primary systems but model them as white Gaussian processes. Therefore, this detector is robust to system variation. Control signals of the primary system are modelled as a zero mean white Gaussian process with transmission period M_C samples with duration N_C samples. To detect control signals, we collect M_C samples and assume this interval contains one control signal period. The detection problem turns out to be

$$H_0 : r(n) = w(n) \qquad\qquad n = 0, 1, \ldots, M_C - 1$$

$$H_1 : r(n) = \begin{cases} w(n) & n \notin Q \\ s_c(n) + w(n) & n \in Q \end{cases}$$

where Q denotes the sampling interval, containing the control signal, or $Q = \{\phi, \phi + 1, \ldots, \phi + N_C - 1\}$, where ϕ denotes timing offset of control signal. Note that frequency offset is ignored because it contributes phase shift and only the magnitude of received signal provides information in the likelihood function. In addition, the statistics of the received signal under hypotheses are shown as follows:

$$E[r(n)r^*(n+m)|H_1] = \begin{cases} \sigma_s^2 + \sigma_w^2 & m = 0, n \in Q \\ \sigma_w^2 & m = 0, n \notin Q \\ 0 & \text{otherwise} \end{cases}$$

and

$$E[r(n)r^*(n+m)|H_0] = \begin{cases} \sigma_w^2 & m = 0 \\ 0 & \text{otherwise} \end{cases}$$

The GLRT becomes

$$M(\mathbf{r}) = \max_{\phi} \left\{ \sum_{n=\phi}^{\phi+N_C-1} |r(n)|^2 \right\} \underset{H_0}{\overset{H_1}{\underset{<}{\gtrless}}} \tau_C$$

This detector first finds the timing offset and measures the energy of the received signal in the interval of control signal. Next, we release the assumption of knowing noise power, which occurs in detecting superposition of the control signal of the primary system and non-collaborative interference or background noise with unknown power. In addition to noise power, signal power is another unknown parameter under H_1. Then, we have

$$M(\mathbf{r}) = \max_{\phi} \left\{ M_C \ln\left(\sum_n |r(n)|^2\right) - N_C \ln\left(\sum_{n \in Q} |r(n)|^2\right) - (M_C - N_C) \ln\left(\sum_{n \notin Q} |r(n)|^2\right) \right\} \underset{H_0}{\overset{H_1}{\underset{<}{\gtrless}}} \tau_C$$

Finally, consider a situation where a non-target system exists and broadcasts control signals as well. If cognitive radios only measure received energy, they fail to discriminate between this system and the primary system. However, it is reasonable to assume that periods of control signal from different systems are distinct because each standard is designed for specific purposes (e.g., coverage and data rate), which result in individual timing parameters. Therefore, cognitive radios can track the period of the control signal from the primary system to detect the control signal. We rewrite the control signal as follows:

$$r(n) = s_c(n) \sum_{m=-\infty}^{\infty} g(n - mM_C) + w(n) \quad n = 0, 1, \ldots, LM_C - 1$$

where

$$g(n) = \begin{cases} 1 & n = 0, 1, \ldots, N_C - 1 \\ 0 & \text{otherwise} \end{cases}$$

and L denotes number of periods we observe. To track the period of the control signal, we adopt an approach similar to fundamental symbol rate tracking, that is, squaring the received signal and then extracting the magnitude at fundamental frequency of control signal $1/M_C$ by Fourier transform. It is easy to show that

$$E[|r(n)|^2] = \sigma_s^2 \sum_{m=-\infty}^{\infty} g(n - mM_C) + \sigma_w^2 \quad n = 0, 1, \ldots, LM_C - 1$$

We observe that this equation is a periodic function with period M_C, or specifically the magnitude at frequency $1/M_C$ is

$$L \frac{\sin(\pi N_C/M_C)}{\sin(\pi/M_C)} \sigma_s^2$$

To provide high throughput, the control signal is not transmitted frequently and thus M_C is large, which leads to long sensing duration. However, since the primary BSs are reasonably assumed fixed, this process is only taken initially. Furthermore, this detector can also be used to detect OFDM-based signal with zero-padding in the prefix, such as multi-band OFDM. However, in detecting the control signal by received energy, we assumed the observation interval contains one control signal period, which is true if $M_C \gg N_C$. For OFDM-based signals, $M_C = N_{FFT} + N_{CP}$, $N_C = N_{FFT}$ and the above condition is not satisfied. To solve this problem, we could collect $2N_{FFT} + N_{CP}$ samples as we did for the CP detector, and the optimal detector has similar structure as [21]. Alternatively, we could also detect such OFDM-based signals by tracking their symbol periods similar to the method in detecting fundamental symbol rate.

7.2.2.4 Available Radio Resource to Secondary System(s)

To take advantage of the rate-distance nature, we reconsider distance relation in Figure 7.10 and the primary communication link uses some channel, which is acquired by analysing the frame header. Let $SINR_{\min}$ be the minimum SINR at which the primary receiver (Rx^1) maintains current link quality. Note that this value shall depend on modulation scheme and FEC type and rate. Thus, we obtain an inequality

$$SINR_1 = \frac{G_{11}P_1}{G_{21}P_2 + N_1} \geq SINR_{\min}$$

where P_i denotes transmission power at Tx^i, N_i denotes noise power at Rx^i and G_{ij} denotes power loss from Tx^i to Rx^i. Then we have

$$P_{2,\max} = \frac{1}{G_{21}} \left(\frac{G_{11}P_1}{SINR_{\min}} - N_1 \right) = \frac{I_{CR}}{G_{21}} \qquad (7.20)$$

where I_{CR} denotes maximum tolerant interference of the primary system at the channel. Since I_{CR} depends on signal quality of the primary communication link, we can roughly infer it from the transmission rate and assume it as known. In addition, since 'distance' is actually a measure of received signal power, we have to specify the relation between received power and propagation distance. Without loss of generality, we adopt large-scale path loss including log-distance path loss and log-normal shadowing [28], as spectrum sensing does not treat instantaneous signal reception. The power loss between two nodes G is given by

$$G = Kd^{-\alpha}10^{\beta/10}$$

where K is a normalisation constant, d is distance between these two nodes, α is path loss exponent and β is the shadowing parameter, which is modelled as a Gaussian random variable with zero mean and variance σ_β^2. In the following, we determine that radio resource depends on whether information about distance between Tx^2 and $Rx^1 r$ is available or not, which corresponds to the uplink and downlink environment.

7.2.2.5 Radio Resource – Uplink

In uplink, Rx^1 is the primary BS and cognitive radios can trace its location by received signal strength of preamble. Then, we get the following inequality

$$KP_2 r^{-\alpha}10^{\beta/10} \leq I_{CR}.$$

Due to shadowing factor β, interference level is a random variable and hence can be only guaranteed in a probability sense. The optimal criterion is maximum transmission power such that probability of interfering with the primary system is less than ξ, i.e.,

$$P_{2,\max} = \arg \max_{P_2} \{ \Pr(KP_2 r^{-\alpha}10^{\beta/10} > I_{CR}) \leq \xi \}$$

This criterion has the same concept as the *Neyman-Pearson* criterion and we have

$$P_{2,\max}^{UL} = \frac{I_{CR}r^{\alpha}}{K} 10^{-\sigma_{\beta}Q^{-1}(\xi)/10}$$

Note that $P_{2,\max}^{UL}$ is proportional to I_{CR}.

7.2.2.6 Radio Resource – Downlink

In downlink, Rx[1] becomes the primary mobile station (MS) and it is hard for cognitive radios to estimate the location of the MS. Since it is reasonable to assume uniformly distributed MSs around coverage region, by Bayesian approach, we have

$$\Pr\{P_2 G_{21} > I_{CR}\} = \int_0^{\infty} Q\left(\frac{1}{\sigma_{\beta}} 10 \log_{10}\left(\frac{I_{CR}r^{\alpha}}{P_2 K}\right)\right) dF(r)$$

where

$$F(r) = \begin{cases} \dfrac{A_H(r,c,D)}{\pi D^2} & \text{high-rate region} \\[3mm] \dfrac{A_L(r,c,D,R)}{\pi(R^2 - D^2)} & \text{low-rate region} \end{cases} \tag{7.21}$$

and

$$A_H(r,c,D) = \begin{cases} \pi r^2 U(D-c) & 0 \le r < |D-c| \\ \theta_2 r^2 + \theta_1 D^2 - cv & |D-c| \le r < D+c \\ \pi D^2 & D+c \le r \end{cases}$$

$$A_L(r,c,D,R) = A_H(r,c,R) - A_H(r,c,D)$$

and $U(x)$ is the unit step function,

$$u = \frac{c^2 + D^2 - r^2}{2c},$$

$$v = \sqrt{D^2 - u^2},$$

$$\theta_1 = \cos^{-1}(u/D),$$

$$\theta_2 = \cos^{-1}((c - u)/r)$$

and $\theta_1, \theta_2 \in [0,\pi]$. Then,

$$P_{2,\max}^{DL} = \arg_{P_2}\left\{\int_0^{\infty} Q\left(\frac{1}{\sigma_{\beta}} 10 \log_{10}\left(\frac{I_{CR}r^{\alpha}}{P_2 K}\right)\right) dF(r) = \xi\right\}$$

and can be obtained numerically. By the similar procedures, we can easily generalise to multi-level data transmission rates.

We now acquire a rough range of I_{CR} from *a priori* information; however it could be dynamically calculated according to the instantaneous link quality of the primary system. If cognitive radios can exchange information with the primary system, that is, *collaborative co-existence*, after achieving synchronisation with the primary BS, they can squeeze radio resource more efficiently and further achieve channel capacity.

7.2.3 Spectrum Sensing Procedure

The spectrum sensing algorithm senses the whole spectrum of interests serially, sub-band by sub-band. To acquire spectrum utilisation within a sub-band quickly, we discriminate traffic signal of the primary system from noise and interference caused by the received energy and extracting signal feature, including the fundamental symbol rate and CP, and then ensure the existence of the active primary (OFDMA) system. We then consider the existence of the inactive primary system by means of the control signal under three conditions: background noise (C_0), interfering traffic signal (C_3 and C_4) and interfering control signal (C_5). When an active primary system exists, we acquire the utilisation status of each channel along the time axis to further increase spectrum efficiency by decoding the frame header, and then determine the radio resource to secondary system(s). This general sensing algorithm that can be applied to IEEE 802.16 and other OFDMA systems by adjusting system parameters (T_s, N_{FFT}, N_{CP}, M_C, N_C) is summarised as follows:

1. Initially, set channel state table as S_0 and reset radio resource table.
2. For $n = 1$ to N_F (sub-band), do the following sub-steps. Then, go to Step 3.
   2.1 Measure RSSI and distinguish the hypothesis test
       H_0: traffic signals do not exist
       H_1: traffic signals exist
       If H_0 is true, go to Step 2.6, else go to Step 2.2.
   2.2 Track fundamental symbol rate of the primary system
       H_0: traffic signals with fundamental symbol rate of the primary system do not exist
       H_1: traffic signals with fundamental symbol rate of the primary system exist
       If H_0 is true, go to Step 2.7, else go to Step 2.3.
   2.3 By CP of OFDMA signal, separate non-collaborative interference and traffic signal from the primary system
       H_0: traffic signals with CP property of the primary system do not exist
       H_1: traffic signals with CP property of the primary system exist
       If H_0 is true, go to Step 2.7, else go to Step 2.4.
   2.4 Synchronise with the primary base station, including carrier and timing synchronisation, and channel estimation by preamble. Then, go to Step 2.5.
   2.5 Decode frame header (DL_MAP and UL_MAP in this case) and obtain transmission parameters of the primary system, including FEC rate, modulation scheme and resource allocation. These parameters are used to set nth row of channel state table from S_2 to S_4. Then, go to Step 2.9.
   2.6 Measure energy of the control signal and detect the hypothesis test
       H_0: control signals do not exist
       H_1: control signals exist
       If H_0 is true, go to Step 2.9, else go to Step 2.8.
   2.7 Extract control signal of the primary system from non-collaborative interference
       H_0: control signals of the primary system do not exist
       H_1: control signals of the primary system exist
       If H_1 is true, set the nth row of channel state table as S_1. Then, go to Step 2.9.
   2.8 Track period of the control signal and discriminate between control signals from the primary system and other systems
       H_0: control signals of the primary system do not exist
       H_1: control signals of the primary system exist
       If H_1 is true, set the nth row of channel state table as S_1. Then, go to Step 2.9.

2.9 (optional) Identify system by fundamental symbol rate, cyclic properties of OFDMA systems and control signal. This is critical information in cognitive and cooperative networking. Then, go to Step 2.

3. According to channel state table, set radio resource table. Then, end of sensing.

In steps 2.4 and 2.5, if synchronisation or packet decoding cannot be achieved due to lack of frame information, this leads to a degenerative case, in which S_2, S_3 and S_4 merge to one state and we only need two $N_F \times 1$ tables. To illustrate the spectrum sensing procedure, we establish a sensing tree as shown in Figure 7.12. Here, we assume that control signal and traffic signal of a system cannot exist at the same time. Then there are 15 possible combinations to be received signals at cognitive radios as follows:

$$\Pi_1 = C_0;$$

$$\Pi_2 = C_1 \cap C_0; \ \Pi_3 = C_2 \cap C_0; \ \Pi_4 = C_3 \cap C_0; \ \Pi_5 = C_4 \cap C_0; \ \Pi_6 = C_5 \cap C_0;$$

$$\Pi_7 = C_1 \cap C_3 \cap C_0; \ \Pi_8 = C_1 \cap C_4 \cap C_0; \ \Pi_9 = C_1 \cap C_5 \cap C_0; \ \Pi_{10} = C_2 \cap C_3 \cap C_0;$$

$$\Pi_{11} = C_2 \cap C_4 \cap C_0; \ \Pi_{12} = C_2 \cap C_5 \cap C_0; \ \Pi_{13} = C_3 \cap C_4 \cap C_0;$$

$$\Pi_{14} = C_1 \cap C_3 \cap C_4 \cap C_0; \ \Pi_{15} = C_2 \cap C_3 \cap C_4 \cap C_0$$

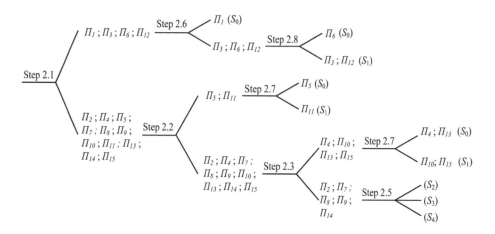

Figure 7.12   Sensing tree

As shown in the figure, there may be several possible signals belonging to the same state. Since transmission of other potential systems should be honoured, a similar sensing procedure can be applied to other potential systems as well. Then, combining these channel state tables and radio resource tables, we obtain complete information of spectrum usage and corresponding maximum transmission power. Finally, to fulfil this sensing/cognition cycle, a spectrum sensing block diagram is as shown in Figure 7.13. The functionalities are labelled in corresponding blocks.

To achieve effective spectrum sensing for CRNs (beyond traditional cognitive radio links) by taking advantage of rate-distance nature, we propose to acquire a set of cognitive information for OFDMA systems. To avoid interfering with the overlay primary system, CRs could determine the operation states of the primary system, including frequency band, existence, activity, resource allocation and data transmission rate. Spectrum sensing has been therefore generalised to multi-state discrimination and illustrated in the sensing tree in Figure 7.12. Elaborating existing energy detection and cyclostationary feature detection, we designed the optimal detectors to recognise the operation state of the primary system in the generalised system model. Simulation results show that this design

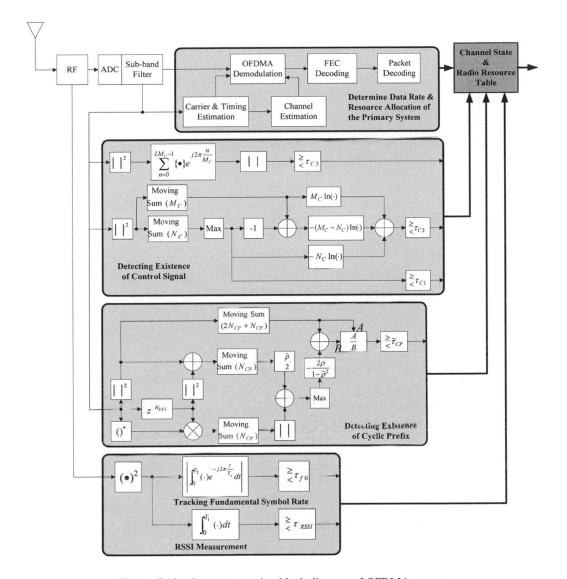

Figure 7.13   Spectrum sensing block diagram of OFDMA systems

can satisfactorily achieve system requirements ($P_F < 0.05$, $P_D > 0.95$, SINR $= 0$ dB) in the AWGN channel and frequency selective fading channel by increasing sensing duration.

After identifying the operation state of the primary system, we generalise conventional cognitive radio and sense the radio resource in busy duration of the primary system by considering the rate-distance nature to improve the opportunity using spectrum. Figure 7.14 shows that the radio resource to secondary system(s) critically depends on the information (e.g., link quality of the primary system and distance information) available to CRs and is proportional to the tolerable interference level of the primary system. This greatly generalises the scope of spectrum sensing for networking purposes.

Furthermore, these methodologies can be applied to identification of multi-radio systems, which is critical information in cognitive and cooperative networking. By adjusting the sensing parameters

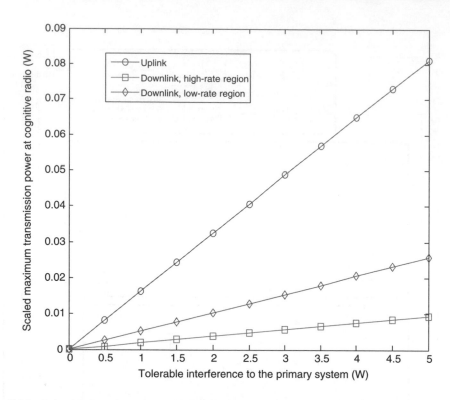

Figure 7.14   Tolerable interference to the primary system versus available radio resource to secondary system

of CR, our proposed sensing procedure can be extended to state-of-the-art OFDMA systems and even to most mobile communication systems, along with MIMO processing technologies.

7.3 Spectrum Sensing for Cognitive Multi-Radio Networks

Spectrum sensing techniques have traditionally been mainly developed for a single target system to either determine whether this system is active or discover the location of the unused spectrums (spectrum holes for link-level transmissions). On the other hand, conventional power scaling power control is also designed for a single target system to determine the maximum transmission power of CRs. Since in CRNs the cooperative routing and network efficiency optimisation are done in multiple-radio environments, the sensing of CRs should also be established with multiple co-existing radio systems. In such multiple-radios scenarios, these conventional spectrum sensing and power control techniques are not enough due to the inter-system interference. The energy approach can only distinguish whether any of the communication systems are active, not identify which they are. Since the matched filters are needed for every kind of target systems, the complexity becomes a significant challenge. The identification of a cyclostationary detector for multiple systems encounters the problem that we need to determine the active systems efficiently out of a predetermined library. The wavelet detector is used to identify the locations of unused spectrum and thus it is hard to cope with the inter-system interference. Therefore, a more reliable and general multiple systems sensing algorithm is needed to overcome this challenge. We provide a general multiple-system sensing algorithm in Section 4.4.3.

In the conventional power scaling power control methodology, CRs are passively listening to the signal transmitted from the target system. The transmission power control of CRs for multiple pre-existing systems becomes complex because we need to guarantee that all pre-existing systems are under acceptable interference, especially those suffering from the hidden terminal problem. We will propose an aggressive sensing algorithm to improve the system capacity and extend it to multiple systems in Section 7.3.2, which serves as a critical step toward sensing of general CRNs consisting of multiple co-existing radio systems [11].

7.3.1 Multiple System Sensing

To develop spectrum sensing that is feasible to multi-radio CRN, we describe a general multiple-system sensing algorithm exploiting the system-specific characteristics to identify multiple active systems. As energy detection alone is not reliable, we start by identifying the fundamental frequencies of candidate communication systems periodically filtered by pulse shaping filters. To accomplish the multiple-system sensing, we have to further exploit the unique power spectrum density pattern. If the additive noise is coloured Gaussian with an unknown covariance matrix, we may further make use of high-order statistics (fourth-order cumulant) to ensure the success of our algorithm.

7.3.1.1 Problem Formulation

To achieve a cooperative routing purpose and optimise the overall network efficiency, the CRs in CRNs have to sense the surrounding communication systems/networks. We call it system/network sensing. However, because of inter-system interference, conventional spectrum sensing techniques are not sufficient for a multiple systems environment as described in Section 2.3. Therefore, we now provide a more reliable and general multiple-system sensing algorithm.

Assume that there are Q candidate communication systems, and the transmitted signal of each system is going through a flat uncorrelated Rayleigh fading channel; that is, each has an independent complex amplitude $a_i = |a_i| \cdot e^{j\theta_i}$ where amplitude $|a_i|$ is Rayleigh distributed with $E\{|a_i|^2\} = y_i^2$ and phase θ_i is uniformly distributed over $[0, 2\pi]$. In addition, a white Gaussian noise $w(t)$ with zero mean and variance σ_w^2 is added to the received radio signal. Suppose that the activities of systems are unchanged during the period of each sensing. With the assumption of P active systems ($P \leq Q$), the received radio signal can be expressed as

$$r(t) = Re\left\{ \sum_{i=1}^{P} a_i s_i(t) + w(t) \right\} = Re\{y(t) + w(t)\} \tag{7.22}$$

where $s_i(t)$ is the signal of ith active system and

$$y(t) = \sum_{i=1}^{P} a_i s_i(t) \tag{7.23}$$

The object of the multiple-system sensing is to determine the number of active systems P, and identify them.

7.3.1.2 General Multiple-System Sensing Algorithm

The block diagram of the general multiple-sensing algorithm is shown in Figure 7.15. We summarise our sensing algorithm as follows:

1. Energy detection and carrier locking to initiate the algorithm.

2. Square the received radio signal and filter it by a narrowband bandpass filter containing all potential fundamental frequency. Detect the fundamental frequencies and identify the corresponding systems.
3. Estimate power spectrum density of target spectrum.
4. If the result of step 1 is none, end; otherwise, go to step 4.
5. If the covariance matrix of noise is known, go to step 5; otherwise, go to step 6.
6. Perform singular value decomposition (SVD) of the spectrum estimation result and identify systems. End.
7. Estimate the trispectrum matrix of the target spectrum. Perform EVD of the trispectrum matrix and identify systems by MUSIC algorithm. End.

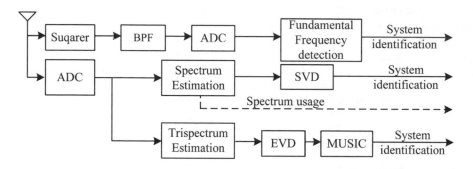

Figure 7.15 Block diagram of general multiple-system sensing algorithm

7.3.1.3 Fundamental Frequency

Because of pulse shaping in digital communication systems, there are energy peaks in the reciprocal of the symbol period (baud rate) and in its harmonics. We define the lowest frequency (baud rate) with energy peak as the fundamental frequency. Suppose that these P active systems are digital communication systems with a fundamental frequency characteristic. The transmitted signal of ith system can be written in the form

$$s_i(t) = \sum_{n=-\infty}^{\infty} x_{i,n} h_i(t - nT_i - \tau_i) e^{j(2\pi f_{c,i} t + \alpha_i)} \tag{7.24}$$

for $I = 1, 2, \ldots P$ where $x_{i,n}$ is the data sequence, $h_i(t)$ is the impulse response of the pulse-shaping filter with frequency response $H_i(j\omega)$, T_i is the symbol duration, $\tau_i \in [0, T_i)$ and $\alpha_i \in [0, 2\pi)$ are the time offset and the phase offset, respectively (both are regarded as constants during sensing) and $f_{c,i}$ is the carrier frequency. To be specific, we assume that $\{x_{i,n}\}$ are zero mean variance σ_i^2 stationary sequences with statistically independent and identically distributed elements.

There are various methods to extract the fundamental frequency information and here we adopt the nonlinear spectral line method with squared magnitude followed by a narrowband bandpass filter. Assume $y(t)$ with equal variance in real and imaginary parts and independent of $w(t)$. It is easy to show that $E\{r^2(t)\} = 1/2 \cdot E\{|y(t)|^2 + |w(t)|^2\}$. Thus, we can obtain

$$E\left\{r^2(t)\right\} = \frac{1}{2} \sum_{i=1}^{P} \frac{\sigma_i^2 \gamma_i^2}{T_i} Z_{i,1} \cos\left[\frac{2\pi(t - \tau_i)}{T_i}\right] + \frac{1}{2} \sum_{i=1}^{P} \frac{\sigma_i^2 \gamma_i^2 Z_{i,0}}{T_i} + \frac{1}{2} \sigma_w^2 \tag{7.25}$$

where

$$Z_{i,m} = \frac{1}{2\pi} \int_{-\infty}^{\infty} H_i(j\theta) \cdot H_i^* \left(-j \left(\frac{2\pi m}{T_i} - \theta \right) \right) d\theta \tag{7.26}$$

Therefore the squared signal $r^2(t)$ can be decomposed as

$$r^2(t) = E\{r^2(t)\} + \varepsilon(t)$$
$$= \frac{1}{2} \sum_{i=1}^{P} \frac{\sigma_i^2 \gamma_i^2}{T_i} Z_{i,1} \cos \left[\frac{2\pi(t - \tau_l)}{T_i} \right] + \frac{1}{2} \sum_{i=1}^{P} \frac{\sigma_i^2 \gamma_i^2 Z_{i,0}}{T_i} + \frac{1}{2} \sigma_w^2 + \varepsilon(t) \tag{7.27}$$

where $\varepsilon(t)$ is the disturbance term with zero mean.

We can observe spectral lines at frequency $\{1/T_i\}$ in Equation (7.27). After filtering the signal with a narrowband bandpass filter containing all potential fundamental frequencies, we can detect these tones to identify the corresponding systems. Figure 7.16 shows spectrum magnitude after squarer with 802.11b, 802.11g and Bluetooth co-existing under SNR $= 10$ dB (assuming these three systems have equal signal power). Three tones at 1, 11 and 20 MHz are observed, which correspond to the fundamental frequencies of these systems. Note that because the spectrum magnitude is mainly determined by $Z_{i,1}$, under the same signal power condition, the system with a higher fundamental frequency has a lower spectrum magnitude after squarer (assuming the same roll-off factor). On the other hand, we can observe that the disturbance is less in higher spectrum.

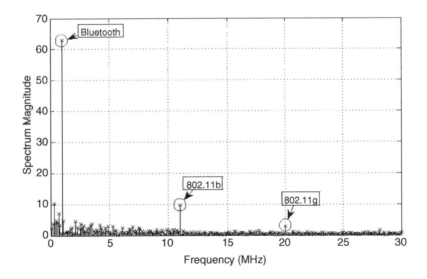

Figure 7.16   Spectrum magnitude after squarer with SNR $= 10$ dB and roll-off factor $= 0.5$

7.3.1.4 Power Spectrum Density Pattern

Assume the power spectrum density of each candidate system in the target spectrum is known and portioned into M equal-bandwidth sub-bands. When only the power spectrum density shape is concerned, we can model the transmitted signal in discrete-time as

$$s_i[n] = u_i[n] * h_i[n] \qquad i = 1, 2, \dots, P \tag{7.28}$$

where $u_i[n]$ is from white noise with variance σ_i^2 and $h_i[n]$ has a frequency response $H_i(e^{j\omega})$ so that $\left\{|H_i(e^{j2\pi k/M})|^2 \equiv P_i(e^{j2\pi k/M})\right\}_{k=0}^{M-1}$ are real and known. Thus, the received signal in discrete time can be written as

$$r[n] = \sum_{i=1}^{P} a_i s_i[n] + w[n] = \sum_{i=1}^{P} a_i u_i[n] * h_i[n] + w[n] \tag{7.29}$$

Let

$$\mathbf{p}_i \equiv [P_i(1)\ P_i(e^{j2\pi/M})\ \cdots\ P_i(e^{j2\pi(M-1)/M})]^T \tag{7.30}$$

be the power spectrum density pattern of the ith candidate system and assume that these power spectrum density patterns $\{\mathbf{p}_i\}_{i=1}^{Q}$ are linear independent. Many well-known spectrum estimation methods such as periodogram, Blackman-Tukey method, Barlett-Welch method, multitaper, etc, are suitable for different spectrum estimation requirements. Assume a large observation length is available and thus the power spectrum estimation error can be neglected. The power spectrum estimation result can expressed in vector form

$$\hat{p} = \sum_{i=1}^{P} \gamma_i^2 \sigma_i^2 \mathbf{p}_i + \mathbf{w} = \mathbf{S} \cdot \mathbf{h} + \mathbf{w} \tag{7.31}$$

where

$$\hat{P} \equiv [\hat{P}(1)\hat{P}(e^{j2\pi/M})\ \cdots\ \hat{P}\left(e^{j2\pi(M-1)/M}\right)]^T \tag{7.32}$$

and $\hat{P}(e^{j2\pi(M-1)/M})$ is the estimation at frequency $\omega_k = 2\pi k/M$,

$$\mathbf{S} \equiv [\mathbf{p}_1\ \mathbf{p}_2 \cdots \mathbf{p}_Q]_{M \times Q} \tag{7.33}$$

is the power spectrum pattern matrix,

$$\mathbf{w} \equiv [W(1)\ W(2)\ \cdots\ W(M-1)]^T \tag{7.34}$$

is the noise contribution to power spectrum (assume that it can be estimated accurately), and

$$\mathbf{h} \equiv [\gamma_1^2 \sigma_1^2\ \gamma_2^2 \sigma_2^2 \cdots \gamma_P^2 \sigma_P^2]^T \tag{7.35}$$

is the received power vector.

Since $\{\mathbf{p}_i\}_{i=1}^{Q}$ are linear independent, $\mathbf{S}$ is a $M \times Q$ matrix of rank Q. If we perform SVD, $\mathbf{S}$ becomes $\mathbf{S} = \mathbf{U}\Lambda\mathbf{V}^T$ where $\mathbf{U}$ and $\mathbf{V}$ are an $M \times M$ orthogonal matrix and an $Q \times Q$ orthogonal matrix, respectively, and Λ is a $M \times Q$ matrix with (i, j)-entry

$$\begin{cases} s_{ii} = s_i & \text{for } i = 1, 2, \ldots, Q \\ s_{ij} = 0 & \text{otherwise} \end{cases} \tag{7.36}$$

where $\{s_i\}_{i=1}^{Q}$ are the singular values of $\mathbf{S}$. Therefore we can solve $\mathbf{h}$ as follows

$$\mathbf{V}\Lambda^+\mathbf{U}^T(\hat{p} - \mathbf{w}) = \mathbf{V}\Lambda^+\mathbf{U}^T\mathbf{U}\Lambda\mathbf{V}^T\mathbf{h} = \mathbf{h} \tag{7.37}$$

where Λ^+ is a $Q \times M$ matrix with (i, j)-entry

$$\begin{cases} q_{ii} = 1/s_i & \text{for } i = 1, 2, \ldots, Q \\ q_{ij} = 0 & \text{otherwise} \end{cases} \tag{7.38}$$

and thus $\Lambda^+\Lambda = \mathbf{I}_Q$.

Ideally **h** contains nonzero elements only when the corresponding systems are active. In practice, **h** is always nonzero because of estimation errors. A heuristically serial search can be applied here. Arrange the elements of **h** in decreasing order, $k_1 \geq k_2 \geq \cdots \geq k_Q$. Compute the ratio $\sum_{i=1}^{\hat{P}} k_i / \sum_{i=1}^{Q} k_i$ from $\hat{P} = 1$ and stop when the ratio exceeds a predetermined threshold. When the search stops, the number of active systems is determined as $\hat{P}$ and the corresponding elements and systems are identified.

Figure 7.17 shows the estimation of power spectrum density by Welch's method. Here we assume that the power spectrum density pattern of the microwave oven is a rectangular shape from 2415 to 2465 MHz and the 802.11b is active with central frequency 2437 MHz. We can see that when the spectrums of these systems overlap, the estimation results are summed by the power spectrum density patterns of these systems and thus we can perform the SVD to detect the activity of systems.

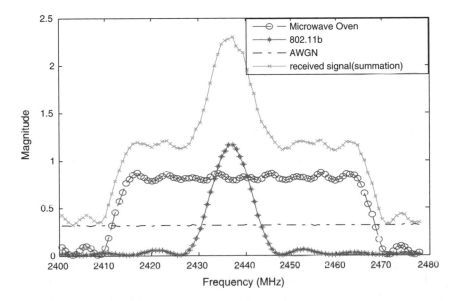

Figure 7.17 Estimation of power spectrum density with microwave oven and 802.11b (carrier frequency = 2437 MHz)

7.3.1.5 Fourth-Order Cumulant

The earlier discussion is based on the additive Gaussian noise with known covariance matrix. However, if the additive noise is coloured Gaussian with unknown covariance matrix, it is not possible to express the equation as a linear combination of known spectrum vectors. This implies that second-order statistics of the measurements (the power spectrum density pattern methodology) are not sufficient to detect and identify the systems. In this case, *high-order statistics*, in which cumulants are blind to any kind of a Gaussian process, becomes a useful characteristic to ensure the success of our multi-system sensing. Suppose that the transmitted signal $S_i[n]$ is modelled as the form of Equation (7.28) where $u_i[n]$ becomes a stationary, white, non-Gaussian random process with fourth-order cumulant ρ_i and $h_i[n]$ is with the same assumption that $\mathbf{p}_i$ is known and $\{\mathbf{p}_i\}_{i=1}^{Q}$ are linear independent. Define the elements in the $M \times M$ trispectrum matrix **C** as

$$c_{ij} = T(\omega_i, -\omega_i, \omega_j) \quad 1 \leq i,j \leq M \quad \omega_i = \frac{2\pi(i-1)}{M} \tag{7.39}$$

where

$$T(\omega_1, \omega_2, \omega_3) = \sum_{i=1}^{P} \gamma_1^4 \rho_i H_i(\omega_1) H_i(\omega_2) H_i(\omega_3) \tag{7.40}$$
$$\times H_i(-\omega_1 - \omega_2 - \omega_3)$$

is the received trispectrum. Therefore we can express $\mathbf{C}$ as

$$\mathbf{C} = \sum_{i=1}^{P} \gamma_1^4 \rho_i \mathbf{p}_i \mathbf{p}_i^T = \mathbf{S}\Gamma\mathbf{S}^T \tag{7.41}$$

where $\mathbf{S}$ is the source spectrum matrix and Γ is a $Q \times Q$ diagonal matrix with only P nonzero diagonal elements. Ideally $\mathbf{C}$ is a real symmetric matrix with rank P and the number of active systems can be determined by computing the rank of $\mathbf{C}$. Unfortunately, in practice the estimation of $\mathbf{C}$ is always full rank because of estimation error. By performing an eigendecomposition on the estimation of the trispectrum matrix, $\hat{\mathbf{C}}$ can be written as

$$\hat{\mathbf{C}} = \sum_{m=1}^{M} \lambda_m \mathbf{g}_m \mathbf{g}_m^H = G\Sigma G^H \tag{7.42}$$

with the eigenvalues arranged in decreasing order, $\lambda_1 \geq \lambda_2 \geq \cdots \geq \lambda_M$.

The signal subspace is spanned by the vectors $\{\mathbf{g}_i\}_{i=1}^{P}$ and the noise subspace is spanned by the vectors $\{\mathbf{g}_i\}_{i=P}^{M+1}$. Therefore the number of active systems and system identification can be carried out by serial search and the MUSIC algorithm, respectively. We can compute the MUSIC pseudospectrum

$$\overline{R_{music}}(i) = \frac{\mathbf{p}_i^H \mathbf{p}_i}{\sum_{m=Q+1}^{M} |\mathbf{p}_i^H \mathbf{g}_m|^2} \tag{7.43}$$

and the systems with the corresponding largest P values are selected and identified as active ones.

Ideally the trispectrum matrix in Equation (7.39) is not affected by the power of additive Gaussian noise σ_w^2, which means that the fourth-order cumulant methodology is expected to perform well in a low SNR environment. On the other hand, to estimate the trispectrum matrix $\mathbf{C}$ accurately, a large convergence time is required, which means that we need a longer sensing time.

7.3.1.6   2.4G Hz ISM Band Illustration

The most common environment with multiple co-existing systems may well be the 2.4 GHz ISM (industrial, scientific, medical) band, which is likely to include WLAN (802.11b and 11g), Bluetooth and the microwave oven as potential active systems. The system parameters are listed in Table 7.2.

A flow chart of multiple systems sensing for this case is shown in Figure 7.18. When the energy detection indicates there are active systems in the spectrum, we try to lock the carrier frequencies 2412, 2437 and 2462 MHz. If some carrier frequencies are locked, we say that 802.11b or/and 802.11g systems exist(s) in the corresponding channels. If none are locked, we conclude there is no active 802.11b or 802.11g system. Next we apply the fundamental frequency methodology. Since squaring a broadband RF spectrum in analogue fashion needs very efficient nonlinear devices, which might be not realistic, we can divide the frequency band into several parts to proceed with the sensing. The fundamental frequency methodology can help us to determine the existence of 802.11b, 802.11g and Bluetooth systems since they have with different fundamental frequencies.

After spectrum estimation, we can obtain the spectrum usage status. If we know the power spectrum density of the microwave oven, the power spectrum density pattern methodology can be used. On the other hand, if the power spectrum density of the microwave oven is unknown, assuming that the

Table 7.2   System parameters

System	Carrier frequency	Fundamental frequency
802.11b	2412 MHz, 2437 MHz, 2462 MHz	11 MHz
802.11g	2412 MHz, 2437 MHz, 2462 MHz	20 MHz
Bluetooth	Not Fixed	1 MHz
Microwave oven	None	None

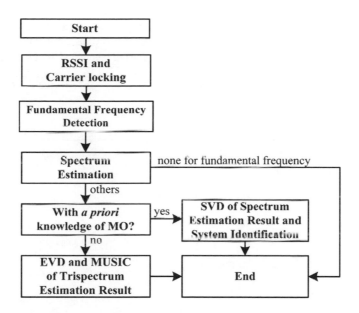

Figure 7.18   Flow chart of general multiple-system sensing algorithm

microwave oven interference is an additive Gaussian noise, the fourth-order cumulant methodology can be applied to identify other systems with non-Gaussian signals. The RSSI and fundamental frequency detection are usually faster than the other methods. Thus no matter whether *a priori* knowledge of the microwave oven is known or unknown in advance, the flow chart still holds because the sensing time is regarded as one critical parameter.

Figure 7.19 shows the receiver operating characteristic (ROC) of fundamental frequency detection in $f1 = 10$ MHz and $f2 = 20$ MHz co-existing environments. The data length is 10,000 with the signal sampled at rate 100 MHz after squarer and band-pass filter. In Figures 7.20 and 7.21, we plot probability of detection versus SNR of the power spectrum density pattern methodology. We consider the three main channels to be as shown in Table 7.2, where 802.11b channel 1 (has carrier frequency = 2412 MHz), channel 2 (has carrier frequency = 2437 MHz) and channel 3 (has carrier frequency = 2462 MHz). The microwave oven is also operating.

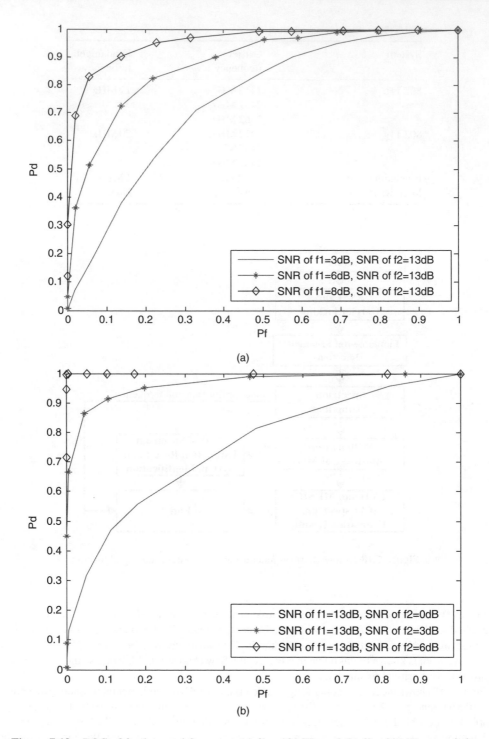

Figure 7.19 ROC of fundamental frequency (a) f1 = 10 MHz and (b) f2 = 20 MHz co-existing

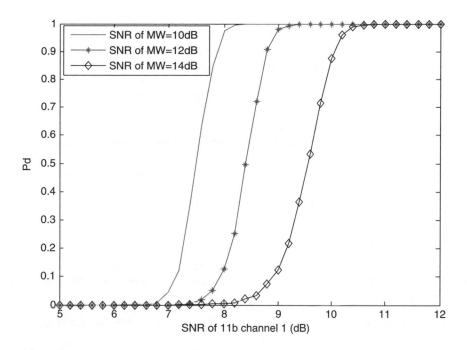

Figure 7.20 Probability of detection versus SNR of power spectrum density pattern with 11b channel 2 = 10 dB and channel 3 = 10 dB

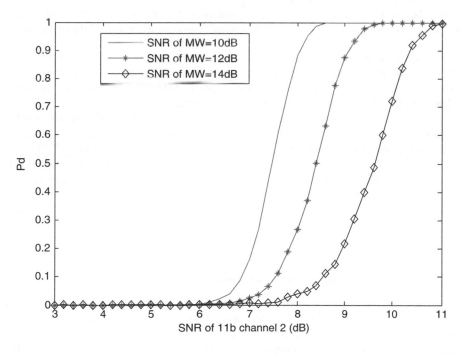

Figure 7.21 Probability of detection versus SNR of power spectrum density pattern with 11b channel 1 at 10 dB and channel 3 at 10 dB

7.3.2 Radio Resource Sensing

Since the purpose of conventional spectrum sensing is to indentify CR transmission opportunities at link level, it is not enough for CRN to operate by obtaining information from passive operating techniques. To explore multi-radio system availability further, we borrow a concept from CSMA networks and propose the *aggressive sensing* algorithm with cooperative adaptive modulation coding (AMC). This algorithm can determine whether the operation of the secondary system is feasible and detect the available radio resource under the tolerable interference constraint. CR aggressively transmits a *pseudo carrier* subject to an acceptable interference level to the PS. Through dynamical change of the pseudo carrier power and the cooperative adjustment of modulation and coding parameters (MCP) of the PS, CR can determine the optimal utilisation of system power (and thus the possible rate). Consequently, both the outage performance in time duration and outage probability of the PS can be maintained, and the available radio resource for the secondary system can be utilised with greater efficiency.

7.3.2.1 Passive Sensing Problem

The power scaling power control as illustrated earlier is an intuitive approach. However, since it senses the signal from PS-Tx passively and then infers the transmission power control, there may be several problems as follows:

- The inference result of maximum transmission power may be inaccurate because of the reliance on the assumption of the channel parameter α.
- Since the margin for shadowing is reserved in an ensemble average sense (when PS-Rx and CR-Tx do not move), the quality of service of the PS in time duration is not guaranteed. Thus a vulnerable and undetectable period of PS-Rx may be induced.
- The secondary system should reserve a huge transmission power margin to protect the worst case of PS-Rx. It may waste the radio resource and thus reduce the system capacity of the secondary system.

Although cooperative sensing proves helpful in protecting the hidden terminal, the nature of passive sensing is not changed so that the sensing behaviour focuses on the transmission of PS-Tx and not actually on the interference in PS-Rx. Thus the worst case is still considered to conservatively prevent PS-Rx from the harmful interference, which results in a significant reduction of the radio resource. Relying on passive listening to the channel inherently has these problems and is not sufficient for radio resource sensing, owing to the fact that passive sensing is totally carried out at the CR side. Here we introduce a novel collaborative mechanism of sensing for CRs: aggressive sensing. Collaboration is accomplished in both CRs (we only consider one cognitive radio pair) and the pre-existing systems to sense the available radio resource of the secondary system suffering from the hidden terminal problem. Through the cooperation the CRs can actually 'test' the channel and thus know the impact to the pre-existing systems. We expect aggressive sensing to alleviate the problems of passive sensing. Moreover, the aggressive sensing algorithm should be extendable to the multiple-system environment for CRNs; that is, the transmission power should not interfere with any pre-existing systems.

7.3.2.2 Problem Formulation of Active Probing

Consider a simple interference channel model with two pairs, a pair of PS-Tx and PS-Rx with the feedback channel and a pair of CR-Tx and cognitive radio receiver (CR-Rx), as shown in Figure 7.22. The feedback channel is used for the PS to communicate the channel quality and adjust the transmission parameters, e.g., modulation type and coding rate, to optimise system capacity. The SINR of PS-Rx is given by

$$\gamma_{PS} \equiv \frac{P_{PS} \cdot G_{PP}}{P_{CR} \cdot G_{CP} + P_{N,PS}} \tag{7.44}$$

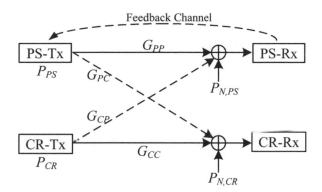

Figure 7.22 Interference channel model with the feedback channel of the primary system in CRNs

and the SINR of CR-Rx is given by

$$\gamma_{CR} \equiv \frac{P_{CR} \cdot G_{CC}}{P_{PS} \cdot G_{PC} + P_{N,CR}} \tag{7.45}$$

where P_{PS} and P_{CR} are the transmission power of the PS-Tx and the CR-Tx, respectively, G_{PP}, G_{PC}, G_{CP} and G_{CC} are the channel power gains, and $P_{N,PS}$ and $P_{N,CR}$ are the noise power received by the PS-Rx and the CR-Rx, respectively.

We are considering slow time-varying channels, and hence the channel power gains G_{PP}, G_{PC}, G_{CP} and G_{CC} are assumed to be unchanged during the sensing period. On the secondary system side, the transmission power of CR-Tx P_{CR} is inherently known and the transmission power of PS-Tx P_{PS} may be obtained through the cooperation between the primary system and the secondary (CR) system. Furthermore, we do not consider the power control of PS-Tx to treat P_{PS} as a fixed constant. The channel power gains G_{PC} and G_{CC} can be estimated and are thus regarded as known. Note that as a common condition in CRNs the channel power gains G_{PP} and G_{CP} are unknown to the secondary CR system.

To incorporate AMC, we suppose that the PS-Tx sends a packet of symbols with modulation coding mode M_j. For example, M_j corresponds to several possibilities: no transmission, BPSK with rate 1/2 convolutional code, QPSK with rate 1/2 convolutional code, QPSK with rate 3/4 convolutional code, 16-QAM with rate 1/2 convolutional code, 16-QAM with rate 3/4 convolutional code, 64-QAM with rate 2/3 convolutional code or 64-QAM with rate 3/4 convolutional code, donated as $M_0, M_1, \ldots$ and M_8, respectively. The adaptive modulation coding scheme switches the modulation of PS-Tx to M_0 as $\gamma_{PS} < l_1$, to M_1 as $l_1 < \gamma_{PS} < l_2$, etc. Note that when $\gamma_{PS} < l_1$, the PS-Tx does not transmit data and outage of the PS occurs. The theoretical SINR required to satisfy a requested BER under the AWGN channel for the only modulation type is listed in Table 7.3.

Table 7.3 The theoretical SINR required to satisfy a requested BER under the AWGN channel where GRAY mapping is employed.

BER_{req}	BPSK (dB)	QPSK (dB)	16QAM (dB)	64QAM (dB)
10^{-2}	4.3	7.3	13.9	19.6
10^{-3}	6.8	9.8	16.5	22.6
10^{-4}	8.4	11.4	18.2	24.3
10^{-5}	9.6	12.6	19.4	25.6

We consider two operation constraints for the secondary system:

- The SINR of the PS-Rx γ_{PS} should be above l_1 during the transmission of CR-Tx to not cause outage of the primary system.
- The SINR of the CR-Rx γ_{CR} should be above a threshold η in order for received signals to be decodable.

The constraints can be written as

$$
\begin{cases}
P_{CR} < \dfrac{G_{PP}}{G_{GP}} \cdot \dfrac{1}{l_1} \cdot P_{PS} - \dfrac{P_{N,PS}}{G_{CP}} \\
P_{CR} > \dfrac{G_{PC}}{G_{CC}} \cdot \eta \cdot P_{PS} + \dfrac{\eta}{G_{CC}} \cdot P_{N,CR}
\end{cases}
\tag{7.46}
$$

and are illustrated in Figure 7.23. There exists a feasible region for the secondary system operation when

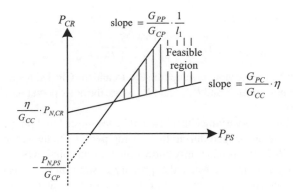

Figure 7.23   The feasible region for the secondary system operation

$$
\frac{G_{PP}}{G_{CP}} \cdot \frac{1}{l_1} > \frac{G_{PC}}{G_{CC}} \cdot \eta
\tag{7.47}
$$

Now we want to determine whether the operation of the secondary CR system is feasible and the corresponding maximum allowable transmission power of the secondary system. Note that verifying Equation (7.44) actually means to evaluate the channel impact on PS-Rx from CR-Tx. Since G_{PC} and G_{CC} are known, and the threshold values l_1 and η are predetermined and known, the feasible region existence problem becomes to find the ratio G_{PP}/G_{CP}.

7.3.2.3 Aggressive Sensing Algorithm with Cooperative AMC

Borrowing the concept from CSMA of 'testing' whether the channel is clear before transmission, CR-Tx can sense the channel through testing to evaluate the interference in PS-Rx. This kind of spectrum sensing concept is called *aggressive sensing* (or *active sensing*), which acts as a contrast to conventional passive sensing. In aggressive sensing, CR-Tx senses the channel by: (i) transmitting a pseudo carrier with dynamically controlled power; and (ii) observing the transmission behaviour variation of the PS bearing a feedback channel. Since the variation of transmission behaviour is due to the dynamically controlled interference of the pseudo carrier, we can actually infer the impact on PS-Rx from the transmission of CR-Tx.

Cooperative AMC involves three steps:

1. PS-Tx adjusts the modulation and coding parameter (MCP) according to SINR of PS-Rx to maintain the requested bit error rate.
2. PS-Tx and the secondary system cooperate so that transmission parameters of PS-Tx such as transmission power and MCP are provided to the secondary system.
3. The secondary system adjusts its transmission power P_{PS} and adapts the modulation and coding scheme to control the interference in PS-Rx and optimally utilise the radio resource. Because of the cooperation between PS-Tx and the secondary system, the MCP of the primary system and P_{PS} are available for the secondary system. This cooperation may not need to modify the primary systems and can implement on CRs by decoding the control message, e.g., the headers of frames, or the detection process from received signals.

Combining the aggressive sensing and the cooperative AMC concepts, we propose an algorithm to deal with the hidden terminal problem. The algorithm is summarised as follows with initial values $k = 1$ and $I_{CR}[1] = \Delta I_{CR}[0]$:

1. Detect MCP. If the MCP corresponds to the SINR level only higher than no transmission, end; otherwise, go to step 2.
2. Transmit the pseudo carrier with power $I_{CR}[k]$.
3. Wait for a period τ.
4. Detect MCP. If the corresponding SINR level to MCP becomes one level lower compared to step 1, go to step 6. If the descent is more than one level, end. Otherwise, increase the power of the pseudo carrier by $\Delta I_{CR}[k]$.
5. If $I_{CR}[k] \geq P_{regulation}$, end; otherwise increase k by 1 and go to step 3.
6. Denote the corresponding switching SINR level as γ_{switch}. If

$$\frac{\gamma_{switch}}{l_1} \cdot \frac{\Delta I_{CR}[k]}{P_{PS}} > \frac{G_{PC}}{G_{CC}} \cdot \eta \tag{7.48}$$

holds, the operation of the secondary system is feasible; otherwise, not. End.

A flow chart is shown in Figure 7.24. In step 1, the SINR level corresponding to the lowest transmission rate leads to the end of the algorithm to protect the outage of the PS due to pseudo carrier transmission. In step 5, When the power of the pseudo carrier exceeds the transmission power of the secondary system of the regulation $P_{regulation}$, it means the secondary system can communicate in the maximum allowable transmission power $P_{regulation}$ without causing intolerable interference in PS-Rx. In Step 6, the switching SINR γ_{switch} can be expressed as

$$\gamma_{switch} = \frac{P_{PS} \cdot G_{PP}}{I_{CR} \cdot G_{CP} + P_{M,PS}} \tag{7.49}$$

and with the assumption that interference power dominates γ_{switch} can be approximated as

$$\gamma_{switch} \approx \frac{P_{PS} \cdot G_{PP}}{I_{CR} \cdot G_{CP}} \tag{7.50}$$

and thus the power gain ratio can be calculated as

$$\frac{G_{PP}}{G_{CP}} \approx \gamma_{switch} \cdot \frac{I_{CR}}{P_{PS}} \tag{7.51}$$

By substitution Equation (7.51) into Equation (7.47), Equation (7.48) is derived. Therefore when Equation (7.48) holds, the operation of the secondary system is feasible. According to Equation (7.51),

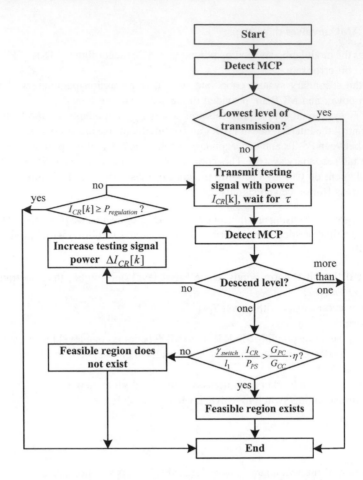

Figure 7.24 Flow chart of the aggressive sensing algorithm

the maximum allowable transmission power of CR-Tx $P_{max,CR}$ can be expressed as

$$P_{max,CR} = \frac{G_{PP}}{G_{CP}} \cdot \frac{P_{PS}}{l_1} \tag{7.52}$$

Note that in this case it is assumed that P_{PS} is known by the CRs through cooperative AMC. If P_{PS} is unknown, the aggressive sensing algorithm can still work by detecting twice the MCP (modulation and coding parameters) variations of PS-Tx.

7.3.2.4 Modulation and Coding Parameters Detection

In general, detection of the MCP can be achieved in one of two ways: (i) MCP information can be directly decoded from the pilot or the header of frames; or (ii) MCP information can be detected by sensing the received traffic signal. In the first method, the traditional signal decoding procedure of the PS can be applied. Here, however, we consider the second method to ensure the reliability and generality of the aggressive sensing algorithm. The MCP are composed of two parts: modulation type and coding rate. The detection of modulation type has been well-studied as part of the modulation classification issue, which is an intermediate step between signal detection/interception and signal demodulation. On the other hand, after demodulation, the detection of the coding rate from the traffic signal does not have a well-known solution. We can therefore introduce a detection methodology using the frame structure characteristic for the frames with synchronisation words.

Generally speaking, there are two approaches for the modulation classification algorithm, including the decision-theoretic/maximum likelihood (ML) approach, and the feature based approach. We can express the received signal as

$$r(t) = s(\boldsymbol{\mu}_i; t) + n(t) \quad 0 \le t \le KT_s \tag{7.53}$$

whereas T_s is the symbol duration, K is the μ number of observed symbols, $n(t)$ is the two-sided WGN with psd $N_0/2$, with receiver noise, interference and jamming included. Specifically,

$$s(\boldsymbol{\mu}_i; t) = e^{j\theta} e^{j2\pi\Delta f t} \sum_{k=1}^{K} s_k^{(i)} e^{j\varphi_k} g(t - (k-1)T_s - \varepsilon T_s) \tag{7.54}$$

is the noise-free baseband complex envelope of the received signal with θ the carrier phase, Δf the frequency offset, $s_k^{(i)}$ the transmitted data symbol in $[(k-1)T_s, kT_s]$, φ_k is the phase jitter, ε the timing offset and $g(t)$ the convolution of the transmitter pulse shape and the channel impulse response. Here the usage of $\boldsymbol{\mu}_i$ stands for the dependence of the received signal on the unknown symbols and signal parameters. In the following, we will take into consideration only part of these unknown quantities, while assuming that the others are known to the receiver, and hence simplify the complexity of receiver.

In the decision-theoretic approach, the modulation classification is formulated as a *multiple composite hypothesis-testing* problem. The decision is made by comparison of a likelihood ratio with predefined threshold. With different assumptions of the unknown quantities, the variety of the *likelihood ratio test* (LRT) is developed. The *average likelihood ratio test* (ALRT) treats unknown quantities as random variables and their probability density function (pdf) is assumed to be already known. As long as our presumed pdf coincides with the actual one, the optimum solution is guaranteed by ALRT. However, for most cases, ALRT usually has much complexity and thus is simplified to quasi log-likelihood ratio (qLLR). The generalised likelihood ratio test (GLRT) treats unknown quantities as deterministic variables which can be estimated assuming each hypothesis is true. GLRT provides smaller error probabilities compared with qLLR in discriminating 16-QAM and V.29. However, GLRT fails in classifying BPSK and QPSK. The hybrid likelihood ratio test (HLRT) is obtained by treating some of the unknown quantities with ALRT and others with GLRT. We can apply HLRT by treating the data sequence as random variables with equal probability, and the carrier phase remains the unknown deterministic variable to classify BPSK and QPSK.

By and large, a feature based system contains two stages: feature extraction and pattern recognition. Features are extracted and used to separate the received signal into different modulation types. Intuitively, the signal itself and its instantaneous amplitude, frequency and phase supply a good set of time domain features, while zero crossing can also be used. The most descriptive characteristic is definitely pdf, but this not easily derived, and therefore moments and cumulants, the variation of moments, are the secondary choice. If the spectrum is not stationary, the wavelet transform can be applied. In our operating environment of the secondary system, the modulation classification is usually to decide the modulation types out of a predetermined library of primary systems. Therefore, appropriate classifiers can be chosen in advance.

The data-link protocols most commonly used for space-to-ground transmission employ known binary sequences or sync words for proper frame synchronisation. This kind of characteristic may be used to determine the details of the channel encoding that was applied. For example, if we apply the Viterbi decoding with rate 1/2 convolutional code to a received signal, and if the decoded sequence has an occurrence of one of these sync words, it is highly likely that the transmitted signal was encoded in this way. For MCP detection, the modulation type and coding rate may be determined jointly by trying all the possibilities. That is, suppose the received signal is modulated by one of the possible modulation types, and encoded with one of the possible coding rates. By demodulating and decoding under such an assumption, we can search each of the sequences for sync word occurrences. If this happens, the modulation type and coding rate is determined.

7.3.2.5 Pseudo Carrier Design

The purpose of the pseudo carrier is to change dynamically the interference in the PS-Rx and thus the MCP of the primary system. When we design the mechanism of the pseudo carrier, the MCP detection period T_D and the increasing power level $\Delta I_{CR}[k]$ are two critical parameters. Define P_{target} as the final target power level of the pseudo carrier to change MCP and $K = \left\lfloor \frac{P_{target}}{\Delta I_{CR}[k]} \right\rfloor$ as the maximum period index for $\Delta I_{CR}[k]$ lower than P_{target} where $\lfloor \cdot \rfloor$ is the floor function. We can express the relation of T_D and $\Delta I_{CR}[k]$ as

$$I_{CR}(t) = I_{CR}[k] \quad \text{for} \quad (k-1)T_D \le t \le kT_D \tag{7.55}$$

and

$$I_{CR}[k+1] = I_{CR}[k] + \Delta I_{CR}[k] \tag{7.56}$$

For $k = K + 1$, the pseudo carrier power exceeds P_{target}, and the transient delay $\tau_{transient}$ from the transmission of the pseudo carrier to the MCP change can be expressed as

$$\tau_{transient} = \tau_{SINR,PS} + \tau_{feedback,PS} + \tau_{adaption,PS} + \tau_{detection,CR} = \tau + \tau_{detection,CR} \tag{7.57}$$

where $\tau_{SINR,PS}$ is the required time for SINR estimation of PS-Rx, $\tau_{feedback,PS}$ is the delay of feedback information delivery, $\tau_{adaption,PS}$ is the MCP adaption of PS-Tx according the feedback information and $\tau_{adaption,CR}$ is the required MCP detection time of the secondary system. We can see that T_D should be longer than the total delay $\tau_{transient}$ to ensure the success of the algorithm. However, T_D should not be too large because then the period remaining at an insufficient pseudo carrier power level becomes long, which results in a large waste of energy. We discuss this aspect a little later.

The increasing power level $\Delta I_{CR}[k]$ can be predetermined or adaptive: the predetermined method sets $\Delta I_{CR}[k]$ by a deterministic function which may be derived by experiments and the adaptive method sets $\Delta I_{CR}[k]$ according to the present environment. To simplify the analysis, here we consider the predetermined method with a constant function as an example. That is, $\Delta I_{CR}[k] = \Delta I_{CR}$ is a constant as shown in Figure 7.25. It is subject to the tradeoff between two factors: granular error and waste of energy. When ΔI_{CR} is too large, the granular error may become serious. Thereby the final power level $\Delta I_{CR}[K+1]$ could have a huge error or cause multi-level descent of the MCP in step 4 of the algorithm, which leads to the estimation error of G_{PP}/G_{CP} or the failure of the algorithm. On the other hand, the waste of energy can be calculated as

$$E_{waste} = \sum_{k=1}^{K} \Delta I_{CR} \cdot kT_D = \frac{K(K-1)}{2} \Delta I_{CR} \cdot T_D \propto \frac{T_D}{\Delta I_{CR}} \tag{7.58}$$

To lower E_{waste}, T_D should be small and $\Delta I_{CR}[k]$ should be large for certain P_{target}. Note that T_D has a lower bound larger than $\tau_{transient}$ and $\Delta I_{CR}[k]$ should not be too large due to the granular error. Thus there is a tradeoff when designing the pseudo carrier.

7.3.2.6 Extension to Multiple-System Environment

Let us turn to our original application scenario: there might be multiple primary systems and the secondary CR system must not cause destructive interference in all receivers of nodes. Assume that

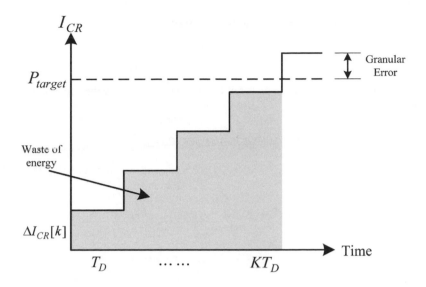

Figure 7.25   Illustration of granular error and waste of energy

there are M primary systems and each has one transmission pair. The SINR of PS-Rx in the mth primary system can be expressed as

$$\gamma_{PS,m} \equiv \frac{P_{PS,m} \cdot G_{Pm,Pm}}{P_{CR} \cdot G_{CR,Pm} + \sum_{i=1,i\neq m}^{M} P_{PS,i} \cdot G_{Pi,Pi} + P_{N,PSm}} \tag{7.59}$$

In the multiple primary systems case, the maximum allowable transmission power $P_{max,CR}$ should be determined under the following constraint: the SINR of each PS-Rx should above l_1. That is,

$$\gamma_{PS,m} > l_1 \quad \text{for } m = 1, \ldots, M \tag{7.60}$$

We can aggressively sense the channel and target one specific primary system (assumed to be ith primary system) to observe its transmission parameters variation, and then calculate the maximum allowable transmission power under the constraint of this primary system, denoted as $P_{max,CR}^{i}$. The procedure continues until $P_{max,CR}^{i}$ is determined for all primary systems and the maximum allowable transmission power is

$$P_{max,CR} = \min\{P_{max,CR}^{1}, \ldots, P_{max,CR}^{M}\} \tag{7.61}$$

However, the aggressive sensing process is slightly different from the one primary system case: the variation of transmission parameters should be observed twice. We can express the first time switching SINR targeting at the mth primary system as

$$\begin{aligned}
\gamma_{switch}^{1} &= \frac{P_{PS,m} \cdot G_{Pm,Pm}}{P_{CR}^{1} \cdot G_{CR,Pm} + \sum_{i=1,i\neq m}^{M} P_{PS,i} \cdot G_{Pi,Pi} + P_{N,PSm}} \\
&= \frac{P_{signal}}{P_{CR}^{1} \cdot G_{CR,Pm} + P_{I,PS} + P_{N,PSm}}
\end{aligned} \tag{7.62}$$

where $P_{PS,m} \cdot G_{Pm,Pm} = P_{signal}$ and $\sum_{i=1,i\neq m}^{M} P_{PS,i} \cdot G_{Pi,Pi} = P_{I,PS}$.

The second time switching SINR can also be expressed as

$$\gamma_{switch}^2 = \frac{P_{signal}}{P_{CR}^2 \cdot G_{CR,Pm} + P_{I,PS} + P_{M,PSm}} \tag{7.63}$$

We want to determine the transmission power of CR-Tx that will achieve the constraint of the primary system. That is,

$$\frac{P_{signal}}{P_{max,CR}^m \cdot G_{CR,Pm} + P_{I,PS} + P_{N,PSm}} = l_1 \tag{7.64}$$

The known parameters include γ_{switch}^1, P_{CR}^1, γ_{switch}^1, P_{CR}^2, l_1 and we want to solve P_{CR}^t. After some derivation, we can obtain

$$P_{max,CR}^m = \frac{\gamma_{switch}^2 P_{CR}^2 (\gamma_{switch}^1 - l_1) + \gamma_{switch}^1 P_{CR}^1 (l_1 - \gamma_{switch}^2)}{l_1 \cdot (\gamma_{switch}^1 - \gamma_{switch}^2)} \tag{7.65}$$

7.3.2.7 Numerical Example

To illustrate such aggressive sensing (or active probing), we consider the ideal condition of the secondary system has several assumptions as follows (see Figure 7.26):

- The feedback channel between PS-Tx and PS-Rx is perfect.
- Through the cooperation of the primary system and the secondary system, P_{PS} and the MCP can be accurately known.
- G_{PC} and G_{CC} are accurately estimated by the secondary system.
- P_{targer} can be perfectly reached without granular error.
- The noise power is small compared with the interference power so it is ignored when we consider the SINR.

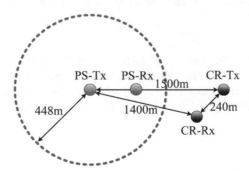

Figure 7.26 Simulation environments

Under these assumptions, the dominating error in determining $P_{max,CR}$ is the SINR estimation of PS-Rx. Without loss of generality, the SINR estimation result (in dB scale) of the PS-Rx is modelled as a random variable $\hat{\gamma}$ with Gaussian distribution where $E\{\hat{\gamma}\} = \gamma$ is the actual SINR value and σ_{SINR} is the standard deviation. We can rewrite $P_{max,CR}$ as

$$P_{max,CR}(\text{dBm}) = \hat{\gamma}_{switch}(\text{dB}) - \gamma_{switch}(\text{dB}) + [G_{PP}(\text{dB}) - G_{PC}(\text{dB})] + P_{PS}(\text{dBm}) - l_1(\text{dB}) \tag{7.66}$$

where $\hat{\gamma}_{switch}$ is the switching SINR estimation result of PS-Rx.

Define the outage probability of PS as

$$
\begin{aligned}
P_{outage,PS} &= Pr\{\gamma_{PS} < l_1(\mathrm{dB})\} \\
&= Pr\{P_{PS}(\mathrm{dBm}) - P_{CR}(\mathrm{dBm}) + [G_{PP}(\mathrm{dB}) - G_{PC}(\mathrm{dB})] < l_1\}
\end{aligned}
\tag{7.67}
$$

In our algorithm, $P_{CR} = \delta P_{max,CR}$ (in linear scale) where δ is used o reserve the margin for the constraint on $P_{outage,PS}$. We can get

$$
\begin{aligned}
P_{outage,PS} &= Pr\{\gamma_{switch}(\mathrm{dB}) - \hat{\gamma}_{switch}(\mathrm{dB}) < 10 \log \delta\} \\
&= Q\left(\frac{10 \log \delta}{\sigma_{SINR}}\right)
\end{aligned}
\tag{7.68}
$$

Please note that $P_{outage,PS}$ is affected by δ and the accuracy of SINR estimation of PS-Rx but not by the parameters of propagation model because our algorithm is actually sensing the interference at PS-Rx.

For a comparison, please recall the optimal case of the conventional power scaling power control technique. Suppose that the distance between PS-Tx and CR-Tx can be estimated perfectly and thus CR can determine the maximum transmission power. Consider the propagation model as a path-loss channel model so that the channel power gain can be written as

$$
G(r) = r^{-\alpha}
\tag{7.69}
$$

where r is the distance between the transmitter and the receiver and α is the path-loss exponent. Assume that the system parameters are $P_{PS} = 10\,\mathrm{dBm}$, $\Delta = 71.6\,\mathrm{dB}$, $\mu = 1\,\mathrm{dB}$, $\sigma^2 = -70\,\mathrm{dBm}$, $d - 1500\,\mathrm{m}$ and $\gamma_{dec} = l_1 = 8.4\,\mathrm{dB}$. Figure 7.26 shows the SINR of PS-Rx versus the transmission power of CR-Tx. We can find that under the SINR constraint of PS-Rx, the maximum transmission power of CR-Tx is different in different environments. Thus a harmful interference with the primary system may occur when the assumption of an environment does not coincide with the actual one. Table 7.4 summarises α-values for different environments. We can transform the result in Figure 7.27 to the outage probability of the primary system (Figure 7.28). If a PS-Rx is in the service range of the primary system (SINR $\geq \gamma_{dec}$) when the secondary system signal is absent and SINR $\geq \gamma_{dec}$ when the secondary system signal is present, the PS-Rx incurs outage. We define the percentage of outage due to the secondary system as the outage probability (assume the PS-Rx is uniformly distributed). Figure 4.29 shows this equivalent outage probability of the primary system for conventional spectrum sensing.

Table 7.4 Typical path-loss exponents

Environment	α range
Urban macrocells	3.7–6.5
Urban microcells	2.7–3.5
Office building (same floor)	1.6–3.5
Office building (multiple floors)	2–6
Store	1.8–2.2
Factory	1.6–3.3
Home	3

Now we consider $\alpha = 2.7$ in Figure 7.28. We can find $P_{CR} = 6.120.7\,\mathrm{dBm}$ in conventional sensing under the constrain $\gamma_{dec} = 8.4$. The corresponding outage probability in Figure 7.29 is $P_{outage,PS} = 0.02$. With the assumption $\sigma_{SINR} = 0.3\,\mathrm{dB}$ and $P_{outage,PS} = 0.02$, we can get the corresponding $\delta = 0.8675$ in the aggressive sensing algorithm. Consider a specific simulation environment as shown in Figure 7.30.

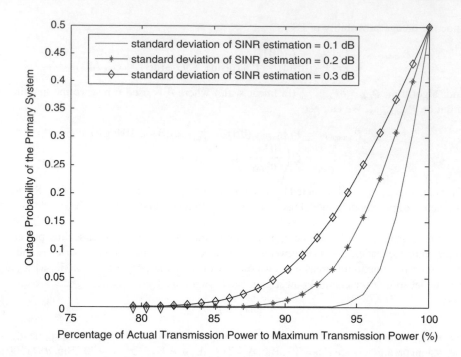

Figure 7.27   Outage probability of the primary system for the aggressive sensing algorithm

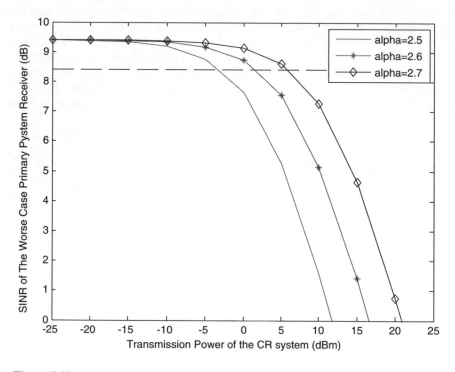

Figure 7.28   SINR of the worse case primary system receiver for conventional sensing

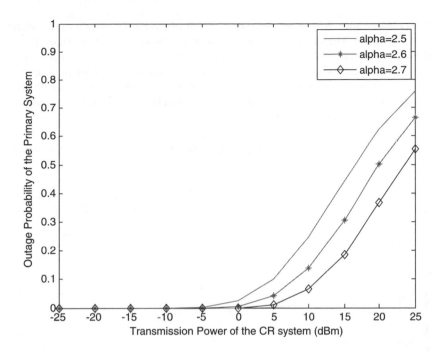

Figure 7.29 Equivalent outage probability of the primary system for conventional sensing

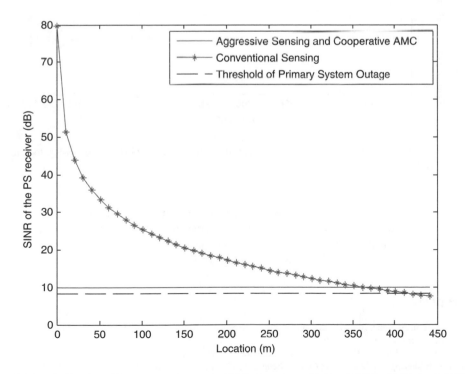

Figure 7.30 SINR of the PS-Rx according to the PS-Rx location with $\delta = 0.8675$ and $P_{CR} = 6.1207\,\text{dBm}$

From on PS-Rx side, Figure 7.30 shows the average SINR of the PS-Rx when it moves along the line from the PS-Tx to CR-Tx. The location 0 m corresponds to the PS-Tx. According to these parameters, in Figure 7.30 we can observe that when the location is larger than about 410 m, outage of the PS occurs in conventional sensing and does not in our algorithm. As long as the PS-Rx does not move towards the PS-Tx, the outage remains in conventional sensing. This shows the long time communication outage problem and the improvement of our algorithm. Also, as shown in Figure 7.31, the normalised capacity of the secondary system can be derived as

$$\frac{C}{B} = \log_2(1 + \text{SINR}) \tag{7.70}$$

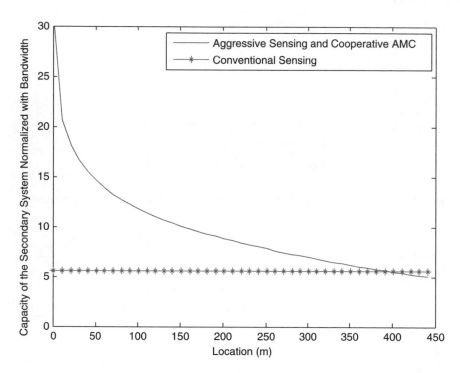

Figure 7.31 The normalised capacity of the secondary system according to the PS-Rx location with $\delta = 0.8675$ and $P_{CR} = 6.1207$ dBm

A significant improvement of the capacity is achieved in the aggressive sensing algorithm. This is because the aggressive sensing algorithm actually senses the channel to PS-Tx rather than considering the worse case. Note that the higher capacity of conventional sensing with locations larger than 410 m is because of an intolerant P_{CR} which causes the outage of PS.

References

[1] S.M. Mishra, A. Sahai, R.W. Brodersen, 'Cooperative Sensing Among Cognitive Radios', in *Proc. IEEE Int. Conf. Communications*, **4**, pp. 1658–1663.
[2] E. Visotsky, S. Kuffner, R. Peterson, 'On Collaborative Detection of TV Transmissions in Support of Dynamic Spectrum Sharing', in *Proc. 1st IEEE Int. Symp. New Frontiers in Dynamic Spectrum Access Networks DySPAN*, 2005, pp. 338–345.

[3] M. Gudmundson, 'Correlation Model for Shadow Fading in Mobile Radio Systems', *Electron. Lett.*, **27**(23), 1991, 2145–2146.

[4] C-X Wang, H-H Chen, X. Hong, M. Guizani, 'Cognitive Radio Network Management', *IEEE Vehicular Technology Magazine*, **3**(1), 2008, 28–35.

[5] J. Unnikrishnan, V.V. Veeravalli, 'Cooperative Sensing for Primary Detection in Cognitive Radio', *IEEE Journal on Selected Topics in Signal Processing*, **2**(1), 2008, 18–27.

[6] Z. Quan, S. Cui, A.H. Sayed, 'Optimal Linear Cooperation for Spectrum Sensing in Cognitive Radio Networks', *IEEE Journal on Selected Topics in Signal Processing*, **2**(1), 2008, 28–40.

[7] A. Ghasemi, E.S. Sousa, 'Interference Aggregation in Spectrum Sensing Cognitive Wireless Networks', *IEEE Journal on Selected Topics in Signal Processing*, **2**(1), 2008, 41–56.

[8] G. Ganesan, J. Li, 'Cooperative Spectrum Sensing in Cognitive Radio, Part I: Two-User Networks', *IEEE Transactions on Wireless Communications*, **6**(6), 2007, 2204–2213.

[9] G. Ganesan, J. Li, 'Cooperative Spectrum Sensing in Cognitive Radio, Part II: Multiuser Networks', *IEEE Transactions on Wireless Communications*, **6**(6), 2007, 2214–2222.

[10]   S.Y. Tu, K.C. Chen, R. Prasad, 'Spectrum Sensing of OFDMA Systems for Cognitive Radio Networks', to appear in the *IEEE Transactions on Vehicular Technology*, 2009.

[11] C.K. Yu, K.C. Chen, 'Multiple Systems Sensing for Cognitive Radio Networks over Rayleigh Fading Channel', *Proceedings IEEE VTC Spring, Singapore*, 2008.

[12]   C.K. Yu, Multiple System Sensing for Cognitive Radio Networks, Master Thesis, National Taiwan University, 2008.

8

Medium Access Control

The simplest concept for CR operation (the lower two layers for CRN) is to detect the opportunity to transmit by spectrum sensing and to access the spectrum dynamically (known as DSA). Having discussed general spectrum sensing for CRN, we are going to introduce more information regarding medium access control (MAC) for CRN, to coordinate multiple CRs accessing the shared medium in an opportunistic way, as opposed to simple spectrum utilisation for opportunistic transmission at the link level.

8.1 MAC for Cognitive Radios

We first introduce the general mathematical model for MAC operation; note that Lai, El Gamal, Jiang and Poor [35] developed derivations from statistical communication theory to achieve optimal medium access.

Let us consider a set of frequency bands to represent the general case, although more dimensional radio resources can be considered. Suppose the frequency bands that we are interested in (typically PS operating) are a set of numbered bands, $\mathcal{M} = \{1, \ldots, M\}$. At time t_n, CRN operation allows an update of spectrum utilisation. The nth observation (or allocation) time interval is $[t_n, t_{n+1})$. Due to the opportunistic nature of each link (and thus frequency band) modelled as a Markov chain, the ith frequency band is available following a Bernoulli process with probability π_i available, and is invariant to time.

We define the following indicator function (just like the clear channel indicator defined in Chapter 5):

$$\mathbf{1}_i[n] = \begin{cases} 1, & \text{channel i available in } [t_n, t_{n+1}) \\ 0, & \text{otherwise} \end{cases} \tag{8.1}$$

For perfect spectrum sensing, we can determine Equation (8.1) in a reliable way. However, any spectrum sensing has some vulnerable situations, and thus we need to consider more to decide medium access control. Following Equation (8.1), the probability mass function (pmf) of Bernoulli random variable at the ith frequency band is

$$f_{\mathbf{1}_i[n]}(x|\pi_i) = \pi_i x + (1 - \pi_i)\delta(x) \tag{8.2}$$

It is reasonable to assume $\{\mathbf{1}_i[n]_{i=1}^{M}\}$ independent, $n = 1, \ldots, L$ where L implies the observation interval depth. Denote

$$\pi = [\pi_1, \ldots, \pi_M] $$

Cognitive Radio Networks   Kwang-Cheng Chen and Ramjee Prasad
© 2009 John Wiley & Sons, Ltd

For reliable CR operation, spectrum sensing is necessary, so that CR-Tx can have information about the availability of each frequency band. However, for network operations on top of CR links, the strategy would be highly related to π.

Case 1: π is known.
Case 2: π is unknown.
Case 3: π can be detected or estimated via some CRN sensing or tomography (network tomography is introduced in Chapter 9) methods.

Traditional CR functions are as follows: At time t_n, CR learns the availability of a selected frequency band s_n (typically via spectrum sensing). If $\mathbf{1}_{s_n}(n) = 1$, information amount B can be successfully transmitted. For L time durations, the overall throughput is

$$W = \sum_{n=1}^{L} \mathbf{1}_{s_n}(n) \tag{8.3}$$

When π is known, the spectrum sensing strategy for a CR is simply to select channel $\hat{i} = \arg\max_{i \in M} \pi_i$ to sense. Then, the access decision is optimally or sub-optimally made based on certain decision criteria and conditions, such as the partially observed Markovian decision process [27]. Instead of pursuing mathematical and theoretic solutions, in this chapter we are going to explore some fundamental medium access (or DSA) in CRN; that is, when there are multiple CRs functioning.

8.2 Multichannel MAC

Under the co-existing multi-radio systems scenario, the MAC of CRs might contend for multiple channels, which makes multi-channel MAC (McMAC) attractive, although almost no literature on McMAC for CRN exists. After the medium access control problem first appeared in the packet switch system in 1969, the idea of using parallel channels to communicate first appeared in the paper 'Multichannel Local Area Network Protocols' in 1983 by Marsan and Roffinela [1]. The traditional problem formulation of medium access control, as in the Chapter 3 on MAC, considers multiple terminals sharing a medium. Instead of using all the medium, the multichannel MAC protocols separate the original channel into several sub-divided channels and use them. The performance of the multichannel MAC protocol is proved to outperform a single channel protocol under same radio resource. With the development of wireless communications, new paradigms appeared in medium access control. Because of limited radio transmission ranges, a node (i.e., a wireless device) may not be accessible to all other nodes. The design of MAC is a considerably more difficult task in wireless than in wired networks. This is due to the fact that in the wireless medium, the signal strength decays (path loss, fading) causing the medium characteristics to be highly location-dependent. Hence, traditional listen-before-transmit mechanisms such as CSMA do not work very well because the channel state might be different at the receiver from what is estimated at the transmitter. This gives rise to the well-known *hidden terminal problem*.

To overcome the problem, random channel access protocols in wireless LANs often use channel reservation techniques by exchanging short request-to-send (RTS) and clear-to-send (CTS) control packets before the data packet is sent [18]. This effectively performs a *virtual carrier sensing* at the receiver by letting the receiver centralise the controlling of the medium access [18]. In addition, the channel is temporarily reserved for the data packet transmission since neighbours of the receiver who receive the CTS may defer transmission at least for the duration of data transmission. The duration can be included in the RTS/CTS packets. IEEE Standard 802.11 for wireless LAN essentially uses this handshaking technique for asynchronous data transfer (i.e., *distributed coordination function*, or DCF).

The above access protocol on a single channel is prone to inefficiencies at heavy loads, and collisions can occur among the control packets. With possible unsynchronised backoff delays, the medium can be idle if all contending nodes are in backoff. Furthermore, any device hearing RTS or CTS must defer at least until the end of the entire exchange (i.e., until end of ACK), which means that concurrent transmissions cannot take place when two senders hear each other, even though the respective receivers do not hear any device other than their respective senders (the so called *exposed terminal problem*). Collision may happen, and the concurrent transmission is limited. The use of multichannel consequently provides performance advantages. Multichannel MAC protocols reduce collisions and enable more concurrent transmissions and thus better bandwidth usage even with the same aggregate capacity. Multichannel MAC protocols allow a number of nodes in the same neighbourhood to transmit concurrently on different channels without interfering with one another. Carrier sensing can be coupled with an efficient channel selection mechanism to pick the clearest channel for transmission. The development of multichannel MAC protocols now begins a new phase: 'How to select the channel' or 'How to make all devices know where to go?' Most recently, the proposed multichannel MAC research focuses more on how to make a channel access negotiation, that is, how the devices find each other. Please recall the fact that a wireless device (or node) can only use one radio due to its half-duplex nature and is vulnerable to simultaneous transmission or reception. With the development of software-defined radio (SDR) and CR, multi-channel MAC plays a critical role in CRN MAC.

There are two important factors to demonstrate the advantages of McMAC:

- *One wide-channel versus multi-channels:* An intuitive thought is that if we divide the fixed amount of radio resource into many sub-channels, parallel transmission is possible. The immediate drawback is that the channel capacity/throughput is factorised. However, the overall system throughput can increase. Another point of view is that, in future communication networks, devices may be accessible to a wide range of frequency bands; instead of using a wide band that may suffer from frequency selective fading, the protocol is suitable for having multichannel selection capability to select the good channels. Multichannel instead of one wide-channel intuitively makes sense.

  A simple illustration can show this advantage. Consider a scenario as in Figure 8.1. Each node A and node B has a probability p to transmit to node C at each slot; that is, nodes A and B follow i.i.d. Bernomial processes. There is no carrier sensing for ALOHA. The channel capacity is K, that is, for each successful transmission, the throughput is K bytes. Delay and retransmission is neglected.

Figure 8.1 One-channel random access MAC (slotted aloha)

The total throughput is

$$2K(p)(1 - p) = 2Kp - 2Kp^2$$

We fix the aggregate channel capacity K and divide the channel into two identical channels, each with capacity $K/2$. We also enable A and B to randomly choose a divided channel to transmit as in Figure 8.2. Suppose that C is capable of receiving from channel 1 and 2 simultaneously.

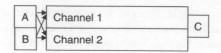

Figure 8.2 Divided-channel random access MAC (slotted aloha)

The total throughput is:

$$
2\frac{K}{2}\left\{ p\cdot\underbrace{\frac{1}{2}\left[(1-p)+p\cdot\frac{1}{2}\right]}_{\text{pick channel 1, successful}} + p\cdot\underbrace{\frac{1}{2}\left[(1-p)+p\cdot\frac{1}{2}\right]}_{\text{pick channel 2, successful}} \right\}
$$

$$
= 2Kp - Kp^2
$$

Even with equally aggregated channel capacity, multichannel MAC protocol outperforms single channel when $p > 0.667$. In a crowded network, the multichannel MAC protocol outperforms even with lower p. A more rigorous result is shown in Figure 8.3 for an M-multichannel, random channel selection, CSMA protocol. From the figure the throughput is increased with the number of channels divided to form a fixed channel.

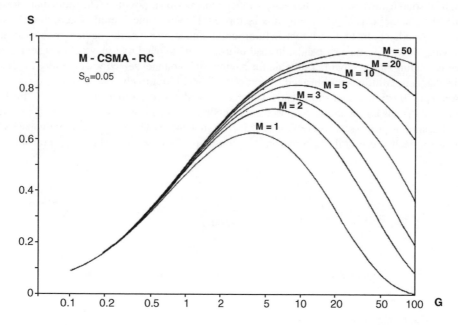

Figure 8.3 Throughput of M-channel multichannel CSMA versus offered traffic, for varying number of channels

- *Divide existing channel versus extension to multi-channels:* The next question arises: how to make multiple channels. Based on the way multiple channels are used, these protocols may be broadly classified into two classes. The first class assumes a separate channel for every device in the network, formed on the basis of individual CDMA codes (or orthogonal signalling). Transmissions may be transmitter-oriented (i.e., each device transmits using its own code/signalling) or receiver-oriented (i.e., all transmissions made to the same receiver using the same code/signalling). In the first class, all channel are dedicated, that is, every device has one and only one unique channel with respect to its code. The second class of multichannel MAC protocols does not assume dedicated channels for every node. Instead, the available bandwidth is assumed to be divided into a number of channels whose number is smaller than the number of devices. A device may transmit and receive on any channel. There are some advantages of using a smaller number of shared channels over unique channels for every device in the mobile ad-hoc network. For example, this does not require every device to know the whole set of codes used in the network that can increase the design complexity significantly.

8.2.1 General Description of Multichannel MAC

Without losing generosity, we start from the original MAC with K devices/nodes in a LAN as shown in Figure 8.4.

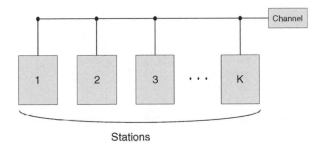

Figure 8.4 Original MAC problem

All nodes/devices share a common medium and try to communicate with each other. If more than two nodes try to access the shared medium, the collision will happen. In other words, there is contention for the limited resource. There have been many protocols that try to resolve this problem, such as CSMA/CD and CSMA/CA, using techniques such as carrier sensing, inter-frame space (IFS) and random backoff, etc. The multichannel scenario differs from the traditional MAC in that it provides a sort of diversity over channels. Assume the number of channels is M and the number of devices/stations is K. Every node is capable of sensing, transmitting and receiving over multiple channels. A collision still happens if any co-channel transmissions exist. This multi-channel scenario is shown in Figure 8.5.

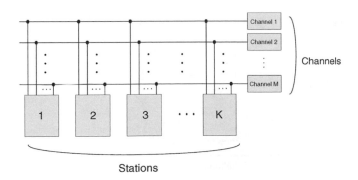

Figure 8.5 Multichannel MAC in LAN

In wireless networks, however, Figure 8.5 does not make sense because nodes are not capable of listening while transmitting. In addition, the number of radios is limited, in many cases, to be one. As a result, every node must tune to a specific channel and perform transmission or reception over it. A collision still occurs if there are more than two transmissions over the same channel. Such a multichannel scenario for wireless (radio) networks is depicted in Figure 8.6.

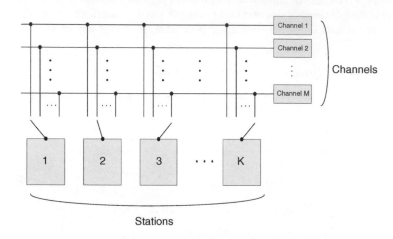

Figure 8.6   Generalised multichannel MAC (for wireless networks)

To transmit a packet successfully, the following conditions must fit:

- transmitter can find the receiver (Access Negotiation);
- no simultaneous transmission on the same channel (Collision Avoidance/Resolution);
- receiver must on the same channel with transmitter (Access Negotiation).

A successful transmission is illustrated in Figure 8.7, in which collisions or errors are typically handled by ARQ.

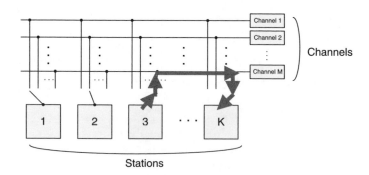

Figure 8.7   A successful transmission of node 3 to K over channel M

The second condition in the above list (Collision Avoidance/Resolution) is inherent from the original MAC problem, as described in Section 6.3.2. Contention must be avoided/resolved when more than two nodes try to access the channel. The first part of the multichannel MAC protocol is again Collision Avoidance/Resolution, which is still within the scope of Gallager's pioneer view in his 1985 *IEEE Transactions on Information Theory* paper [36].

For the first and third conditions (Access Negotiation) with the limited capability of channel access, a node may listen/transmit over one/few selected channels, and a scheme/protocol must be properly designed to access the network. That is, transmitter and receiver must have a mechanism to find each other, which is known as Access Negotiation in multichannel MAC. Consequently, the multichannel MAC consists of

- Collision Avoidance and Collision Resolution;
- Access Negotiation.

We may note that the wireless multichannel MAC structure in Figure 8.8 is equivalent to a mixed network of *packet-switch networking* and *circuit-switch networking*.

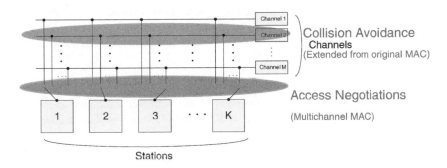

Figure 8.8   Two fundamental parts of multichannel MAC collision avoidance/resolution and access negotiation in wireless networks

Although the collision avoidance/resolution in the multichannel MAC protocol is inherited from the original MAC, it differs in the following aspects:

- The number of competing nodes may not be fixed over one channel, that is, a transmitter may 'leave' the current channel where contention occurs and try to access its receiver on other channels in the next frame.
- The channels may be asymmetric, with different capacity, delay, jitter, etc. Thus, an adaptive structure of the MAC protocol may be required to increase the efficiency over different channels.

The access negotiation in multichannel is certainly a new problem in a wireless multichannel environment, where a node can only listen/transmit over one channel. The protocol must guarantee the transmission/reception accessibility. In other words, nodes must be able to find each other.

As illustrated in Figure 8.8, the generalised multichannel scenario is determined by a set of parameters, $\{M, q, C, N\}$, where M and N respectively denote the number of channel and the number of devices and $\{q_k | k = 1,2...M\}$ and $\{C_k | k = 1,2...M\}$ denote the probabilities of appearance from other interferences and normalised capacities over channel $1,2...M$. These parameters fully determine the multichannel environment. The interference $\{q\}$ can express the probability of the appearance of other systems (e.g., primary systems in CRN).

8.2.2 Multichannel MAC: Collision Avoidance/Resolution

This part of the multichannel MAC is the same as traditional MAC (or multiple access) described in Chapter 3 and Section 6.3.2.

8.2.2.1 Multi-Layer Collision Avoidance/Resolution (MULCAR)

The aim of the fundamental MAC design is to reduce the impact of collisions over the shared medium either by *avoidance* of anticipating traffic or by *resolution* of collisions, which exactly suggests *carrier sensing* and *collision resolution* like Gallager's paper. In the CSMA family, *listen-before-transmit* is used to *avoid* the anticipated collision. The idea behind collision resolution is to *resolve* collision effectively to enhance the efficiency of multiple access through the splitting algorithm of the tree expansion structure. Mathematically, MAC protocol design has the following information:

- *Carrier sensing in generalised form:* a device can sense the channel to be '0' or 't', that is, there exists transmission(s) or not.
- *Collision resolution in generalised form:* a device (mobile or base station) can teach the channel to be '0', '1' or 'e', that is, idle, success or collision.

Originating from the classification of historical multiple access protocols by the tree expansion structure, the first generalised combination version of the collision *avoidance* and *resolution* tree expansion structure, MULCAR, was proposed by K.C. Chen [22] and is elaborated as follows:

- *Collision Anticipation Tree Expansion (CATE):* This tree structure avoids/reduces possible collision, which is the primary reason prohibiting channel efficiency. Such a tree expansion provides a number of subsets (leaves) to split contending users *before* the transmission or identification phase.
- *Collision Resolution Tree Expansion (CRTE):* This tree structure has been well studied in the 1970s and 1980s. In contrast to CATE, CRTE only generates a tree *after* the transmission or identification phase with some collisions being detected. The devices wait for channel response and try to resolve the collisions as the CRP in Chapter 3. It should be note that the splitting tree may not be a binary tree.
- *Multi-Layer Collision Avoidance/Resolution (MULCAR):* Considering a generalised and combined version of CATE and CRTE, the general structure of multi-layer collision avoidance/resolution is depicted in Figure 8.9. The idea is simple: create several independent 'urns' (groups) and then try to avoid/resolve contentions. MULCAR may consist of the following layers of effective expansion:
  - frequency domain expansion;
  - time domain expansion;
  - spatial domain expansion;
  - communication channel/bands;
  - probability domain as random backoff;
  - signalling domain.

In fact, the multichannel MAC problem is a special case of MULCAR. The main difference lies in the splitting algorithm: how to split devices into different groups to avoid/resolve contention. It is another point of view from 'Access Negotiation'. There have been some solutions such as the randomly addressed polling (RAP) family in centralised networks [18, 23], but in distributed networks, this problem remains open.

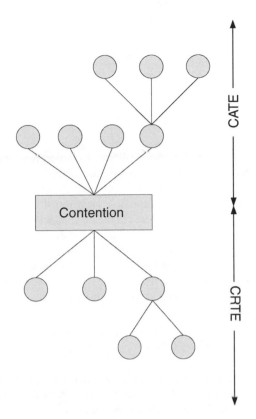

Figure 8.9   Multi-layer collision avoidance/resolution (MULCAR)

8.2.2.2 CSMA-based Multichannel MAC with Random Selection

Please recall the scenario of multichannel MAC depicted in Figure 8.5, which is composed of a set of M parallel broadcast channels to N connected nodes, where each node of M has separate interfaces (one for each channel). These broadcast channels may have different bandwidths; let W_i be the bandwidth of the ith channel. The total available bandwidth is then:

$$W = \sum_{i=1}^{M} W_i$$

Data packets are assumed to be of constant length b bits; thus the packet transmission time on the ith channel T_i is proportional to $1/W_i$. If all channels have the same bandwidth, $W/M, i = 1,2,\ldots,M$, and T_0 is the time needed to transmit a packet on a single broadcast channel with bandwidth W, then we have $T = MT_0$. The channel end-to-end propagation delay is assumed to be D sec. The normalised propagation delays $d_0 = D_0/T_0$ and $a_i = D/T_i$.

Assuming an infinite population model, new data packets are generated according to a Poisson process with rate λ packet/s. In equilibrium conditions, the total system throughput, normalising time to T_0 units, is defined as $S = \lambda T_0$. Define λ_i, as the number of packet/s in steady state that are successfully transmitted on the ith channel. The throughput of channel i is defined as $S_i = \lambda_i T$, and

$$W_i = \frac{W}{M}; \; T_i = MT_0; \; d_i = \frac{d_0}{M} \tag{8.4}$$

$$\lambda = \sum_{i=1}^{M} \lambda_i \tag{8.5}$$

We can further obtain

$$S = T_0 \sum_{i=1}^{M} \frac{S_i}{T_i} \tag{8.6}$$

The total data traffic offered to the broadcast system, including new and rescheduled packets, is assumed to be Poisson with rate γ packet/s. Therefore, the offered traffic is $G = \gamma T_0$. The stochastic properties of the traffic offered to each channel depend on the policy followed by each station in the selection of a channel on which the packet transmission is scheduled. If channels are randomly chosen, independently of their state, then the traffic offered to the ith channel is Poisson with rate γP_i where P_i is the probability of choosing channel i. If a channel is randomly selected, then the offered traffic on each channel is *Poisson*. This symmetric channel property is important in the analysis. If the channels are not identical Poisson, or symmetric, the analysis becomes tedious.

The evaluation of S_i is greatly simplified if we can assume that the traffic offered to channel i is Poisson. In this case S_i is evaluated independently of any other channel, using the single channel expression with offered traffic:

$$G_i = \gamma PL_i T_i = \frac{GP_i T_i}{T_0} = \frac{GP_i W}{W_i} \tag{8.7}$$

And with the other parameters pertinent to channel i (such as the propagation delay normalised to T_i: $d_i = d_0 T_0 / T_i$, etc.). This Poisson assumption is used to approximate the performance of some cases in which the traffic offered to each channel is obviously not Poisson.

8.2.2.3 Non-persistent CSMA Multichannel MAC Protocol with Random Channel Selection

In the single channel non-persistent CSMA protocol, when a node has a packet ready for transmission, whether new or rescheduled, it senses the channel. If the channel is sensed idle, the packet is immediately transmitted. If a carrier is detected (the channel is being used by some other station) the packet is rescheduled for transmission after a random delay. Packets may collide due to the nonzero propagation delay between any two nodes. Collided packets are scheduled for retransmission after a random delay. The extension to multiple channel systems implies that a node must select a channel where the packet is transmitted. Several ways of selecting the channel can be envisioned and we introduce the simplest case of random channel selection.

In multichannel CSMA-RC (random channel selection), a node with a ready packet randomly chooses a channel before sensing it; this procedure is named random channel selection (RC) and thus the resulting protocol is named M-SCMA-RC. Channel i is randomly chosen with probability P_i; thus the traffic offered to the ith channel is Poisson with rate G_i in Equation (8.7) per packets per T_i. The throughput of channel i is obtained from the non-persistent CSMA single channel expression as in the original CSMA, substituting G_i, for G, and a_i, for a. The total throughput S then can be express as:

$$S = G \sum_{i=1}^{M} \frac{P_i e^{-\frac{a_i GPW_i}{W_i}}}{GP_i(2a_i+1)\frac{W}{W_i} + e^{-a_i GP_i \frac{W}{W_i}}} \tag{8.8}$$

In the case of two channels ($M=2$), the probabilities P_1 and $P_2=1-P_l$ that maximise the total throughput can be obtained by setting the partial derivative of S with respect to P_l equal to zero. The optimal of P_1 is shown in Figure 8.10. From the result we can observe that the optimal channel selection is approximately equal to the ratio of channel bandwidth when the offered traffic G is low. When the fraction of channel bandwidth is assigned equal, i.e., $W_1=W_2=W/2$, the optimal probability is $P_1=P_2=0.5$. It can further be proved that the equally probable to achieve maximum in M equally factored channels. In the two channel case, that is $P_1=P_2=0.5$.

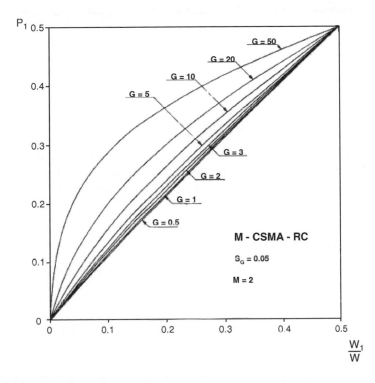

Figure 8.10   Optimal probability of selecting channel 1 to maximise total throughput

For P_i the value is $1/M$. The optimised throughput can further being simplified as

$$S = \frac{Ge^{-\frac{a_0 G}{M}}}{G(2\frac{a_0}{M}+1)+e^{-\frac{a_0 G}{M}}}$$

(8.9)

Average packet delay analysis is estimated as follows. We first observe that every time a packet is scheduled for transmission it is transmitted successfully with probability S/G. The number of scheduling of a packet is thus a geometrically distributed random variable with mean G/S. Define P_{li} as the probability of idle while sensing the chosen channel. Then P_{li} can be derived as

$$P_{li} = \frac{1+a_i G_i}{G_i(2a_i+1)+e^{-a_i G_i}}$$

(8.10)

The average number of scheduling that results in no transmission is $G(1-P_{li})/S$. The average number of collisions of a packet is $GP_{li}/S-1$, where 1 denotes the last success transmission.

The expression of the average packet delay normalised to T_0 is thus

$$D_0 = \underbrace{\left(\frac{G}{S}P_{li} - 1\right)(M + X_0 + \tau_0 + a_0 + a_0)}_{\substack{\text{\# of collision} \\ \text{Time required for re-transmissions}}} + \underbrace{\left(\frac{G}{S}(1 - P_{li})\right)X_0 + M + a_0}_{\substack{\text{\# no trnamission} \\ \text{Time required for a successful transmission}}} \qquad (8.11)$$

where X_0 is the average re-transmission delay normalised to T_0, τ_0 is the time required for ACK transmission time normalised to T_0 and P_{li} is the probability that a station senses the chosen channel to be idle.

The splitting of a single CSMA channel into M parallel links accessed according to M-CSMA-RC yields a throughput increase with the reduction of the normalised propagation delay on each channel. Results show that the average normalised packet delay can also be improved, provided that the operating throughput level is not too low.

8.2.3 Multichannel MAC: Access Negotiation

The second part of multichannel MAC protocol is *access negotiation*, that is, how nodes commit right access negotiation, and it is equivalent to how nodes find each other. First we introduce a new hidden terminal problem of multichannel MAC that must be resolved when designing a multichannel MAC; then we introduce the recent multichannel MAC protocols.

8.2.3.1 Multichannel Hidden Terminal Problem

Hereafter, we describe a new type of hidden terminal problem pertaining to multi-channel environment, which we call the *multi-channel hidden terminal problem*. For illustration, we start with a simple multichannel MAC protocol that does not address this problem.

The protocol is similar to MAC of 802.11 with DCF using one transceiver. Suppose there are N channels available. One channel is dedicated for exchanging control messages (control channel), and all the other channels are for data. When a node is neither transmitting nor receiving, it listens to the control channel. When node A wants to transmit a packet to node B, A and B exchange RTS and CTS messages to reserve the channel as in IEEE 802.11 DCF. RTS and CTS messages are sent on the control channel. When sending an RTS, node A includes a list of channels it is willing to use. Upon receiving the RTS, B selects a channel and includes the selected channel in the CTS. After that, node A and B switch their channels to the agreed data channel and exchange the DATA and ACK packets. When this handshake is done, node A and B immediately switch to the control channel.

Now consider the scenario in Figure 8.11. Node A has a packet for B, so A sends an RTS on Channel 1, which is the control channel. B selects Channel 2 for data communication and sends a CTS back to A. The RTS and CTS messages should reserve Channel 2 in the transmission ranges of A and B; no collision occurs. However, when node B sends the CTS to node A, node C is busy receiving on another channel, and thus it does not hear the CTS. Not knowing that node B is receiving on Channel 2, node C might initiate a communication with node D, and end up selecting Channel 2 for communication. This results in a collision at node B.

The above problem occurs due to the fact that nodes may listen to different channels, which makes it difficult to use *virtual carrier sensing* to avoid the hidden terminal problem. If each node listens to only one channel, node C would have heard the CTS and thus deferred its transmission. This is why we call the above problem the *multi-channel hidden terminal problem*.

The multichannel hidden terminal problem must be resolved during the design of the access negotiation part of the multichannel MAC protocol. This problem is severe in distributed networks

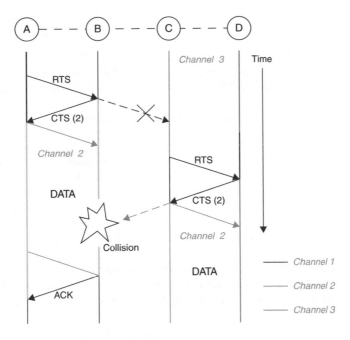

Figure 8.11   Multichannel hidden terminal problem

such as the ad-hoc mode of multichannel MAC. Different multichannel protocols are therefore proposed to relief the multichannel hidden terminal problem.

8.2.3.2 Access Negotiation Mechanisms

Access negotiation mechanisms in multichannel MAC protocols fallen into two categories: narrow band approach and wide band approach. With a narrowband transmitter and receiver, the frequency band to be used for transmission can be predetermined, or dynamically chosen. With a wideband system, the transmitter can transmit over multiple frequency bands that are detected to be unoccupied. Note that in wide band systems, each device may need more than two radios to achieve the parallel transmissions over multiple bands/channel, whereas it may require less than two radios in narrow band systems.

8.2.3.3 Narrow Band Access Negotiation Mechanisms

Dedicated Control Channels
Wu *et al.* [7] proposed a protocol that assigns channels dynamically, in an on-demand style. In this protocol, called Dynamic Channel Assignment (DCA), they maintain one dedicated channel for control messages and other channels for data. Each host has two transceivers, so that it can listen on the control channel and the data channel simultaneously. RTS/CTS packets are exchanged on the control channel, and data packets are transmitted on the data channel. In RTS packets, the sender includes a list of preferred channels. On receiving the RTS, the receiver decides on a channel and includes the channel information in the CTS packet. Then, DATA and ACK packets are exchanged on the agreed data channel. Since one of the two transceivers is always listening on the control channel, the multichannel hidden terminal problem does not occur. Figure 8.12 illustrates the idea of the dedicated control channel protocol.

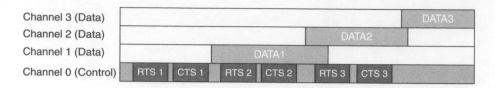

Figure 8.12   Dedicated control channel

This protocol does not need synchronisation and can use multiple channels with little control message overhead. But it does not perform well in an environment where all channels have the same bandwidth. When the number of channels is small, one channel dedicated for control messages can be costly. In the case of IEEE 802.11b, only three channels are available, and so having one control channel results in 33% of the total bandwidth as the control overhead. On the other hand, if the number of channels is large, the control channel can become a bottleneck and prevent data channels from being fully utilised.

Jain *et al.* [6] propose a similar protocol that has one control channel and *N* data channels, but selects the best channel according to the channel condition at the receiver side. The protocol achieves throughput improvements by intelligently selecting the data channel, but still has the same disadvantages as DCA.

Common Hopping

The idea of hopping protocols (Figure 8.13) is to use the hopping nature to separate the traffic into multiple channels/bands. *Hop Reservation Multiple Access* is a multichannel protocol for networks using the slow frequency hopping spread spectrum (FHSS). The hosts hop from one channel to another according to a predefined hopping pattern. When two hosts agree to exchange data by an RTS/CTS handshake, they stay in a frequency hop for communication. Other hosts continue hopping, and more than one communication can take place on different frequency hops. Receiver Initiated Channel-Hopping with Dual Polling takes a similar approach, except that the receiver initiates the collision avoidance handshake instead of the sender. These schemes can be implemented using only one transceiver for each host, but they only apply to frequency hopping networks, and cannot be used in systems using other mechanisms such as direct sequence spread spectrum (DSSS) communication.

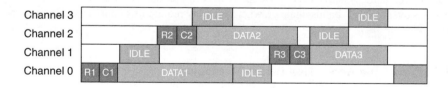

Figure 8.13   Common hopping protocol

Split Phase

In this approach, devices use a single radio. Time is divided into an alternating sequence of control and data exchange phases. During a control phase, all devices tune to the control channel and attempt to make agreements for channels to be used during the following data exchange phase. If device A has some data to send to device B, it sends a packet to B on the control channel with the ID of the lowest numbered idle channel, say, *i*. Device B then returns a confirmation packet to A. At this point, A and B have agreed to use channel *i* in the upcoming data phase. Once committed, a device cannot accept other agreements that conflict with earlier agreements. (Note that, when hidden nodes are prevalent, the sender and the receiver might have very different views of which channels are free. A more sophisticated agreement protocol is then needed, as proposed in MMAC.) Figure 8.14 illustrates the idea of split phase.

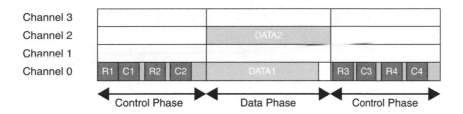

Figure 8.14   Split phase protocols

In the second phase, devices tune to the agreed channel and start data transfer. The protocol allows multiple pairs of devices to choose the same channel because each pair might not have enough data to use up the entire data phase. As a result, the different pairs must either schedule themselves or contend during the data phase. The advantage of this approach is that it requires only one radio per device. However, it requires time synchronisation among all devices, although the synchronisation can be looser than in Common Hopping because devices hop less frequently. Examples of this approach are MMAC and Multichannel Access Protocol (MAP). Their main difference is that the duration of the data phase is fixed in MMAC, whereas it is variable in MAP and depends on the agreements made during the control phase.

Hybrid
Hybrid protocols consider the multichannel multi-radio channel assignment and interfaces assignment in a multi-layer approach. Assuming the routing information is widely known, each node can dynamically build its channel assignment table and inform the whole network. In the MAC layer, the approach can be dynamically selected or in the hybrid version of existing multichannel protocols. Note that the pre-assumption for stable routing information in the hybrid protocols is the wireless mesh network, which is not widely considered in the more generalised ad-hoc mode.

SSCH
There are as many hopping sequences that each node can follow as the number of channels. Each sequence is uniquely determined by the seed of a pseudorandom generator. Each node picks multiple sequences and follows them in a time-multiplexed manner. When node A wants to talk to node B, A waits until on the same channel as B. If A frequently wants to talk to B, A adopts one or more of B's sequences, thereby increasing the time they spend on the same channel. For this mechanism to work, the sender learns the receiver's current sequences via a seed broadcast mechanism.

MCMAC
Each node picks a seed to generate a different pseudorandom hopping sequence. When a node is idle, it follows its 'home' hopping sequence. Each node puts its seed in every packet that it sends, so its neighbours eventually learn its hopping sequence. However, nodes are not required to align their hopping boundaries in practice. When node A has data to send to node B, A flips a coin and transmits with some probability p during each time slot. If it decides to transmit, then it tunes to the current channel of B and sends an RTS. If B replies with a CTS, both A and B stop hopping to exchange data. Data exchange normally takes place over several time slots. After the data exchange is over, A and B

return to their original hopping sequence, as if no pause in hopping had happened. Figure 8.15 depicts the idea of McMAC.

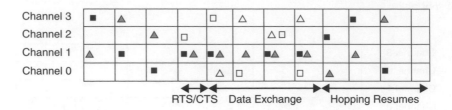

Figure 8.15 McMAC

SSCH and McMAC are similar in that they allow nodes to rendezvous simultaneously on different channels. However, there are subtle differences. In SSCH, each node selects four different hopping sequences and time multiplexes them to form a single hopping sequence. Nodes adapt their hopping sequences over time to the traffic but are not allowed to deviate from their hopping sequences. In McMAC, each node has one hopping sequence, which never changes. However, nodes are allowed to deviate from their default hopping sequence temporarily to accommodate sending and receiving. In SSCH, a sender must wait until its current channel overlaps with that of the receiver before it can send data. In McMAC, the sender can temporarily deviate from its sequence to jump to the receiver's channel to send.

Cognitive Multichannel MAC

Following the idea of cognitive radio, nodes should be able to sense the environment and reconfigure its structure. The cognitive multichannel MAC appears as an enhanced version of McMAC. Every protocol operating frame is divided into three phases: the spectrum sensing (SS) phase, the CSMA contention phase and the data transmission phase. In the cognitive Multichannel MAC, devices are able to sense the outer environment, the information of the network and the information of other interference resources (i.e., primary systems). After gathering and exchanging such information, nodes are able to set transmission probabilities to maximum the system throughput. Figure 8.16 illustrates the idea of cognitive multichannel MAC.

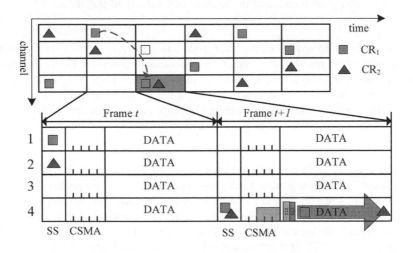

Figure 8.16 Cognitive multichannel MAC

8.2.3.4 Wide Band Access Negotiation Mechanisms

CDMA

The idea of using the CDMA approach in the multichannel system is simple. Instead of using divided bands, these kind of protocols utilise the orthogonal property of CDMA as 'multiple channels'. By different spreading codes, nodes are able to transmit in different rates, that is, a shorter code represents a larger bandwidth just like wideband CDMA cellular. However, the main drawback of this approach is that it is almost impossible to have the global codeword information and the efficiency is low.

M-channel selection

Nasipuri *et al.* proposed a multi-channel CSMA protocol with 'soft' channel reservation. If there are N channels, the protocol assumes that each host can listen to all N channels concurrently. A host wanting to transmit a packet searches for an idle channel and transmits on that idle channel. Among the idle channels, the one that was used for the last successful transmission is preferred. The protocol can be extended to select the best channel based on the signal power observed at the sender. These protocols require N transceivers for each host, which may be a concern.

Frequency Coding

This is an enhanced version of the M-channel selection protocol. Within a wideband system, the transmitter can transmit over multiple frequency bands that are detected to be unoccupied, i.e., signals/codewords are sent over several frequencies. The receiver monitors all the frequencies that are detected to be available at its end. Please note that such a frequency coding scheme, unlike frequency hopping, requires the channel availabilities in all the different frequency bands before every transmission. The main drawback of the N-transceiver requirement remains, as in M-channel selection.

Frequency Cooperation (Cooperative Multichannel MAC)

This is a new multichannel MAC protocol combining two prospective ideas: multichannel and cooperative communication. One of the open questions in the multichannel problem is the asymmetric nature: some channels are better for transmission. However, another question lies in the routing: what if the destination is not the neighbour? The two questions can be considered to be the same question: which channel/device to send? Cooperative communication/networks have been proven that will improve the throughput performance by repeat/relay of the information from the transmitter. The general MAC design follows the dedicated control channel scheme as illustrated in Figure 8.17 where each device is assumed to have radio/interfaces. Note that the data transmits on different channels with different rates. The multichannel hidden terminal problem is solved because every node continuously listens to the same control channel.

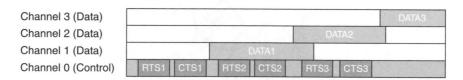

Figure 8.17   Multichannel cooperative MAC

In the practical MAC design, there are two types of transmissions: cooperative channel selection or direct transmission. The selection of different types is determined by the minimum transmission time/delay or the maximum throughput. Figure 8.18 illustrates the direct transmission. The different circles indicate the different transmission channel/systems. In this case, the lighter-coloured channel is used as the selected channel. If there is more communication in the lighter-coloured channel/system, then the dark-coloured channel/system is selected as the communication channel/system.

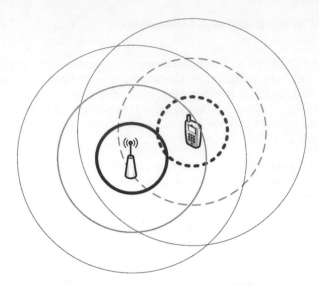

Figure 8.18   Direct transmission in the cooperative multichannel MAC

Figures 8.19 and 8.20 illustrate the cooperative transmission type in the cooperative multichannel MAC. The transmitter may calculate the expected transmission time. If there is a cooperative node in a high transmission rate channel/system, it sends a request for help. If the cooperative node replies with accessibility of cooperation, the transmitter will send the data to the cooperative node, and the cooperative node will help the relay. The locations/status of these nodes result in different types of cooperation. For example, in Figure 8.19, the overall time is saved. In Figure 8.20, the transmitter is able to access the node out of its transmission range.

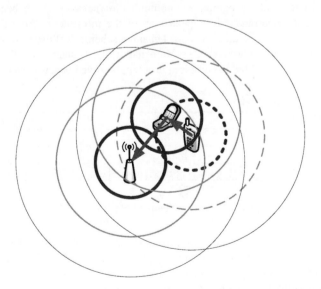

Figure 8.19   Cooperative multichannel MAC (overall time saved)

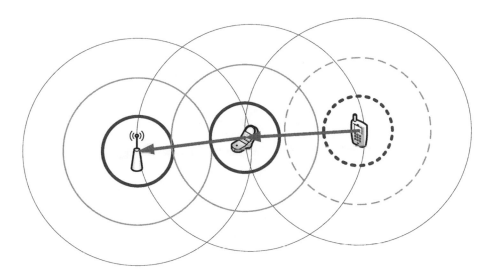

Figure 8.20   Cooperative multichannel MAC (transmitter accesses node out of its transmission range)

8.2.3.5 Analysis and Comparison

In this section, we describe and analyse the simplified model of the multichannel MAC protocols. Our goal is numerically to compare the performance of different multichannel protocols under identical conditions. Several assumptions in our environment are as follows:

- It is a fully distributed network where all devices can transmit and receive from one another. Each device contains single radio. It can tune to access one specific potential channel in one time slot. For simplicity, we assume the network is fully distributed and there is no hidden/exposed terminal. The medium access control is therefore simplified to 1-hop problem.
- Time is perfectly synchronised and divided into frames/slots within the network.
- Devices always have a packet to transmit to other devices, but with probability p they try to transmit at the beginning of every time slot. The receiver of the transmitter is randomly chosen by the transmitter with equal probability among all devices except the transmitter.
- The channel is perfect and there will be no transmission error, however, the collision may happen while more than one transmitter starts to transmit over the same channel at the same time. There is no ARQ and no receivers can receive the packet if a collision happens.
- There will be no re-transmission over time slots. If a transmitter decides to transmit to a specific receiver, somehow the transmitter fails (collision or busy channel). It will not persist in transmission to fix the past failures. It will just flip a coin, and with probability p to transmit, the receiver is randomly chosen again taking no regard of the previous receiver.

The number of potential channels is set as M. To start, we define a Markov chain $\{Xt\}$ as shown in Figure 8.21 with state Xt denoting the number of communicating pairs within CRN at time t. When $Xt = k$, there are $2k$ devices transmitting/receiving over different channels within the CRN, and other $N - 2k$ devices remain silent and follow their original hopping sequence. The maximum number of transmission pairs is bound by 0 and the minimum number between potential channels and the half of

CRN devices. The state space of Xt is:

$$S \triangleq \left\{ 0, 1, 2, \ldots \min\left(\left\lfloor \frac{N}{2} \right\rfloor, M_D\right) \right\} \tag{8.12}$$

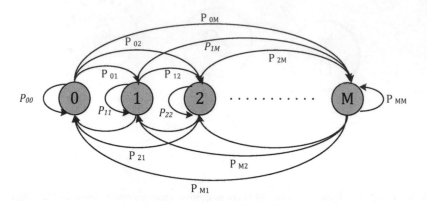

Figure 8.21 Markov chain of multichannel protocols

A state transmission will happen when new agreements are made or the existing transmission terminates. Let $S_k^{(i)}$ and $T_k^{(j)}$ respectively denote the probability that i new agreements are made and j agreements terminate in the next slot when state $Xt = k$. Then the transition probability from state k at time t to state l at time $t + 1$ can be expressed as

$$p_{kl} = \begin{cases} \displaystyle\sum_{m=0}^{k} S_k^{m+l-k} T_k^m, \ \text{for } l \geq k \\[4mm] \displaystyle\sum_{m=k-l}^{k} S_k^{m+l-k} T_k^m, \ \text{for } l < k \end{cases} \tag{8.13}$$

$$= \sum_{m=(k-l)^+}^{k} S_k^{m+l-k} T_k^m$$

where m is the number of transfers that terminate at time t to $t + 1$ and thus its value is between $k - l$ to k. In this expression, m is the number of transfers that terminate and its value is between $(k-l)$ and k. At least $(k-l)$ transfers must terminate to have l pairs in the next slot and k is the maximum number of terminating transfers. Also, the probability $T_k^{(j)}$ that i transfers terminate when the system is in state k is given by the following:

$$T_k^{(j)} = Pr[j \text{ transfers terminate at time } t | X_t = k]$$
$$= \binom{k}{j} q^j (1 - q)^{k-j}$$

where S^l the key of the multichannel MAC protocol design. It describes how a transmission is initialised. Since the transition probability of the Markov chain of McMAC in CRN can be known, we can further compute the limiting probabilities on each state by solving the balancing equations of the Markov chain. We can have the average utilisation as

$$\rho = \frac{\sum_{i \in S} i \cdot \pi_i}{M}$$

where π_i is the limiting probability that the system is in state i and S is the state of the Markov chain. Then we can further derive the total system throughput by multiplying ρ by the channel transmission rate and number of potential channels. The system throughput is

$$R_d = M \cdot C \cdot \rho_d$$

We summarise the notations in Table 8.1.

Table 8.1 Summary of notations

Symbol	Meaning of symbol
N	Number of devices
M	Number of potential channels
$T_k^{(j)}$	Probability that j transfers terminate in the next slot
$S_k^{(i)}$	Probability that i agreements are made in the next slot
P_{kl}	Transition probability from state k to state i
p	Probability of attempt to transmit for each CR device
q	Probability of appearance of primary
C	Channel capacity of each channel in bits/sec
ρ	Channel utilisation between [0,1]

By comparing S^l and the transition probabilities of the multichannel MAC protocol, we can derive and compare their performances. Here we neglect the detailed result but in Table 8.2 list the representative results of the four protocols: dedicated control channel, common hopping, split phase, McMAC.

Figure 8.22 illustrates the performance of different multichannel protocols under 802.11a/b. We can see that the McMAC outperforms the other protocols, and that the dedicated control channel is the second. It is noteworthy that when the number of the channel is large, the dedicated control channel may outperform McMAC.

It is worth noting that McMAC is not the best solution to the multichannel hidden terminal problem (see Figure 8.23). A feasible solution may be the dedicated control channel, and the latter can also scale as the number of channel increased.

8.3 Slotted-ALOHA with Rate-Distance Adaptability

In wireless communication networks, wireless nodes are in general assumed to be randomly and uniformly distributed over the service area. Thus, the distances between nodes are generally random. Recalling the *rate-distance nature* of Section 5.4.3, we may consider the distances between nodes as a measure of the received signal power, rather than the Euclidean distance or the propagation distance. Note that the propagation distance is regarded as the Euclidean distance between transmitter and receiver to have the same received power. Consequently, distance D means a permissible set of any possible location point with received signal power as propagation Euclidean distance D under certain long-term fading. Rate-distance nature usually comes together with AMC, and we can observe that high spectral efficiency modulations are generally more sensitive to interference and noise. In other words, systems operating at higher rates are more vulnerable to interference from other co-existing systems or the primary/secondary systems in CRNs. To achieve the spectrum efficiency, after sensing idle radio resources (in time and/or frequency domain) of the primary user system(s), CR nodes as the secondary system(s) adapt themselves (both in PHY and MAC layers) to access optimally the available radio resources. In this section, we employ the rate-distance nature to study the adaptability of the

Table 8.2 Analytic model of multichannel MAC protocols

Protocol	$S_k^{(i)}$ and p_{kl}
Dedicated Control Channel	$S_k^{(i)} = \begin{cases} (N = 2k)p(1-p)^{(N-2k-1)} & \text{if } i = 1 \\ 1 - S_k^{(1)} & \text{if } i = 0 \\ 0 & \text{otherwise.} \end{cases}$
	$p_{kl} = \begin{cases} 0 & \text{if } l > k+1 \\ T_k^{(0)} S_k^{(1)} & \text{if } l = k+1 \\ T_k^{(k-l)} S_k^{(0)} + T_k^{(k-l+1)} S_k^{(1)} & \text{if } 0 < l \leq k \\ T_k^{(k)} S_k^{(0)} & \text{if } l = 0. \end{cases}$
Common Hopping	$S_k^{(i)} = \begin{cases} (N - 2k)p(1-p)^{(N-2k-1)} \times \dfrac{N-2k-1}{N-1} \times \dfrac{M-k}{M} & \text{if } i = 1 \\ 1 - S_k^{(1)} & \text{if } i = 0 \\ 0 & \text{otherwise.} \end{cases}$
	$p_{kl} = \begin{cases} 0 & \text{if } l > k+1 \\ T_k^{(0)} S_k^{(1)} & \text{if } l = k+1 \\ T_k^{(k-l)} S_k^{(0)} + T_k^{(k-l+1)} S_k^{(1)} & \text{if } 0 < l \leq k \\ T_k^{(k)} S_k^{(0)} & \text{if } l = 0. \end{cases}$
Split Phase	Details in Reference 35
McMAC	$S_k^{(j)} = \sum_{o,i,a} P[A = a] \times P[O = o\|A = a]$ $\times P[I = i\|O = o, A = a]P[J = j\|I = i, A = a, O = o],$
	Details in Reference 35
	$p_{kl} = \begin{cases} 0 & \text{if } l > k+1 \\ T_k^{(0)} S_k^{(1)} & \text{if } l = k+1 \\ T_k^{(k-l)} S_k^{(0)} + T_k^{(k-l+1)} S_k^{(1)} & \text{if } 0 < l \leq k \\ T_k^{(k)} S_k^{(0)} & \text{if } l = 0. \end{cases}$

multiple access protocols. More specifically, a two-rate slotted Aloha is selected as a pilot example to demonstrate the rate-distance feature in the multiple access protocol.

8.3.1 System Model

We assume that the number of nodes in the system is N. All nodes are randomly and uniformly distributed over the service area. The service area is partitioned into i regions. For simplicity, the service area and each region are assumed to be circles. The radius of region i is d_i. The number of newly arrived packets for each node in a slot is assumed to be a Poisson random variable. The number of nodes in the region i is n_i and $\sum n_i = N$. Nodes with newly arrived packets or collided packets are called active nodes (in the active state) or backlogged nodes (in the backlogged state) respectively. The number of active and backlogged nodes in the region i is $n_{a,i}$ and $n_{b,i}$ respectively. Therefore, we have

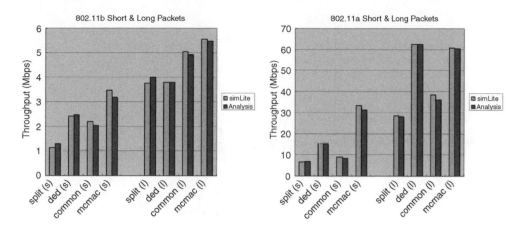

Figure 8.22 Performance of different multichannel protocols

$n_{a,i} + n_{b,i} = n_i$. The packet arrival rate (or accessible rate) of an active node in the region i is r_i. The probability of an active node and a backlogged node in the region i sending a packet in a given slot is $q_{a,i}$ and $q_{b,i}$ respectively. If r_i is small, the transmission probability of an active node can be approximated to $q_{a,i} = 1 - e^{-r_i} \approx r_i$. In other words, we can use the transmission probability of an active node in a given slot to represent the packet arrival rate. To simplify our studies, we assume $l = 2$, i.e., the service area is partitioned into two regions.

An example to demonstrate the rate-distance feature in a two-region services area of the cognitive radio scenario is illustrated in Figure 8.24 (the same as Figure 5.13). The base station (BS) and mobile station (MS) in the primary system are communicating through the primary communication link (PCL). Due to their effective distance, the low-rate is selected. Near the boundary of the service area of the BS, there are two cognitive radio devices wishing to establish a secondary communication link (SCL) under the region of low-level interference from the primary system. As Figure 8.24 shows, high-rate

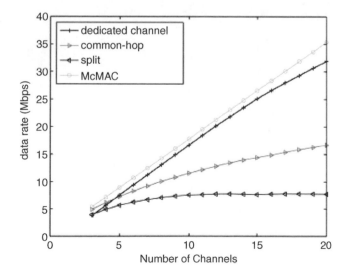

Figure 8.23 Performance of different multichannel protocols via a number of channels

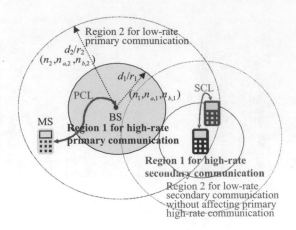

Figure 8.24 Rate-distance nature of co-existing PS and CR systems

communication might be possible between these two cognitive radios without affecting the primary system, and interference from the active primary system nodes to the CRs can be tolerated.

To analyse the rate-distance characteristics, as shown in Figure 8.24, a 2-D Markov chain is used to model the number of backlogged nodes in region 1 and 2. The probability that there are u active users among the $(n_i - v)$ unbacklogged nodes in region i transmit packets at a given slot when there are v backlogged nodes is

$$P_{a,i}(u, v) = \binom{n_i - v}{u}(q_{a,i})^u(1 - q_{a,i})^{n_i - v - u} \tag{8.14}$$

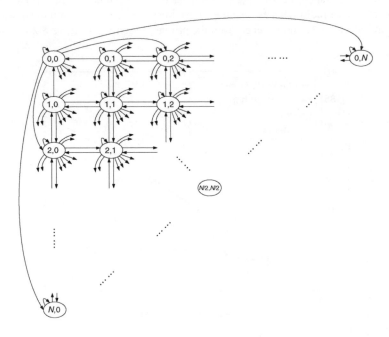

Figure 8.25 2-D Markov chain model

The probability that there are u backlogged nodes among the v backlogged nodes in region i transmit packets at a given slot is

$$P_{b,i}(u, v) = \binom{v}{u}(q_{b,i})^u (1 - q_{b,i})^{v-u} \tag{8.15}$$

The steady-state probability can be computed from the following two equations:

$$
\begin{aligned}
&P(n_{b,1}, n_{b,2}) \\
&- \sum_{i=0}^{n_1} \sum_{j=0}^{n_2} \Big[P(i,j) P_{(n_{b,1},n_{b,2})}^{(i,j)} + P(n_{b,1}+1, n_{b,2}) P_{(n_{b,1},n_{b,2})}^{(n_{b,1}+1,n_{b,2})} \\
&\qquad + P(n_{b,1}, n_{b,2}+1) P_{(n_{b,1},n_{b,2})}^{(n_{b,1},n_{b,2}+1)} \Big]
\end{aligned}
\tag{8.16}
$$

$$\sum_{n_{b,1}=0}^{n_1} \sum_{n_{b,2}=0}^{n_2} P(n_{b,1}, n_{b,2}) = 1 \tag{8.17}$$

where

$$P_{(n_{b,1}+i,n_{b,2}+j)}^{(n_{b,1},n_{b,2})}$$

is the state transition probability from state $(n_{b,1}, n_{b,2})$ to state $(n_{b,1} + i, n_{b,2} + j)$ and is derived in the following. Then, the probability of a successful transmission when the number of nodes in regions 1 and 2 are n_1 and n_2 (or $N - n_1$) and the number of backlogged nodes in regions 1 and 2 are $n_{b,1}$ and $n_{b,2}$ is

$$
P_{(n_{b,1}+i,n_{b,2}+j)}^{(n_{b,1},n_{b,2})}
$$

$$
= \begin{cases}
\begin{aligned}
&P_{a,2}(0,n_{b,2})P_{a,1}(1,n_{b,1})P_{b,1}(0,n_{b,1})P_{b,2}(0,n_{b,2}) \\
&+P_{a,2}(1,n_{b,2})P_{a,1}(0,n_{b,1})P_{b,1}(0,n_{b,1})P_{b,2}(0,n_{b,2}) \\
&+P_{a,1}(0,n_{b,1})P_{a,2}(0,n_{b,2})[(1-P_{b,1}(1,n_{b,1})-P_{b,1}(0,n_{b,1})) \\
&+(1-P_{b,2}(1,n_{b,2})-P_{b,2}(0,n_{b,2})) \\
&+P_{b,1}(0,n_{b,1})P_{b,2}(0,n_{b,2})+P_{b,1}(1,n_{b,1})P_{b,2}(1,n_{b,2})], & i=0,\ j=0 \\
\end{aligned} \\
P_{a,1}(1,n_{b,1})P_{a,2}(0,n_{b,2})[(1-P_{b,1}(0,n_{b,1}))+P_{b,1}(0,n_{b,1})(1-P_{b,2}(0,n_{b,2}))], & i=1,\ j=0 \\
P_{a,1}(0,n_{b,1})P_{a,2}(1,n_{b,2})[(1-P_{b,2}(0,n_{b,2}))+P_{b,2}(0,n_{b,2})(1-P_{b,1}(0,n_{b,1}))], & i=0,\ j=1 \\
P_{a,1}(0,n_{b,1})P_{b,1}(1,n_{b,1})P_{a,2}(0,n_{b,2})P_{b,2}(0,n_{b,2}), & i=-1,\ j=0 \\
P_{a,1}(0,n_{b,1})P_{b,1}(0,n_{b,1})P_{a,2}(0,n_{b,2})P_{b,2}(1,n_{b,2}), & i=0,\ j=-1 \\
P_{a,1}(i,n_{b,1})P_{a,2}(j,n_{b,2}), & 2\le i+j\le n_1+n_2-n_{b,1}-n_{b,1}
\end{cases}
\tag{8.18}
$$

$$
\begin{aligned}
P_{succ}(n_{b,1},n_{b,2},n_1,n_2) =\ & P_{a,1}(1,n_{b,1})P_{b,1}(0,n_{b,1})P_{a,2}(0,n_{b,2})P_{b,2}(0,n_{b,2}) \\
&+P_{a,1}(0,n_{b,1})P_{b,1}(1,n_{b,1})P_{a,2}(0,n_{b,2})P_{b,2}(0,n_{b,2}) \\
&+P_{a,1}(0,n_{b,1})P_{b,1}(0,n_{b,1})P_{a,2}(1,n_{b,2})P_{b,2}(0,n_{b,2}) \\
&+P_{a,1}(0,n_{b,1})P_{b,1}(0,n_{b,1})P_{a,2}(0,n_{b,2})P_{b,2}(1,n_{b,2})
\end{aligned}
\tag{8.19}
$$

Let the attempt rate $\Lambda(n_{b,1}, n_{b,2}, n_1, n_2)$ be the number of attempted transmissions in a slot when the number of nodes in regions 1 and 2 are n_1 and n_2 (or $N - n_1$) and the number of backlogged nodes in regions 1 and 2 are $n_{b,1}$ and $n_{b,2}$. Thus,

$$
\begin{aligned}
\Lambda(n_{b,1}, n_{b,2}, n_1, n_2) =\ & (n_1 - n_{b,1})q_{a,1} + n_{b,1}q_{b,1} \\
&+ (n_2 - n_{b,2})q_{a,2} + n_{b,2}q_{b,2}
\end{aligned}
\tag{8.20}
$$

Now, by using $(1-x)^y \approx e^{-xy}$ for small x, Equation (8.19) can be approximated to

$$P_{succ}(n_{b,1}, n_{b,2}, n_1, n_2) = \Lambda(n_{b,1}, n_{b,2}, n_1, n_2)e^{-\Lambda(n_{b,1}, n_{b,2}, n_1, n_2)} \tag{8.21}$$

This is also regarded as the throughput of the two-rate slotted Aloha. Let $\bar{n}_{b,i}$ be the average number of backlogged nodes in region i. Then, the average attempt rate $\bar{\Lambda}(\bar{n}_{b,1}, \bar{n}_{b,2}, n_1, n_2)$ is given by

$$\bar{\Lambda}(\bar{n}_{b,1}, \bar{n}_{b,2}, n_1, n_2) = (n_1 - \bar{n}_{b,1})q_{a,1} + \bar{n}_{b,1}q_{b,1} + (n_2 - \bar{n}_{b,2})q_{a,2} + \bar{n}_{b,2}q_{b,2}, \tag{8.22}$$

where $\bar{n}_{b,i} = \sum_{n_{b,1}=0}^{n_1} \sum_{n_{b,2}=0}^{n_2} n_{b,i}P(n_{b,1}, n_{b,2})$.

Based on the stability analyses in Chapter 3, for the slotted Aloha to be stable, the average attempt rate in Equation (8.22) must be smaller than or equal to 1. Furthermore, the maximal throughput of the slotted Aloha (i.e., $1/e$) can also be achieved when the average attempt rate is equal to 1.

Consider the situation when the system is stable and in the steady state. Assume that all backlogged nodes have the same retransmission probability $q_{b,1} = q_{b,2} = q_b$. Let $k_r = q_{a,2}/q_{a,1}$ and $k_d = d_2/d_1$. To apply the rate-distance slotted Aloha to the cognitive radio networks as shown, we let $q_{a,2} \le q_{a,1}$ and $d_2 > d_1$. In this case, we have $0 \le k_r \le 1$ and $k_d \ge 1$. Then, Equation (8.22) can be rewritten as

$$\bar{\Lambda}(\bar{n}_{b,1}, \bar{n}_{b,2}, n_1, n_2) = [N(k_d^{-1})^2 - Nk_d^{-1}k_r + k_r(N - \bar{n}_{b,2}) \tag{8.23}$$
$$- \bar{n}_{b,1}]q_{a,1} + (\bar{n}_{b,1} + \bar{n}_{b,2})q_b.$$

Let $N = 40$, $d_2 = 80$, $q_{a,1} = 0.02$ and $q_b = 0.05$. First, we maintain the attempt rate to be less than 1 (i.e., $\bar{\Lambda}(\bar{n}_{b,1}, \bar{n}_{b,2}, N_1, N_2) < 1$) so that the system is stably operated. Then, we let the attempt rate approach to 1 (i.e., $\bar{\Lambda}(\bar{n}_{b,1}, \bar{n}_{b,2}, N_1, N_2) \to 1$) so that the throughput can be maximised. The rate-distance relationship is shown in Figure 8.26, where we can see that if k_d is small (e.g., $k_d < 1.2$), the difference of the packet arrival rates between regions 1 and 2 can be very large. On the contrary, if k_d is large (e.g., $k_d > 3$), the packet arrival rates of nodes in regions 1 and 2 are around the same level.

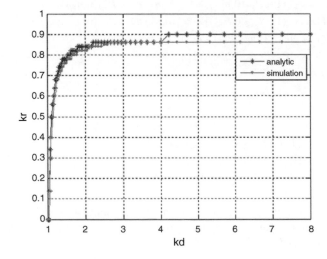

Figure 8.26 Rate-distance relationship under stable and maximal throughput constraints ($N = 40$, $q_{b,1} = q_{b,2} = q_b = 0.05$, $q_{a,1} = 0.02$, $d_2 = 80$)

Delay analysis is critical to the MAC protocol performance. By assuming $q_{b,1} = q_{b,2} = q_b$ and the number of backlogged nodes in regions 1 and 2 are $n_{b,1}$ and $n_{b,2}$, the probabilities of successful transmission for the nodes in regions 1 and 2 are

$$P_{succ,1} = P_{a,1}(1, n_{b,1})P_{a,2}(0, n_{b,2})P_{b,1}(0, n_{b,1})P_{b,2}(0, n_{b,2})$$
$$+ P_{a,1}(0, n_{b,1})P_{a,2}(0, n_{b,2})P_{b,1}(1, n_{b,1})P_{b,2}(0, n_{b,2}) \tag{8.24}$$

and

$$P_{succ,2} = P_{a1}(0, n_{b,1})P_{a2}(1, n_{b,2})P_{b,1}(0, n_{b,1})P_{b,2}(0, n_{b,2})$$
$$+ P_{a1}(0, n_{b,1})P_{u2}(0, n_{b,2})P_{b,1}(0, n_{b,1})P_{b,2}(1, n_{b,2}). \tag{8.25}$$

According to Equations (8.24) and (8.25), when q_b is small, the probability of successful transmission of nodes in region i can be approximated to

$$P_{succ,i} \approx [(n_i - n_{b,i})q_{a,i} + n_{b,i}q_b]e^{-\Lambda(n_{b,1}, n_{b,2}, n_1, n_2)}. \tag{8.26}$$

Let the increase of the number of backlogged nodes in region i be $D_i = (n_i - n_{b,i})q_{a,i} - P_{succ,i}$. For the slotted Aloha to be stable, the increase of the average number of backlogged nodes in region i must be zero, i.e., $E[(n_i - n_{b,i})q_{a,i}] = E[P_{succ,i}]$. Thus, the average throughput in the region i is

$$E[P_{succ,i}] = E[(n_i - n_{b,i})q_{a,i}] = (n_i - \bar{n}_{b,i})q_{a,i}. \tag{8.27}$$

From Little's formula, we know that the 'delay (or waiting time)' = 'average system time' – 'service time'. Thus, the average delay of nodes in the region i is

$$\bar{W}_i = \frac{(n_i - \bar{n}_{b,i})q_{a,i} + \bar{n}_{b,i}}{(n_i - \bar{n}_{b,i})q_{a,i}} - 1 = \frac{\bar{n}_{b,i}}{(n_i - \bar{n}_{b,i})q_{a,i}}. \tag{8.28}$$

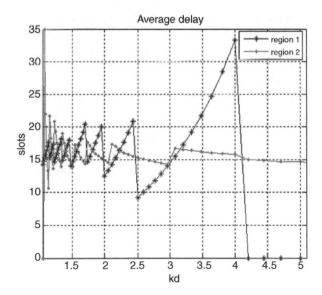

Figure 8.27   The average delays incurred in region 1 and 2 for $q_{b,1} = q_{b,2} = 0.05$

As expected, the average delays incurred in regions 1 and 2 for $N = 40$, $d_2 = 80$, $q_{a,1} = 0.02$ and $q_b = 0.05$, as shown in Figure 8.27. Note that we only focus on the situation that satisfies the stable and maximal throughput constraints. In Figure 8.27, when k_d approaches 1, the packet arrival rate for nodes in region 2 should be set to very small in order not to interfere with nodes in region 1. As a consequence, the throughput for nodes in region 2 is small. In this case, if a backlogged packet occurs in region 2, it will have a rare chance of being transmitted. Therefore, the average delay for nodes in region 2 increases.

On the contrary, when k_d is large, the packet arrival rate for nodes in region 2 increases. Thus, the average delay for nodes in region 2 decreases. The zigzag phenomenon in Figure 8.27 is because we round off the expected number of backlogged nodes to the integer. In addition, as the k_d increased, the size of region 1 is decreased and thus the number of nodes within region 1 is decreased. Under this circumstance, the waiting time of nodes in region 1 can be very large due to the denominator in Equation (8.28) being decreased. Once the k_d exceeds 4, there are no nodes in region 1 and so the waiting time becomes zero. The same reason can be used to explain the waiting time of nodes in region 2.

To complete the analysis of the ALOHA system with the rate-distance adaptation, we now explore the case for an infinite numbers of nodes. Assume that the new arrived packets in regions 1 and 2 are regarded as backlogged immediately on arrival and are transmitted in the next slot. In this case, the probability of a successful transmission is

$$
\begin{aligned}
P_{succ} = \ & n_{b,1} q_{b,1} (1 - q_{b,1})^{n_{b,1} - 1} (1 - q_{b,2})^{n_{b,2}} \\
& + n_{b,2} q_{b,2} (1 - q_{b,1})^{n_{b,1}} (1 - q_{b,2})^{n_{b,2} - 1},
\end{aligned}
\tag{8.29}
$$

Let $\partial P_{succ}/\partial q_{b,1}$ and $\partial P_{succ}/\partial q_{b,2}$ be zero to maximise P_{succ}. We can then obtain $q_{b,1} = q_{b,2} = 1/(n_{b,1} + n_{b,2})$. This means that the number of backlogged users, and which region they are in, is not important. All we have to do is to estimate the total number of backlogged users in the system. The total system arrival rate is $\lambda = \lambda_1 + \lambda_2$, where λ_i is the arrival rate in region i. The system is stable if $\lambda < 1/e$, and a pseudo-Bayesian algorithm can be used to estimate the total number of backlogged users:

$$
\hat{n}_{b,k+1} = \begin{cases} \max[\lambda_1 + \lambda_2, \hat{n}_{b,k} + \lambda_1 + \lambda_2 - 1], & \text{for idle or success} \\ \hat{n}_{b,k} + \lambda_1 + \lambda_2 + (e - 2)^{-1}, & \text{for collision,} \end{cases}
\tag{8.30}
$$

where $\hat{n}_{b,k}$ is the estimation of the total number of backlogged packets in the kth time slot.

Let n_i' be the number of packets that have been transmitted in region i. Then, $E[n_{b,1} + n_1'] : E[n_{b,2} + n_2'] = \lambda_1 : \lambda_2$. Since all backlogged packets are transmitted with the same probability $q_b(k) = 1/\hat{n}_k$ in the kth slot, $E[n_1'(k) - n_1'(k - 1)|n_{b,1}(k) + n_1'(k) - n_1'(k - 1)]: E[n_2'(k)n_2' - (k - 1)|n_{b,2}(k) + n_2'(k) - n_2'(k - 1)] = (n_{b,1}(k) + n_1'(k) - n_1'(k - 1)) : (n_{b,2}(k) + n_2'(k) - n_2'(k - 1))$. Taking the expectation of all values, we have

$$
\begin{aligned}
& E[n_1'(k) - n_1'(k - 1)] : E[n_2'(k) - n_2'(k - 1)] \\
& = E[n_{b,1}(k) + n_1'(k) - n_1'(k - 1)] : E[n_{b,2}(k) + n_2'(k) - n_2'(k - 1)]
\end{aligned}
\tag{8.31}
$$

Since,

$$
E[n_1'(k)] : E[n_2'(k)] = E[n_1(k) + n_1'(k)] : E[n_2(k) + n_2'(k)] = \lambda_1 : \lambda_2,
\tag{8.32}
$$

we have

$$
E[n_{b,1}] : E[n_{b,2}] = \lambda_1 : \lambda_2
\tag{8.33}
$$

According to Little's formula, $E[n_{b,1}] = \lambda_1 E[W_1]$ and $E[n_{b,2}] = \lambda_2 E[W_2]$, we have

$$
E[W_1] = E[W_2]
\tag{8.34}
$$

Let $\lambda_2 = 0.05$ packets/slot, and the expected number of backlogged in region 1 (i.e., user 1) and region 2 (i.e., user 2) are as shown in Figure 8.28. In this figure, consistent with the above derivations, we can find that the expected backlogged packets in regions 1 and 2 can be obtained according to the ratio of arrival rate (i.e., $\lambda_1 : \lambda_2$), and the total number of backlogged packets can be estimated by the pseudo-Bayesian algorithm. The average delays in regions 1 and 2 are shown in Figure 8.28. We can also find that the average delays in regions 1 and 2 are almost the same and can be obtained from the average time of the pseudo-Bayesian algorithm, which demonstrates greater throughput-delay performance than just ALOHA without using rate-distance nature.

If the total system arrival rate λ is known, the arrival rates for the nodes in regions 1 and 2 are $\lambda_1 = \lambda \cdot d_1^2/d_2^2$ and $\lambda_2 = \lambda(1 - d_1^2/d_2^2)$ respectively when all nodes are uniformly and fixed-distributed over the entire system area. Since the system throughput is $G(n)e^{-G(n)}$ and $G(n) = n_{b,1}q_r + n_{b,2}q_r$ where q_r is the retransmission probability, the throughput is influenced by the number of backlogged packets in regions 1 and 2 and the backlogged retransmission probability. If $q_{b,1} = q_{b,2} = q_r = 1/(n_{b,1}+n_{b,2})$, the throughput can be maximised. Otherwise, if $q_{b,1} = q_{b,2} \neq 1/(n_{b,1} + n_{b,2})$, the only parameter that influences the throughput is the retransmission probability q_r.

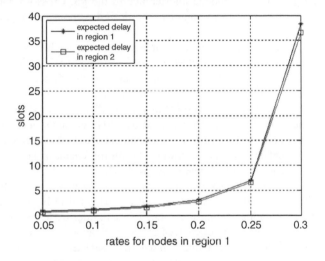

Figure 8.28   Average delay performance

8.4 CSMA with AMC

The MAC protocol to support DSA among multiple CRs is obviously a critical issue in the cognitive radio network (CRN). Here, we consider the CRN formed only by multiple CRs that cognise communication parameters to co-exist with legacy PSs. After treating ALOHA in overlay CRN, it is straightforward to explore the CSMA protocols. For this major goal, various multiple access protocols for the CRN have been proposed. Among them, the CSMA has been shown to be a simple and effective multiple access scheme for the CRN to provide a high throughput, without, however, challenging how to leverage the spatial reuse. According to the inspired observation in Chapter 5 (and earlier sections of this chapter) regarding the rate-distance nature of the wireless communications, communications of CRs can possibly proceed concurrently during transmissions of the PS with an acceptable interference to each other for the spatial reuse. Thus, an effective multiple access protocol of the CRN can incorporate the spatial reuse abilities to exploit fully the radio resource, in both the time and spatial domains.

CSMA combining with concurrent transmissions has received considerable attention in ad-hoc and multi-hop networks, mainly as homogeneous networks. There exists a challenge in the CRN, which is that the modifying operation of the legacy PS is usually unacceptable. Around the same time as Wang *et al.* [30] proved that CRs concurrently transmitting packets during the transmission of the PS do indeed improve the overall throughput, if precise location information between each station is available for the CRN. It can be achieved by analysing uplink SINR and downlink SINR, then installing

CSMA/CA with precise location information. In the following, we extend the scope based on the rate-distance nature, and propose a *CSMA* with *spatial-reuse* transmissions (named as CSMA-ST) as the multiple access protocol for the CRN. Instead of precise location information that is usually unavailable for the CRN even using the globe position system (GPS), CRs only require the signal to interference and noise ratio (SINR) to adapt to the legacy PS with an already decided operation. The proposed CSMA-ST therefore generally allows possible simultaneous/concurrent packet transmissions of CRs during the transmissions of the PS-Tx without location information. Concurrent transmissions may imply higher interference, which induces another critical task of the CRN: to control properly the interference to the PS. The multiple access protocol alone is not sufficient to achieve the optimum protocol capacity and prevent unacceptable interference to the PS. Cross-layer design with power control and AMC is therefore needed to complete the design without the requirement of location information.

For realistic operation, PS just functions as a sort of legacy device and there is no need to make nodes of PS as CR and running the newly developed MAC protocols. The MAC for the CRN shall be able to accommodate the existence of legacy devices. In the mean time, the ultimate goal of the CRN MAC should be to allow each CR simultaneously to transmit the packet with the maximal data rate, subject to the (interference) constraint that the outage probability of the PS is restricted below a pre-determined value, and maximises the throughput of the CRN and the overall throughput (of PS and CRN). Recalling the interaction among transmission rate, distance, tolerable interference at each node and interference level to other nodes, a channel inversion based power-rate control scheme and thus a cross-layer power-rate control procedure is required to facilitate the CSMA-ST.

The rate-distance nature of wireless communications is described in Section 5.4.3. We are considering a co-existence scenario of a CRN and a PS as depicted in Figure 8.29, which is pretty similar to Figure 8.24. Modulation schemes and thus the data rates between the BS and the mobile stations (MSs) of the PS are based on their received SINR. Without loss of generalisation, we consider two rates: high and low. A high data rate modulation scheme is selected for the MS near to the BS of the PS and a low data rate modulation scheme is selected for the MS far from the BS of the PS. Near the coverage boundary of the BS of the PS, a communication link of CRs can be established. A high data rate modulation scheme might be possible between these two CRs to resist the interference from the

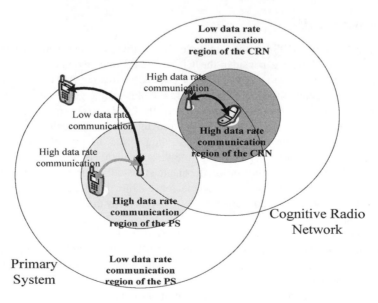

Figure 8.29   Networking scenario of CSMA-ST in CRN (here the CRN consists only of CRs)

transmitter station (TS) of the PS. The interference from the CR to the receiver station (RS) of the PS might also be acceptable for the modulation scheme adopted by the RSs of the PS.

8.4.1 Carrier Sense Multiple Access with Spatial-Reuse Transmissions

To alleviate the hidden terminal problem and to improve the throughput, the proposed CSMA-ST adopts the *four-way handshaking* procedure [18]. The CRN is composed of a BS or the message sink of the ad-hoc network and multiple CRs that attempt to transmit packets to the BS. The legacy PS adopts the conventional CSMA family protocols (e.g., CSMA with collision avoidance, CSMA/CA). Different from conventional CSMA, which considers the channel as either the occupied state (a TS of the PS is transmitting packets) or the idle state (all TSs of the PS do not transmit packets), the proposed CSMA-ST considers the channel as: (i) the occupied state (a TS of the PS is transmitting packets and there are no transmission opportunities for the CR); (ii) the busy state (a TS of the PS is transmitting packets but the transmission opportunity can be found by the CR); or (iii) the idle state.

A transmission opportunity means that the CR can identify a feasible power and rate for the transmission to achieve the following:

- The interference from the CR transmitter to the RS of the PS can be controlled at an acceptable level.
- The interference from the TS of the PS can be overcome; that is, the bit error rate (BER) of the CR is maintained.

When the channel is sensed idle by the CRs and the TSs of the PS, CRs waits a period, τ_s, so that the TSs of the PS can access the channel prior to the channel accesses of CRs. Two statistical parameters are used to characterise the behaviour of the TSs of the PS (shown in Figure 8.30(a)):

- the aggregated successful data transmission probability of all TSs of the PS before the end of the waiting period of CRs, q_p;
- the remaining data transmission time, τ_x, of a TS after the end of the CRs' waiting period.

When CRs attempt to transmit packets and the channel is considered as the busy state at the end of the waiting period, the *request-to-send* (RTS) messages are sent to the BS of the CRN simultaneously during the transmission of the TS of the PS to contend the channel. If the channel contention is successful, the simultaneous packet transmission with the feasible power and data rate proceeds as illustrated in Figure 8.30(a). Otherwise, the CRs wait until the channel becomes idle as shown in Figure 8.30(b), where these data packets are considered as backlogged.

There is one simultaneous transmission opportunity and only one packet can be transmitted for each successful channel contention for a CR. To more efficiently use the radio resources, the CSMA-ST treats a long backoff time or continuous packet collisions between multiple TSs of the PS like a 'spectrum hole'. If a successful packet transmission of the PS does not occur before the end of the waiting period of the CRs, they can contend the channel and transmit the packet. Under this circumstance, the TSs of the PS consider the channel as occupied and do not transmit packets until the next channel idle moment as illustrated in Figure 8.30(c). The RTS messages of the CRs can be transmitted when the channel is idle at the end of CRs' waiting period (it is also considered as a spectrum hole) as illustrated in Figure 8.30(d).

The impact of an unacceptable interference from the CR transmitter to the PS is modelled as an outage probability of the PS; that is, the probability that the interference from the CR transmitter to the RS of the PS is larger than I_{acc}. This outage probability will not exceed a pre-determined p_{outage}, and is expressed as

$$\Pr\{I_{ps} > I_{acc}\} \le p_{outage} \tag{8.35}$$

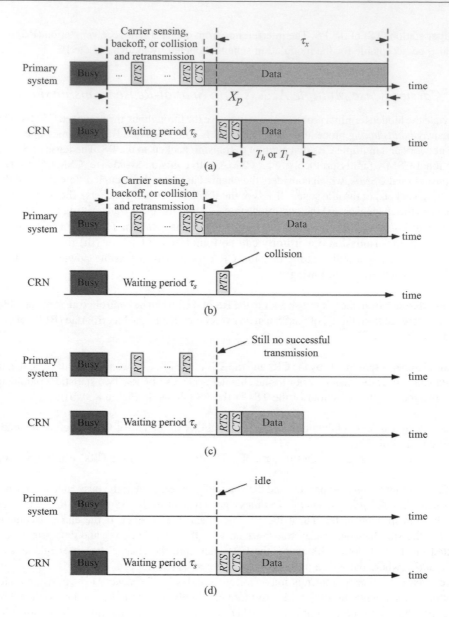

Figure 8.30 The procedure of the CSMA-ST protocol for the CRN (ACK messages are included at the end of the data transmission durations)

where I_{ps} is the interference power from the CR transmitter to the RS of the PS. This is the interference constraint for the transmissions of CRs.

Based on the features described above, the detailed operations of the proposed CSMA-ST protocol are stated below.

1. For each CR transmitter, when a CR transmitter attempts to transmit a packet during the waiting period, it senses the channel at the end of the waiting period and goes to Step 2a. If it is not the waiting period, the CR transmitter waits until the channel becomes idle.

2. When the channel becomes idle, the CR transmitter waits for a period, τ_s, and senses the channel at the end of this waiting period.

    a) If the interference power from the TS of the PS to the CR transmitter is less than a pre-determined threshold at the end of the waiting period, the CR transmitter sends a RTS message to the BS of the CRN to contend the channel.

    b) Otherwise, the CR transmitter shall wait until the channel becomes idle and repeats step 1 and step 2 to resend the RTS with an appropriate retransmission probability.

3. If a RTS is correctly received by the CR receiver, the CR receiver computes the feasible transmission power and rate based on the SINR of the received RTS. If a feasible transmission power and rate can be obtained, the information is carried back to the CR transmitter through the responded CTS. Otherwise, the CTS does not respond.

4. If the corresponding CTS is received by the CR transmitter, it transmits the data packet with the power and rate according to the embedded information in the CTS. Otherwise, the CR transmitter shall wait until the channel becomes idle and repeats step 1 and step 2 to resend the RTS with an appropriate retransmission probability.

5. When the data packet is correctly received by the CR receiver, an acknowledgement (ACK) is replied, due to the dynamic link availability for CRs.

8.4.2 Analysis of CSMA-ST

To analyse CSMA-ST, we first make the following assumptions:

A1. Without loss of generality, two data rates, high and low, are considered for CRs. CRs near to and far from the BS of the CRN have a high and a low data rate, respectively.

A2. Let packets arrive at each of the N_h high data rate CRs and each of the N_l low data rate CRs, with independent Poisson processes with mean arrival rates λ_h and λ_l, respectively.

A3. If a packet arrives at a non-backlogged CR (that is, the CR without the backlogged packet) during the waiting period, a RTS message can be transmitted to the BS of the CRN to contend the channel. Let $q_{ah} = (1 - e^{-\lambda_h \tau_s})q_{Rh}$ and $q_{al} = (1 - e^{-\lambda_l \tau_s})q_{Rl}$ represent the RTS transmission probabilities for each non-backlogged CR near to and far from the BS of the CRN, respectively. q_{Rh} and q_{Rl} are the probabilities that the interference power from the TS of the PS to the high and low data rate CR transmitter does not exceed the predetermined threshold value, respectively.

A4. If a packet arrives at a non-backlogged CR while the channel is not idle or the CR is transmitting, this packet is considered as a backlogged packet.

A5. No buffer is assumed in each CR, and thus new packets arriving at backlogged CRs (the CRs with a backlogged packet) are dropped.

A6. Each backlogged CR resends the RTS message with probabilities q_{bh} and q_{bl} for high and low data rate CRs, respectively, at the end of the waiting period if the RTS messages are allowed to be transmitted.

A7. The packet transmission time for high and low data rate CRs are T_h and T_l, respectively, $T_h \leq T_l$.

A8. Both T_h and T_l are shorter than the remaining data transmission duration τ_x of the TSs of the PS.

A 2-D finite-state-Markov-chain (FSMC) as shown in Figure 8.31 is adopted for the performance analysis. The state index (n_{bh}, n_{bl}) represents the number of backlogged high and low data rate CRs. When there are neither new packets nor backlogged packets of CRs attempting to be transmitted and all TSs of the PS also have no successful transmissions, the state transmits at the end of the waiting period (a self-state transition). Otherwise, the state transmits at the end of the busy (or occupied) channel duration.

Let τ_{RTS} and τ_{CTS} be the RTS and CTS message receiving times including the worst case packet propagation delay, respectively. We also let $g_h(n_{bh}) = (N_h - n_{bh})q_{ah} + n_{bh}q_{bh}$ and $g_l(n_{bl}) = (N_l - n_{bl})q_{al} + n_{bl}q_{bl}$ be the aggregated RTS transmission probabilities of high and low data rate

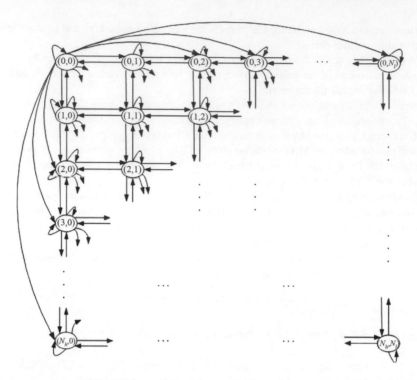

Figure 8.31 2-D finite state Markov chain using (n_{bh}, n_{bl}) as the index of the state used to analyse the number of backlogged high and low data rate CRs

CRs at the end of the waiting period, respectively. If all TSs of the PS have no successful transmission before the end of the waiting period with the probability $(1 - q_p)$, there are four (one step) state transition periods in the 2-D FSMC:

1. τ_s (in the case of neither new packets nor retransmission packets attempting to be transmitted) with the probability $e^{-(g_h(n_{bh}) + g_l(n_{bl}))}$;
2. $\tau_s + \bar{T}_h$ (in the case of a high data rate CR successful transmission, $\bar{T}_h = T_h + \tau_{RTS} + \tau_{CTS}$) with the probability $g_h(n_{bh})e^{-(g_h(n_{bh}) + g_l(n_{bl}))}$;
3. $\tau_s + \bar{T}_l$ (in the case of a low data rate CR successful transmission, $\bar{T}_l = T_l + \tau_{RTS} + \tau_{CTS}$) with the probability $g_l(n_{bl})e^{-(g_h(n_{bh}) + g_l(n_{bl}))}$;
4. $\tau_s + \beta$ (in the case of collisions between CRs, $\beta = \tau_{RTS} + \tau_{CTS}$ with the probability $1 - e^{-(g_h(n_{bh}) + g_l(n_{bl}))} - (g_h(n_{bh}) + g_l(n_{bl}))e^{-(g_h(n_{bh}) + g_l(n_{bl}))})$, or in the case of a failure RTS reception due to the interference from the TS of the PS with the probability q_R.

When the TSs of the PS have a successful transmission before the end of the waiting period of CRs with the probability q_p, the state transition period is X_p, $X_p = \tau_s + \tau_x$. Therefore, the expected duration of the state transition is

$$\Gamma_c(n_{bh}, n_{bl}) = (1 - q_p)[\tau_s(1 + q_R) + (\bar{T}_h g_h(n_{bh}) + \bar{T}_l g_l(n_{bl})) \cdot$$
$$e^{-(g_h(n_{bh}) + g_l(n_{bl}))} + (q_R + 1 - e^{-(g_h(n_{bh}) + g_l(n_{bl}))}$$
$$- (g_h(n_{bh}) + g_l(n_{bl}))e^{-(g_h(n_{bh}) + g_l(n_{bl}))})\beta] + q_p X_p \tag{8.36}$$

8.4.2.1 Throughput of the CSMA-ST

The expected numbers of successful transmissions per state transition of high and low data rate CRs are $g_h(n_{bh})e^{-(g_h(n_{bh})+g_l(n_{bl}))}$ and $g_l(n_{bl})e^{-(g_h(n_{bh})+g_l(n_{bl}))}$, respectively. By obtaining the steady-state distribution of n_{bh} and n_{bl} in the 2-D FSMC, the throughput of the CRN with the proposed CSMA-ST can be expressed as

$$S_c(n_{bh}, n_{bl}) = \frac{(g_h(n_{bh}) + g_l(n_{bl}))e^{-(g_h(n_{bh})+g_l(n_{bl}))}}{\Gamma_c(n_{bh}, n_{bl})} \tag{8.37}$$

which is the number of packets departing per state transition.

8.4.2.2 Average Packet Delays of the CSMA-ST

The average packet delay analysis can be obtained by modifying the average delay analysis for typical CSMA. The average packet delays for high and low data rate CRs can be approximated as

$$W_{high\_rate} \approx \frac{1}{2(1-(\bar\lambda_h+\bar\lambda_l)E[t])}\left[(1-q_p)\left(\frac{\bar\lambda_h(\tau_s+\bar T_h)}{\bar\lambda_h+\bar\lambda_l}+\frac{\bar\lambda_l(\tau_s+\bar T_l)}{\bar\lambda_h+\bar\lambda_l}\right)+q_pX_p]^2+2[E[t]-(\tau_s+\bar T_h)]\right]$$

$$W_{low\_rate} \approx \frac{1}{2(1-(\bar\lambda_h+\bar\lambda_l)E[t])}\left[(1-q_p)\left(\frac{\bar\lambda_h(\tau_s+\bar T_h)}{\bar\lambda_h+\bar\lambda_l}+\frac{\bar\lambda_l(\tau_s+\bar T_l)}{\bar\lambda_h+\bar\lambda_l}\right)+q_pX_p]^2+2[E[t]-(\tau_s+\bar T_l)]\right] \tag{8.38}$$

where $E[t]$ is the reciprocal of the throughput and can be interpreted as the expected required time for each departure packet of the CRN, and $\bar\lambda_h=(N_h-n_{bh})\lambda_h$ and $\bar\lambda_l=(N_l-n_{bl})\lambda_l$ are the aggregated arrival rates of high and low data rate CRs, respectively.

8.4.2.3 Performance of the CSMA-ST under Different Data Rates

Figure 8.32 shows the throughput and average packet delays of the CRN with the proposed CSMA-ST, respectively, under different packet transmission times of low data rate CRs, T_l (as a multiple of T_h).

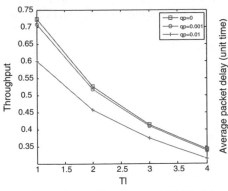
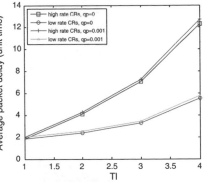

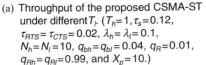

(a) Throughput of the proposed CSMA-ST under different T_l. ($T_h=1, \tau_s=0.12$, $\tau_{RTS}=\tau_{CTS}=0.02$, $\lambda_h=\lambda_l=0.1$, $N_h=N_l=10$, $q_{bh}=q_{bl}=0.04$, $q_R=0.01$, $q_{Rh}=q_{Rl}=0.99$, and $X_p=10$.)

(b) Average packet delays of the proposed CSMA-ST under different T_l. ($T_h=1, \tau_s=0.12$, $\tau_{RTS}=\tau_{CTS}=0.02$, $\lambda_h=\lambda_l=0.1, N_h=N_l=10$, $q_{bh}=q_{bl}=0.04$, $q_R=0.01$, $q_{Rh}=q_{Rl}=0.99$, and $X_p=10$.)

Figure 8.32   Throughput and delay of CSMA-ST

We indicate that reducing the packet transmission time (and thus increasing the data rate) of low data rate CRs increases the throughput and decreases the average packet delays of the CRN. However, a high data rate transmission relies on a high SINR, which also relies on a large transmission power. The data rate cannot increase without limit because CRs will prevent unacceptable interference to the PS. To maximise the throughput and minimise the average packet delay of the CRN subject to the interference constraint, CRs should use the maximal transmission data rate to simultaneously transmit packets subject to the interference constraint. This communication strategy can also be considered to maximise the spectral efficiency subject to the interference constraint.

8.4.2.4 Overall Throughput Analysis

Curious readers may want to know whether the increase on the throughput of the CRN also suggests the increase on the overall throughput. To verify this, we proceed with the following analysis on the overall throughput.

Considering the co-existing network topology shown in Figure 8.33, we denote the CR transmitters near to and far from the CR receiver as CR_{near} and CR_{far}, respectively. The distance vector, $[a, b, c, d, e, f]$, is used to represent the distances between the CRs and the stations of the PS. In this analysis, a path loss channel with the additive white Gaussian noise (AWGN) that has the power spectral density $N_0/2$ is considered. The data rates (normalised to the channel bandwidth) of CR_{near}, CR_{far} and the TS of the PS are expressed as

$$c_{CR_{near}} = \log_2\left(1 + \omega \frac{P_{CR_{near}} a^{-\alpha}}{P_{ps} f^{-\alpha} + N_0 B}\right) \tag{8.39}$$

$$c_{CR_{far}} = \log_2\left(1 + \omega \frac{P_{CR_{far}} b^{-\alpha}}{P_{ps} f^{-\alpha} + N_0 B}\right) \tag{8.40}$$

$$c_{ps} = \log_2\left(1 + \omega \frac{P_{ps} e^{-\alpha}}{(\eta P_{CR_{near}} c^{-\alpha} + (1 - \eta) P_{CR_{far}} d^{-\alpha}) + N_0 B}\right) \tag{8.41}$$

where P_{ps} is the transmission power of the TS of the PS, ω is a constant related to the BER, α is the path loss exponent, $P_{CR_{near}}$ and $P_{CR_{far}}$ are the transmission powers of CR_{near} and CR_{far}, respectively, and B is the channel bandwidth. The successful channel contention probabilities of CR_{near} and CR_{far} are η and $(1 - \eta)$, respectively, where $0 \leq \eta \leq 1$.

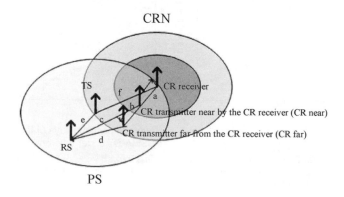

Figure 8.33   A typical co-existing network topology of the CRN and the PS

We approximate the aggregated RTS messages transmission probabilities of high and low data rate CRs at the end of the waiting period as g_h and g_l, respectively. The throughput of the CRN, S_{CRN}, can be approximated as

$$S_{CRN} \approx \frac{(g_h + g_l)e^{-(g_h + g_l)}}{\Gamma_c} \qquad (8.42)$$

where

$$\Gamma_c \approx (1 - q_p)[\tau_s(1 + q_R) + \bar{T}_h g_h e^{-(g_h + g_l)} + \bar{T}_l g_l e^{-(g_h + g_l)} + (1 - e^{-(g_h + g_l)} - (g_h + g_l)e^{-(g_h + g_l)} + q_R)\beta] + q_p X_p$$

The throughput (normalised to the data rate without the interference from the CR transmitter and only suffers the AWGN) of the PS can be approximated as

$$S_{ps} \approx \frac{q_p \tau_x (c_{ps}/c_{ps\_AWGN})}{\Gamma_c} \qquad (8.43)$$

$$c_{ps\_AWGN} = \log_2(1 + \omega P_{ps}e^{-\alpha}/(N_0 B))$$

which can be considered as the throughput under the presence of CRs. Thus, the overall throughput is

$$S_{PS + CRN} \approx \frac{(g_h + g_l)e^{-(g_h + g_l)} + q_p \tau_x(c_{ps}/c_{PS\_AWGN})}{\Gamma_c} \qquad (8.44)$$

8.4.2.5 Capture Effect and Simultaneous Transmission Condition

If more than one transmission simultaneously arrives at the receiver, the stronger signal captures the receiver modem and the weaker signal is rejected as noise or interference. This is known as the *capture effect* in radio communications. We adopt a simple yet widely accepted capture effect model. In this model, the capture effect occurs when

$$P_{CR_{near}} a^{-\alpha} \geq (P_{ps}f^{-\alpha} + N_0 B),$$
$$P_{CR_{near}} b^{-\alpha} \geq (P_{ps}f^{-\alpha} + N_0 B), \qquad (8.45)$$
$$P_{ps}e^{-\alpha} \geq [(\eta P_{CR_{near}}c^{-\alpha} + (1 - \eta)P_{CR_{far}}d^{-\alpha}) + N_0 B]$$

This is the *condition for the simultaneous transmission* (i.e., the SINRs of the CR receiver and the RS of the PS exceed 0 dB).

8.4.2.6 Performance of the Overall Throughput

In Figure 8.34, we numerically plot the throughputs of the CRN, the PS and the overall throughput by considering three representative scenarios. Analysing the overall throughput when the simultaneous transmission condition is satisfied, the distance vector is selected as D_1 in Figure 8.34. Under the assumption that P_{ps} and the noise power are fixed, we gradually decrease $P_{CR_{near}}$ and $P_{CR_{far}}$, and thus the SINR of the RS of the PS increases in Figure 8.34. We observe that, under a given traffic load, CR transmitters filling with the spectrum holes but no simultaneous transmissions results in a 0.22 throughput of the CRN. Under this circumstance, the throughput of the PS is around 0.1 and the overall throughput is around 0.32. The simultaneous packet transmissions of CRs result in a 0.55 overall throughput as maximum.

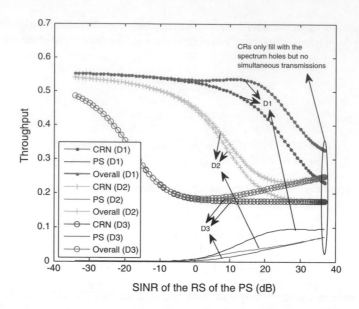

Figure 8.34 Throughputs of the CRN, the PS and the overall throughput ($g_h = gl = 0.4$, $\tau_s = 0.12$, $X_p = \tau_x + \tau_s = 10$, $T_h = 1$, $T_l = T_{h^{c_{CRnear}/c_{CRfar}}}$, $\eta = 1/2$, $B = 200$k Hz, $\omega = 0.28$, $q_R = 0.01$, $q_{Rh} = q_{Rl} = 0.99$, $\alpha = 2$, $D_1 = [20, 40, 300, 240, 50, 260]$, $D_2 = [20, 50, 120, 100, 50, 120]$ $D_3 = [100, 120, 50, 20, 80, 60]$)

We next decrease the distances between the CRN and the PS (and thus the interference between the CRN and the PS increases) while the simultaneous transmission condition still holds. The distance vector is selected as D_2 in Figure 8.34. The CR transmitters filling with the spectrum holes but with no simultaneous transmissions result in a 0.26 overall throughput. However, the simultaneous packet transmissions of CRs result in a 0.47 overall throughput as maximum. When the simultaneous transmission condition is violated, the distance vector is selected as D_3 in Figure 8.34. Under this situation, the simultaneous transmission introduces an unacceptable interference between the PS and the CRN. Thus the overall throughput decreases.

As a summary, we observe that i) a throughput increase of the CRN indeed increases the overall throughput; and ii) a sacrifice on the throughput of the PS gains more on the overall throughput, if the simultaneous transmission condition is satisfied. In practice, the CRN has to guarantee the stricter interference constraint in (i). We can also observe that maximising the throughput of the CRN subject to the interference constraint also maximises the overall throughput subject to this constraint. For this purpose, we propose a cross-layer power-rate control scheme in the following section.

8.4.3 A Cross-Layer Power-Rate Control Scheme

In order to develop a complete design to optimise MAC under the interference constraint, we first introduce the power-rate adjusting scheme to maximise the data rate (and thus the spectrum efficiency). Then, the maximal transmission power subject to the interference constraint of CRs is derived with a cutoff SINR threshold for the power adjustment. Finally, the procedure of the cross-layer power-rate adapting scheme is summarised.

8.4.3.1 Adaptive Power-Rate Adjusting Scheme

A lot of mechanisms concerning maximising the spectral efficiency have been proposed. Under the assumption that there is no restriction on the constellation size of the adaptive M-QAM, the channel inversion scheme [28,29] can be modified and the power-rate adaptation of CRs can be described as

$$R = \max_{\gamma',\gamma > \gamma_0} \log_2 \left(1 + \frac{-1.5}{\ln(5BER)\xi} \right) p(\gamma > \gamma') \quad (8.46)$$

where γ' is the cutoff SINR value such that the power adapting can only be performed when $\gamma > \gamma'$ with the probability $p(\gamma > \gamma')$, γ is the received SINR, γ_0 is a threshold value, R is the data rate normalised to the signal bandwidth, $\xi \triangleq \int_{\gamma'}^{\infty}(1/\gamma)p(\gamma)d\gamma$ is the reciprocal of the channel inversion target SINR that is determined by the average power of CRs, $p(\gamma)$ is the distribution of received SINR, and BER is the required BER. The power adaptation here is slightly different from that in [18,19], where the data rate is maximised by selecting an appropriate cutoff SINR value. Since CRs will prevent an unacceptable interference to the PS, the data rate in Equation (8.46) can only be maximised over the cutoff SINR values exceeding a threshold γ_0. We derive the γ_0 in the following subsection.

8.4.3.2 The Cutoff SINR Threshold

To obtain γ_0, we should first obtain the maximal transmission power of the CRs subject to the interference constraint. This transmission power is referred to the maximal allowable transmission power, Ψ, and is given by

$$\Psi = I_{acc} \left[10\alpha\log_{10}(d_{p_{outage}}/d_0) + q(\gamma)_{p_{outage}} \right] \quad (8.47)$$

where $d_{p_{outage}}$ is the distance such that $\int_0^{d_{p_{outage}}} d(r)dr \leq p_{outage}$ $q(\gamma)_{p_{outage}}$ is the value such that $\int_0^{q(\gamma)_{p_{outage}}} q(\gamma)d\gamma \leq p_{outage}$, $q(\gamma)$ is the distribution of the received SINR, r and $d(r)$ are the Euclidean distance and its distribution between an arbitrary transmitter of the CR and the RS of the PS in Reference [30], respectively, and d_0 is a reference distance. Then, we obtain the cutoff SINR threshold as

$$\gamma_0 = \Psi\rho/(N_0B + I) \quad (8.48)$$

where ρ is the channel gain and I is the interference power from a TS of the PS. To maximise the data rate, the CRs shall search a feasible cutoff SINR that exceeds γ_0 to maximise Equation (8.48).

8.4.3.3 The Procedure of the Cross-Layer Power-Rate Control

The procedure of the cross-layer power-rate control scheme is summarised as the following steps:

1. A CR uses the maximal allowable transmission power according to Equation (8.47) to send the RTS to the CR receiver to contend the channel.
2. When a RTS message is received by the CR receiver, the CR receiver determines γ_0 according to Equation (8.48).
3. If the SINR on the RTS does not exceed γ_0, the CTS message does not reply. Otherwise, the CR receiver searches a feasible cutoff SINR that exceeds γ_0 to maximise Equation (8.46).
   a) If the SINR of the RTS exceeds the cutoff SINR, the transmission power that inverts the channel to the target SINR, $1/\xi$, and the date rate in Equation (8.46), are carried in the CTS message replied to the CR transmitter.
   b) Otherwise, the CTS message is not replied.

Due to the 'hidden terminal problem of the CRN', it is more difficult for a CR to sense an active RS of the PS than sense an active TS of the PS. Therefore, a CR transmitter using the maximal allowable transmission power to send the RTS is necessary, even if it is not for the simultaneous transmission.

8.4.4 Performance Evaluations

To further consider the impact of the unreliability of a wireless channel on performance, we construct the relationship of the BER and the data rate between multiple access schemes with and without adaptive power-rate adjusting. Then, the operations and the theoretical throughputs of four multiple access schemes are described to allow appropriate performance evaluations and comparisons.

We start from the derivation of the approximated relationship between the BER and the data rate of multiple access schemes with and without power-rate adjusting as

$$\frac{BER_2}{BER_1} \approx (5BER_1)^{\frac{2^{c_2}-1}{2^{c_1}-1}} - 1 \tag{8.49}$$

where BER_1 is the BER of the multiple access schemes with the power and rate adjusting and with the data rate c_2. This BER is considered to be harmless to the throughput. The multiple access schemes without power and rate adjusting have the BER, BER_2, according to Equation (8.49), and have the data rate c_1.

Assuming that the CRs near to the BS of the CRN have an acceptable BER but CRs far from the BS of the CRN suffer a transmission error according to Equation (8.49), if the power and rate adapting is not available, we can simply use the BER to approximate the packet error rate. Multiple access schemes 2, 3 and 4 are slotted and packets can only be transmitted at the beginning of each slot. The notation in this section is defined in the same way as in the previous sections unless otherwise specified. We now state the detailed operation of these multiple access schemes.

Scheme 1: CSMA without power adjusting (CSMA non-PRA): CRs adopt the proposed CSMA-ST but only transmit packets via the spectrum hole. The PHY does not adjust the transmission power and rate, thus the BER cannot be maintained to the required value. The throughput is given by

$$S_{CSMA\_nonST} \approx \frac{(1 - q_p)(g_h + g_l \cdot (5BER_1)^{-\left(\frac{2^{c_2}-1}{2^{c_1}-1} - 1\right)})e^{-(g_h+g_l)}}{\Gamma_c} \tag{8.50}$$

where the successful transmission probability of the CRs far from the BS of the CRN is

$$(g_l \cdot (5BER_1)^{-\left(\frac{2^{c_2}-1}{2^{c_1}-1} - 1\right)})e^{-(g_h+g_l)}$$

CRs only have one data rate: $\bar{T}_h = \bar{T}_l$ in Γ_c.

Scheme 2: Random access with power adjusting (RA-PRA): CRs dynamically access the channel in the slotted ALOHA (s-ALOHA) type of manner. That is, CRs near to and far from the BS of the CRN transmit packets with probabilities g_h and g_l, respectively, at the beginning of a slot when the channel is idle. The power-rate adjusting aims to maintain the required BER over the wireless channels, but not for the simultaneous transmissions. We approximate the throughput as

$$S_{CR\_AC} \approx \frac{(g_h + g_l)e^{-(g_h+g_l)}}{\left\{ \begin{array}{l} (1 - e^{-g_l})T_l + (1 - e^{-g_h})e^{-g_l}T_h \\ + q_f T_s e^{-(g_h+g_l)} + e^{-(g_h+g_l)}(1 - q_f)X_p \end{array} \right\}} \tag{8.51}$$

where q_f is the probability that that all TSs of the PS do not have a successful transmission in a slot with the length T_s.

Scheme 3: Random access without power adjusting (RA non-PRA): When all TSs of the PS do not transmit packets, CRs access the channel in the s-ALOHA type of manner. The PHY does not adjust the transmission power and rate. The throughput of this scheme can be expressed as

$$S_{CRnon\_AC} \approx \frac{(g_h + g_l \cdot (5BER_1)^{-\left(\frac{2^{c_2}-1}{2^{c_1}-1} - 1\right)}) e^{-(g_h+g_l)}}{\{T_{CR}(1 - e^{-(g_h+g_l)}) + q_f T_s e^{-(g_h+g_l)} + e^{-(g_h+g_l)}(1 - q_f)X_p\}} \tag{8.52}$$

where T_{CR} is the packet transmission time of the CRs.

Scheme 4: Random access without cognitions (RA non-cognitions): CRs access the channel in the s-ALOHA manner no matter whether a TS of the PS is transmitting packets or not. The PHY does not adjust the transmission power and rate. Under this circumstance, the simultaneously transmitted packets of the CRs interfere with the RS of the PS (packet collisions). The throughput of this scheme is given by

$$S_{MAC\,non\_cognition} \approx \frac{(g_h + g_l \cdot (5BER_1)^{-\left(\frac{2^{c_2}-1}{2^{c_1}-1} - 1\right)}) e^{-(g_h+g_l)}}{\{T_{CR}(1 - e^{-(g_h+g_l)}) + q_f T_s e^{-(g_h+g_l)} + e^{-(g_h+g_l)}(1 - q_f)X_p\}} \tag{8.53}$$

Please note that the RA non-cognition has the same successful packet departure rate as the RA non-AC. Thus they have the same theoretical throughput.

We use the packet transmission time of CRs far from the BS of the CRN, T_l, to capture the interference level from the PS and the channel quality (the received SINR). The parameters we use are assigned as follows: $q_p = 0.01$, $X_p = 10$, $\tau_{RTS} = \tau_{CTS} = 0.02$, $\tau_s = 0.12$, $T_{CR} = T_s = 1$, $T_h = 1$, $BER_1 = 10^{-3}$, $q_R = 0.01$ $q_{RH} = q_{RI} = 0.99$ and $q_f = 0.99$. Figure 8.35 shows the throughputs under a higher received SINR of the BS of the CRN (CR receiver) and thus a higher data rate of the CRs far from the BS of the CRN ($T_l = 2$). The proposed scheme has throughput improvements between 49% and 170% (depending on the traffic load). The throughputs of the CRN under a lower received SINR of the BS of the CRN ($T_l = 3$) are also shown in Figure 8.35. This figure also indicates that the proposed scheme has throughput improvements between 14% and 160%.

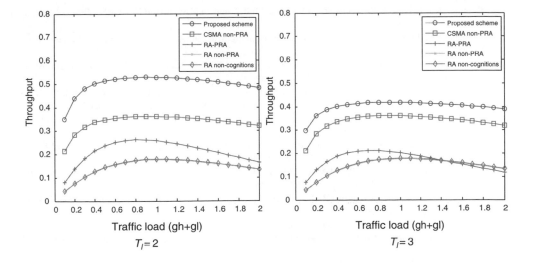

Figure 8.35   Throughputs of the CRN when the PS is present

Without location information to utilise the spatial domain radio resource, CSMA-ST can be adopted as the multiple access protocol for the CRN, with the corresponding cross-layer power-rate control mechanism to track, optimally and simultaneously, the throughput of the CRN and the overall throughput, subject to the pre-determined outage probability of the PS to constrain interference to PS. This lays the foundation for an important revelation about the multiple access protocol design of the CRN: carrier sensing and spatial-reuse simultaneous transmissions are essential to an effective multiple access protocol of the CRN. Numerical results indicate that, under the interference and the channel conditions that T_l adjusts to $2T_h$, the CRN with the CSMA-ST together with the cross-layer power control mechanism reaches the throughput at 0.53, while (i) the multiple access scheme without spatial reuse only has the throughput 0.36 and (ii) the multiple access schemes without carrier sensing or spatial reuse abilities only have throughputs 0.18 to 0.26. This cross-layer CSMA-ST enhances the throughput, although the hidden terminal problem in CRN and the stabilisation of the CSMA-ST remain two unresolved issues.

References

[1] M.A. Marsan, D. Roffinella, 'Multichannel Local Area Network Protocols', *IEEE J. Sel. Areas Commun.*, **1**,1983, 855–897.

[2] H. Okada, Y. Nomura, Y. Nakanishi, 'Multichannel CSMA/CD Method in Broadband-bus Local Area Networks', *Proc. Globecom*, 1984, 642–647.

[3] K.C. Chua, 'Performance Analysis of Multichannel CSMA/CD Network with Noisy Channels', *IEEE International Conference*, **3**, June 1991, pp. 1367–1371.

[4] A.R. Rhaghavan, C.W. Baum, 'Multichannel channel-access schemes for distributed direct-sequence networks', in *Proceedings of the IEEE Military Communications Conference* (MILCOM '98), 1998.

[5] Y.-C. Tseng, S.-L. Wu, C.-Y. Lin, J.-P. Sheu. 'A Multi-Channel Mac Protocol with Power Control for Multi-hop Mobile Ad Hoc Networks', in *Proceedings of 21st International Conference on Distributed Computing Systems Workshops*, April 2001, pp. 419–424.

[6] N. Jain, S.R. Das, A. Nasipuri, 'A Multichannel CSMA MAC Protocol with Receiver-based Channel Selection for Multihop Wireless Networks', *Proceeding of IEEE Vehicular Technology Conference*, 2001.

[7] S.-L. Wu, Y. Lin, Y.-C. Tseng, J.-P. Sheu, 'A New Multi-Channel MAC Protocol with On-Demand Channel Assignment for Mobile Ad Hoc Networks', *Proc. Int'l Symp. Parallel Architectures, Algorithms and Networks* (ISPAN '00), Dec. 2000, pp. 232.

[8] S.-L. Wu, C.-Y. Lin, Y.-C. Tseng, C.-Y. Lin, J.-P. Sheu, 'A Multi-Channel MAC Protocol with Power Control for Multi-Hop Mobile Ad Hoc Networks', *The Computer Journal*, **45**(1), 2002, 101–110.

[9] W.-C. Hung, K.L.E. Law, A. Leon-Garcia, 'A Dynamic Multi-Channel MAC for Ad Hoc LAN', *Proc. 21st Biennial Symp. Comm.*, June 2002, pp. 31–35.

[10] A. Tzamaloukas, J.J. Garcia-Luna-Aceves, 'Channel-Hopping Multiple Access', *Proc. IEEE Int'l Conf. Comm.* (ICC '00), June 2000.

[11] A. Tzamaloukas, J. Garcia-Luna-Aceves, 'Channel-Hopping Multiple Access with Packet Trains for Ad Hoc Networks', *Proc. IEEE Device Multimedia Comm.* (MoMuC '00), October 2000.

[12] J. Chen, S. Sheu, C. Yang, 'A New Multichannel Access Protocol for IEEE 802.11 Ad Hoc Wireless LANs', *Proc. 14th IEEE Int'l Symp. Personal, Indoor and Mobile Radio Comm.* (PIMRC '03), 3, September 2003, 2291–2296.

[13] P. Kyasanur, N.H. Vaidya, 'Routing and Link-layer Protocols for Multichannel Multi-Interface Ad-hoc Wireless Networks', *ACM MC2R*, **10**(1), 2006, 31–43.

[14] P. Bahl, R. Chandra, J. Dunagan, 'SSCH: Slotted Seeded Channel Hopping for Capacity Improvement in IEEE 802.11 Ad Hoc Wireless Networks', *Proc. ACM MobiCom*, September 2004

[15] H.W. So, J. Walrand, J. Mo, 'McMAC: A Multi-Channel MAC Proposal for Ad Hoc Wireless Networks,' *Proc. IEEE Wireless Comm. and Networking Conf.* (WCNC '07), March 2007.

[16] J. Mitola III, G.Q. Maguire, 'Cognitive Radio: Making Software Radios More Personal', *IEEE Personal Communications*, August 1999, 13–18.

[17] J. Mitola III, *Cognitive Radio Architecture*, John Wiley & Sons, Inc. New Jersey, 2006.

[18] K.C. Chen, 'Medium Access Control of Wireless LANs for Mobile Computing', *IEEE Network*, September/October 1994, pp. 50–63.

[19] Z. Tang, J. Garcia-Luna-Aceves, 'Hop Reservation Multiple Access (HRMA) for Multichannel Packet Radio Networks', *Proc. Seventh IEEE Int'l Conf. Computer Comm. and Networks.* (IC3N '98), October 1998.

[20] J. So, N. Vaidya, 'Multi-Channel MAC for Ad Hoc Networks: Handling Multi-Channel Hidden Terminals Using a Single Transceiver', *Proc. ACM MobiHoc*, May 2004.

[21] D. Bertsekas, R.G. Gallager, *Data Networks*, Prentice Hall, 1993.

[22] Y.K. Sun, K.C. Chen, D.C Twu, 'Generalized Tree Multiple Access Protocol for Wireless Communications', *Proceeding IEEE PIMRC*, Helsinki, 1997.

[23] K.C. Chen, C.H. Lee, 'Group Randomly Address Polling for Multicell Wireless Networks', *Proceeding IEEE International Conference on Communications*, 1994.

[24] S. Srinivasa, S.A. Jafar 'The Throughput Potential of Cognitive Radio: A Theoretical Perspective', *IEEE Communications Magazine*, 2007.

[25] J. Mitola III, G.Q. Maguire, 'Cognitive Radio: Making Software Radios More Personal', *IEEE Personal Communications*, August 1999, 13–18.

[26] J. Mo, H.W. So, J. Walrand, 'Comparison of Multichannel MAC Protocols', *IEEE Trans. on Mobile Computing*, 7(1), 2008, 50–65.

[27] S.Y. Lien, C.C. Tseng, K.C. Chen, 'Carrier Sensing Based Multiple Access Protocols for Cognitive Radio Network,' *IEEE International Conference on Communications (ICC)*, Beijing, 2008.

[28] A.J. Goldsmith, S.-G. Chua, 'Variable-Rate Variable-Power MQAM for Fading Channels', *IEEE Trans. Commun.*, 45(10), 1997, 1218–1230.

[29] A. J. Goldsmith, P. P. Varaiya, 'Capacity of Fading Channels with Channel Side Information', *IEEE Trans. Inform. Theory*, 43(6), 1997, 1986–1992.

[30] C.-C. Tseng, K.-C. Chen, 'Layerless Design of A Power-Efficient Clustering Algorithm for Wireless Ad Hoc Networks under Fading', *Journal of Wireless Personal Communications*, 44(1), 2008, 3–26.

[31] C.C. Tseng, S.Y. Lien, K.C. Chen, R. Prasad, 'On the Rate-distance Adaptability of Slotted Aloha,' *IEEE International Conference on Communications (ICC)*, Beijing, 2008.

[32] Q. Zhao, L. Tong, A. Awami, Y. Chen, 'Decentralized Cognitive MAC for Opportunistic Spectrum Access in Ad Hoc Networks: A POMDP Framework', *IEEE Journal on Selected Areas in Communications*, 25(3), April 2007.

[33] H. Kim, K.G. Shin, 'Efficient Discovery of Spectrum Opportunities with MAC-Layer Sensing in Cognitive Radio Networks', *IEEE Tr. on Mobile Computing*, 7(5), May 2008.

[34] L.C. Wang, A. Chen, 'Effects of Location Awareness on Concurrent Transmissions for Cognitive Ad Hoc Networks Overlaying Infrastructure-based Systems', to appear in *IEEE Trans. Mobile Computing*, 2008.

[35] L. Lai, H. El Gamal, H. Jiang, H.V. Poor, 'Optimal Medium Access Protocols for Cognitive Radio Networks', submitted for publication.

[36] R.G. Gallager, 'A Perspective on Multiaccess Channels', *IEEE Tr. on Information Theory*, 31(2), May 1985, pp. 124–142.

9

Network Layer Design

As cognitive radios (CRs) have successfully established the links for opportunistic transmissions, the core function of a cognitive radio network (CRN) lies in network layer design, especially routing, while many other design issues such as flow control, network radio resource management and network mobility management, are based around that routing. A CRN node can be considered to be a node with a dynamic spectrum access capability and programmable multi-radio capability, and in this chapter we construct its capability at the network layer. In other words, a CR node seeks and uses the spectrum hole in multi-radio systems to forward packets in a self-organised way. It is obvious therefore that the central issue among the network layer functions has to be routing in the CRN.

Prior to the routing of any CRN packets/traffic, the very first function of the CRN network layer is *association*, which means a cognitive radio node successfully accessing the general CRN (including the primary system (PS)). In principle, after sensing possible transmission opportunities (i.e., spectrum holes), a CR must complete association, and then execute dynamic spectrum access (DSA) through physical layer transmission and medium access control, to send packet(s) from CR transmitter to CR receiver. Here, the CR receiver can be a CR or a node in PS. In addition to regular association (or registration) to a network/system (typically the PS), the challenge would be quick association for a CR to another node in the CRN (either another CR or a node in the PS or multi-radio system) under a very short available time window, which we will discuss in Chapter 10. In this chapter, we focus on routing and issues related to routing and self-organisation.

9.1 Routing in Mobile Ad-hoc Networks

Before we get into routing of the CRN, we first review what has been done in routing of mobile ad-hoc networks and routing of wireless sensor networks, as per their similarity to the CRN. However, the CRN is generally a heterogeneous wireless network (with part wired network), which fundamentally differs from ad-hoc or sensor networks. Routing in such wireless heterogeneous networks is generally an unresolved problem. To develop a realistic mechanism for routing in the CRN, we will study routing in mobile ad-hoc networks (MANET) and then develop new algorithm fitting the CRN architecture and CRN features.

9.1.1 Routing in Mobile Ad-hoc Networks

We can consider mobile ad-hoc networks (MANET) as a sort of multi-hop packet radio network (mh-PRN). Routing of mh-PRN and therefore mobile ad-hoc networks has been studied for many years. MANET is considered to be a collection of mobile nodes communicating over wireless links without infrastructure. Due to potentially shorter-range radio communications than network coverage, MANET relies on multi-hop concepts to transport the packets and each node acts like a router by itself, with the common assumption of limited resource for routing.

Conventional routing protocols are based on either link-state or distance vector algorithms aimed at identifying optimal routes to every node in the MANET. Topological changes often encountered in MANET are reflected through propagation of periodic updates. To update and maintain the routing consumes tremendous bandwidth and is not practical. For IP-based MANET, routing protocols can be generally categorised as *proactive* or *reactive*, depending on whether the protocol continuously updates the routes or reacts on demand.

Proactive protocols, also known as *table-driven* protocols, continuously determine the network connectivity and already-available routes to forward a packet. This kind of routing protocol is obviously infeasible to frequently re-configurable mobile networks such as CRN, due to the extreme dynamics of links. Reactive protocols, also known as *on-demand* protocols, invoke determination of routes only when it is needed (i.e., on-demand). There are two well known reactive protocols: *dynamic source routing* (DSR) and *ad-hoc on demand distance vector* (AODV). When a route is needed, reactive protocols conduct some sort of global search such as flooding, at the price of delay to determine a route, but reflecting the most updated network topology (i.e., availability of links).

9.1.2 Features of Routing in CRN

The primary differences and challenges between routing of CRNs and routing of wireless ad-hoc (or sensor) networks can be summarised as follows:

- *Link availability:* CRN links are available under idle duration of the primary system(s) so that DSA can effectively fetch such opportunities, after successful spectrum sensing. Consequently, links in CRNs, especially those involving CRs as transmitters and/or receivers, are stochastically available in general, which allows the CRN topology to be random even when all nodes are static, not to mention the mobile nature of CRNs. Although wireless ad-hoc networks and sensor networks have similar phenomena, links in the CRN can vary much more rapidly because link available duration is a only fraction of the inter-arrival time for traffic and control signalling packets. That is, the link available period in the CRN is in the range of milli-seconds, instead of seconds, minutes, hours or even days, as in its wireless networking counterparts.
- *Unidirectional links:* Typical wireless networks have bi-directional links, because radio communication is half-duplex. In typical wireless ad-hoc and sensor networks, unidirectional links might be possible due to the asymmetric transmission power and/or different interference levels at receivers. We may treat unidirectional links as rare in wireless networking. However, in CRNs, unidirectional links are more likely to be due to the fact that a CR node may just have an opportunity to transmit in one time duration and there is no guarantee to allow the opportunity for transmission from the other direction. Another possible situation is that a CR node wants to leverage an existing PS to (cooperatively) relay packets, but the other direction might not be permitted, and *vice versa*. Generally speaking, a link involving a CR node is likely to be uni-directional. This distinguishes CRNs from other wireless networks, especially regarding the network layer functions.
- *Heterogeneous wireless networks:* In contrast to typical wireless ad-hoc or sensor networks, CRNs are generally formed by heterogeneous wireless networks (co-existing primary systems and CR nodes to form ad-hoc networks). Inter-system handover is usually required for routing in such heterogeneous wireless networks. However, CR links might be available for an extremely short duration and the successful networking lies in cooperative relaying among such heterogeneous wireless networks. From the perspective of network security, the enabling of CRNs for spectrum efficiency in wireless networks at the price of losing security is debatable, because there is not enough time for a CR node to

get a secure certificate within the short opportunistic window. A compromise to operate among heterogeneous wireless networks and CR nodes for CRN routing is obviously required.

To ensure a CRN link is available for network layer functioning, we may go back to hardware operation. Assuming that a genie observes CRN operation for both the PS and CR, the CR must use the spectrum hole window to complete transmission of packet(s). Suppose such a spectrum window period is denoted by T_{window}. It is clear that

$$T_{window} \geq T_{sense} + T_{CR - Transmission} + T_{ramp - up} + T_{ramp - down} \tag{9.1}$$

where T_{sense} stands for minimum sensing duration to ensure CR transmission opportunity and acquisition of related communication parameters, $T_{CR - Transmission}$ is the transmission period for CR packets, and $T_{ramp - up/down}$ means the ramping (up or down) period for transmission. This equation ignores propagation delay and processing delay at the transmitter-receiver pair, which can be considered to be a portion of ramp-up/down duration. The maximum duration of spectrum hole (availability) can be considered to be the time duration for beacon signals.

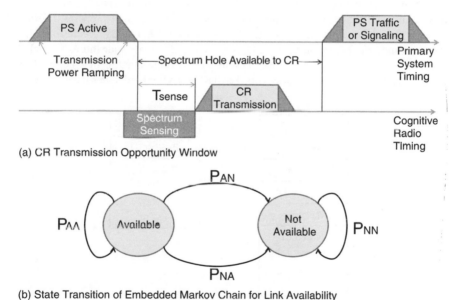

(a) CR Transmission Opportunity Window

(b) State Transition of Embedded Markov Chain for Link Availability

Figure 9.1 CR link availability model: (a) CR transmission opportunity window; (b) state transmission of embedded Markov chain for link availability

It is obvious that we have to model the link availability mathematically in the CRN. Since the link is either available for opportunistic transmission(s) or not available, considering the timings for the change of link availability, we can adopt an embedded continuous-time Markov chain and the rates specifying this continuous-time Markov chain can be obtained from the statistics of spectrum measurement. Figure 9.1(b) illustrates such a 2-state Markov chain with fixed timing (say, the PS's beacon signal time separation), where A stands for 'link available' and N stands for 'link not available'. A link between node X and node Y in the CRN can define two unidirectional links X → Y and Y → X. Using the simple 2-state embedded Markov chain model allows general study of the nature of network layer functions for the CRN, and thus effective design, under the challenges of link availability and unidirectional links.

9.1.3 Dynamic Source Routing in MANET

DSR is an *on-demand* ad-hoc routing protocol that uses a route discovery cycle to search routes. DSR uses source routing and each node maintains a complete topology of routes in its route cache, as shown in Figure 9.2.

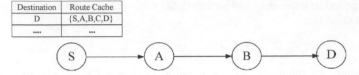

Destination	Route Cache
D	{S,A,B,C,D}
....	...

Figure 9.2   Route caching of source routing schemes

9.1.3.1 Basic Route Discovery

When a source node needs to transmit data packets to a certain destination and has no route in its *route cache*, it will initiate a route discovery procedure. The source node first prepares a *Route Request* (RREQ) packet, which contains the *Source ID*, the *Destination ID*, a unique *Request ID*, the value of *Maximum Hops* and a *Node List* listing the address of each intermediate node through which the RREQ has been forwarded, as shown in Figure 9.3(a). This figure shows the situation in which S has no route to D and thus floods RREQ throughout the network. Each RREQ records all nodes it has passed through in the Node List.

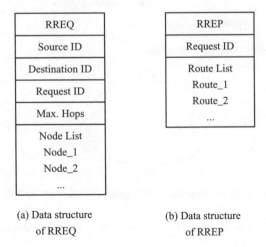

(a) Data structure of RREQ

(b) Data structure of RREP

Figure 9.3   Conceptual data structure of (a) RREQ and (b) RREP of DSR

When a node receives the RREQ, it first checks if it is the destination. If this is not the case, in order to prevent route loops and save bandwidth, it will check the Request ID and the Source ID pair to see if it is receiving a duplicated RREQ. Since the source node increments the Request ID for each new route discovery, a node can distinguish route discoveries from the Source ID and the Request ID of a RREQ. This node can also avoid route loops by simply checking if it is in the Node List. If the receiving RREQ is duplicated or itself is in the Node List, the RREQ will be dropped. For example, as shown in Figure 9.4(a), F has received a RREQ from E (Route 3) and thus drops the RREQ received from node C (Route 2). Finally, if the RREQ is neither duplicated nor forming loops, the node adds its address in the Node List and rebroadcasts the RREQ.

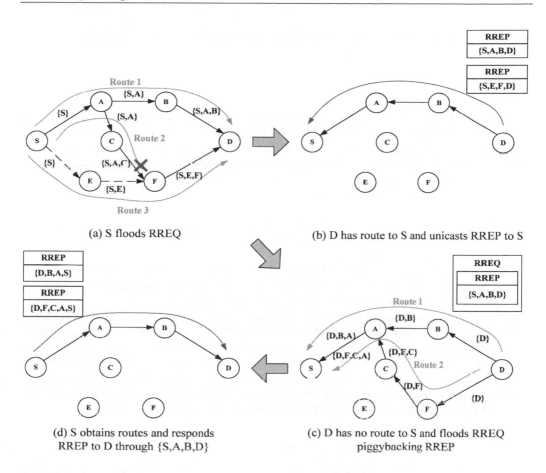

Figure 9.4   Basic DSR route discovery mechanism: (a) S floods RREQ; (b) D has route to S and unicasts RREP to S; (c) D has no route back to S and floods RREQ piggybacking RREPS (d) obtains a route and responds RREP to D through {S,A,B,D}

The task of the destination is simply to tell the source node which route is valid for routing. Thus, it prepares a *Route Reply* (RREP) message, which contains the accumulated route record in the RREQ just received, as shown in Figure 9.4(b). Then it checks its Route Cache to see if it has any route to the source. If so, it unicasts the RREP back to the source, as shown in Figure 9.4(b). If the destination does not have any route in its cache, it will initiate its own route discovery and flood RREQ piggybacking with the RREP to find routes to the source, as shown in Figure 9.4(c). The piggybacked RREP is to prevent an infinite route discovery loop. When the source node receives such a RREQ, it knows that routes to the destination found can reply a RREP to the destination based on these discovered routes, as shown in Figure 9.4(d). DSR assumes that links between nodes can be unidirectional or bidirectional, thus a route only means that unidirectional links towards the destination exist. The destination cannot directly cache the reverse route of the RREQ. Thus the destination cannot reply a RREP by using the reverse route of a RREP directly unless the route is required to be bidirectional. This is why a destination may need to perform a second route discovery after the source. If the lower layers require bidirectional communication, the feedback of RREP *must* follow the reverse order of the route of RREQ. This constraint is used to detect unidirectional links on the discovered route. If any link on the route is unidirectional, the transmission of a RREP might fail and such route would not be discovered by the source. If a route discovery fails, the source node will backoff for a period of time and retry again if the trying times do not exceed the maximum number of route discoveries for the same destination. Otherwise, it informs the upper layers with no route indication.

9.1.3.2 Basic Route Maintenance

The main task of route maintenance is to detect topology changes (e.g., link breaks) and to inform the source node about these changes. DSR detects link breaks only when sending data packets to the next hop node. Each node is responsible for confirming the successful reception of the packets. This can eliminate overheads of neighbour detection packets (e.g., periodic Hello message exchanges) since only when a node needs to forward a data packet through a link does it check the link status. DSR use three kinds of broken link detection mechanism:

- *Link layer acknowledgement:* If the link layer provides feedback for successful reception of data packets, DSR can use such a mechanism to confirm the link availability. This mechanism requires the link between two nodes to be bidirectional.
- *Passive acknowledgement:* If the link layer acknowledgement is not available, a node can passively listen to the next hop's transmission and confirm the successful transmission if the next hop node transmits that packet. This mechanism also requires bidirectional links.
- *Network layer acknowledgement:* If no mechanism is available for determining the reachability of the next hop node, DSR inserts an Acknowledgement Request to the header of the data packet before sending it. The next-hop node receiving such a request prepares an Acknowledgement and uses available routes in its cache to send it back to the node that issued the request. Since this acknowledgement can be routed through a multi-hop path, this mechanism can be used in unidirectional networks. Note that nodes that receive such acknowledgement need not resend another acknowledgement to the node that originally sent the acknowledgement.

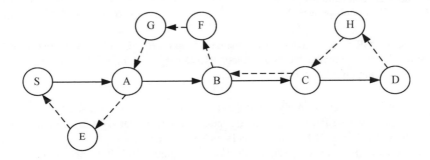

Figure 9.5   An example of reverse routes of unidirectional links

Figure 9.5 shows an example of reverse routes. Links (S, A), (A, B) and (C, D) are unidirectional links and link (B, C) is a bidirectional link. We can see that each unidirectional link has a reverse route for link maintenance. A downstream node of a unidirectional link can feedback an acknowledgement through the reverse route to the upstream node.

If a sender of a data packet detects a link break, which may be due to no reverse route or out of transmission range, it will return a Route Error (RERR) message back to the source. The RRER contains the broken link. The source node will eliminate all routes containing that broken link in its Route Cache. It will not rediscover another new route unless its upper layer has demands and it has no route. Figure 9.6 shows examples of these two reasons.

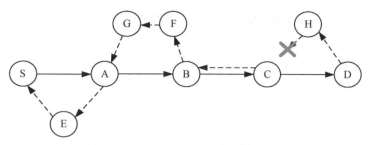

(a) Link (C,D) fails due to no reverse route

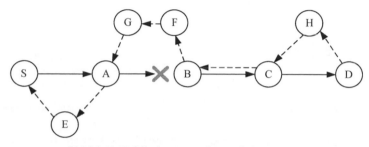

(b) Link (A,B) fails due to out of transmission range

Figure 9.6   Two reasons for link breaks: (a) link (C, D) fails due to no reverse route; (b) link (A, B) fails due to out of transmission range

9.1.3.3 Properties of DSR

We now summarise a few properties regarding DSR, which might be useful in developing routing in CRN.

Property 9.1: In DSR, the RREP is not necessarily forwarded through the reverse order of the path that the RREQ has passed through.

Proof: Since DSR uses source routing, every node that receives a packet only needs to follow the source routes appended to the packet and forward it. Nodes need not maintain a routing table if they just relay packets. After the RREQ reaches the destination, it records all intermediate nodes that it has passed through. The destination only needs to inform the source of the received route by sending the RREP. The path the RREP uses may not be the same as the reverse order of the path of the RREQ.

Property 9.2: In DSR, if each intermediate node only re-broadcasts the first received RREQ of the RREQs that belong to the same route discovery, no route loop can be formed.

Proof: Any RREQ is identified by its Source ID and Request ID. If a node receives two or more RREQs with the same Source ID and Request ID, it can ensure that the following RREQs are duplicated. To form a route loop there must be a certain RREQ that has passed through a node more than once. Since the node can distinguish duplicated RREQs and only re-broadcast the first one received, no route loop will be formed.

Property 9.3: If no duplicated RREQs can be re-broadcast, two routes are *node-disjoint* if they have different first hop nodes.

Proof: Suppose there are two RREQs that have different first hops received by a node. Routes 2 and 3 in Figure 9.7(a) show an example of this situation. Suppose the first intersection of these two routes is that node (e.g., node F). Since only one RREQ can be broadcast, the other received RREQ will be dropped and the corresponding route cannot continue propagating towards the destination. Thus no intersection is allowed between two routes and there can be no common nodes.

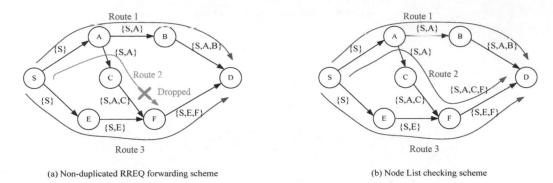

(a) Non-duplicated RREQ forwarding scheme   (b) Node List checking scheme

Figure 9.7   Different route discovery schemes find different kinds of routes: (a) non-duplicated RREQ forwarding scheme; (b) Node List checking scheme

Property 9.4: If no duplicated RREQs can be re-broadcast, two routes can only separate once and have no intersection.

Proof: If two routes have common first hops, it means that they must separate at a certain node, otherwise they are the same route. If we treat the sub-paths from the separating nodes as two paths to the destination with different first hops, they can never intersect again according to Property 9.3. Thus two routes that have separated once can never intersect again.

From Properties 9.3 and 9.4, we know that routes in DSR cannot intersect with each other, which means the source has less chance to discover non-disjoint paths. As shown in Figure 9.7(a), a non-duplicated RREQ forwarding eliminates route 2. If the destination only replies those RREQs that have different first hop nodes, DSR can only discover node-disjoint routes. In DSR, non-duplicated RREQs, re-broadcasting and Node List checking can prevent route loops. However, if we release the constraint of duplicated RREQs re-broadcasting and use Node List checking to avoid loops, a source node may be able to find more non-disjoint routes to a destination, as shown in Figure 9.7(b).

Finding non-disjoint routes can provide more options for the source and intermediate nodes to forward data packets, and thus may increase the packet delivery ratio. For example, in Figure 9.8(b), each intermediate node has more options to forward packets to destination D.

Property 9.5: If re-broadcasting duplicated RREQs is allowed but each node checks the Node List and drop RREQs if they are in the list, no loops can be formed and any route that is within Maximum Hops (maximum *time-to-live* (TTL)) can possibly be discovered.

Proof: Since a RREQ that has passed through a node that it has visited before has that node in its node list, by checking the list the node can easily find a loop and discard that packet. Without the constraint of

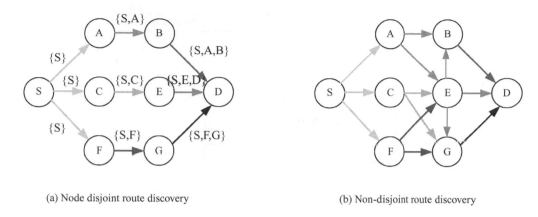

(a) Node disjoint route discovery   (b) Non-disjoint route discovery

Figure 9.8 Benefit of non-disjoint routes for packet forwarding

non-duplicated RREQ re-broadcasting, a route with no loops can reach the destination if and only if it has a smaller length than Maximum Hops. This is because every node only checks the node list and the TTL field to see if there are loops and long path lengths. Since no loop can be found, none exceeding the TTL, all routes can be found.

Property 9.6: If using the Node List scheme, except for those nodes before the first separation of paths, a node that is passed by k different paths needs to re-broadcast duplicated RREQs k times.

Proof: Since these k different paths do not form loops and are with the maximum TTL, their corresponding RREQ must be able to reach the destination. Every time one of the k RREQs passes the node, it has to re-broadcast once. Thus totally it has k broadcasts.

The above two RREQ forwarding schemes are two extreme methods through which non-duplicated RREQ forwarding can discover the least routes and the Node List checking scheme can find all routes that have lengths less than the value of Maximum Hops. A method to control the number of discovered routes it to limit the number of RREQ re-broadcasting of each node.

9.1.4 Ad-hoc On-demand Distance Vector (AODV)

AODV is an on-demand ad-hoc routing protocol and routes are discovered by a route discovery cycle. AODV uses *hop-by-hop* routing, which means routing information is distributed over network nodes. Each node only maintains the next-hop node in its routing table, as shown in Figure 9.9.

9.1.4.1 Basic Route Discovery

When a source node has data packets to send to a certain destination but does not have a route in its routing table, it initiates a *route discovery procedure*. The source node creates a *Route Request* (RREQ) packet which contains its own IP address and sequence number, the destination's IP address and sequence number, and a hop count that is initialised to zero.

When a node receives the RREQ, it increments the hop count value by one and creates a *reverse route entry* in its routing table. The reverse route entry contains the source node and the node from which it received the RREQ. This entry can be used when replying a *Route Reply* (RREP) to the node on the reverse route to the source. After this entry is created, the node checks whether it has a route to

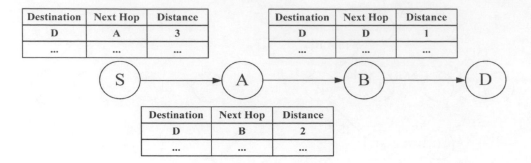

Figure 9.9   Routing table of hop-by-hop routing scheme

the destination or it is the destination. In both cases, it returns a RREP with the destination's IP address and its latest sequence number, and the source's IP address and its sequence number. The RREP also contains the distance (hops) from the replying node to the destination. Finally it unicasts the RREP according to the reverse route entry stored before.

When a node receives the RREP, it increments the hop count value by one and creates a forward route entry which contains the destination node and the node from which it received the reply. This entry can be used to route packets to the destination. Then the node unicasts the RREP to the next node towards the source node according to its reverse route entry. If the source receives the RREP, it can start to send data packets to the node from which it received the RREP. If the source receives multiple RREPs, it would select the route that has the minimum distance (hops) to the destination.

When a node unicasts a RREP, it will expect to receive a Route Reply Acknowledgement (RREP-ACK) from the node to which it sent RREP. This operation is used to prevent using unidirectional links on discovered routes. If no RREP-ACK is replied, that node will be added into a *blacklist* until a predefined timeout value. A node drops a RREQ received from another node that is in the blacklist. This prevents utilisation of unidirectional links in the network.

9.1.4.2 Basic Route Maintenance

The route maintenance mechanism is used to repair *active* routes when they break. An active route is defined to be a route that is currently used for data forwarding. When a node detects that a link break occurs, the upstream node of the link invalidates all routes to those destinations that become unreachable due to the link break. It then creates a *Route Error* (RERR) message which contains each of those lost destinations. It then sends the RERR to all previous hops (so-called *precursors*) that were using the link to destinations. When a node receives the RERR, it checks whether it really has routes to those destinations in the RERR. If so, it eliminates all routes to these destinations and resends this RERR to all of its precursors. Finally when the source receives the RERR, it also invalidates all routes in the RERR. If there is a need for these destinations, it initiates another route discovery procedure.

9.1.4.3 Properties of AODV

Let us summarise some properties of the ad-hoc on-demand distance vector (AODV), which might be useful for routing in CRN (see Figure 9.10).

Property 9.7: In AODV, the RREP *must* be forward through the reverse order of the path that the RREQ has passed through.

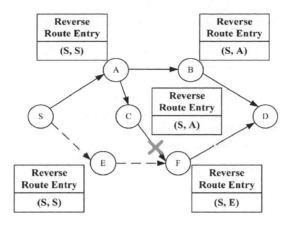

(a) S floods **RREQ** and each node establish the reverse route entry

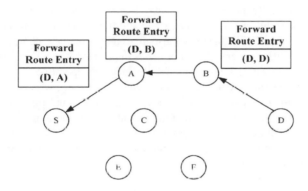

(b) D unicasts **RREP** through the reverse route of the first
received RREQ to S

Figure 9.10 Basic AODV route discovery mechanism: (a) S floods RREQ and each n ode establishes the reverse route entry; (b) D unicasts RREP through the reverse route of the first received RREQ to S

Proof: In AODV, after the RREQ reaches the destination, each intermediate node only knows its upstream node and the corresponding incoming link; so does the destination. Moreover, 'only' the downstream node knows the existence of the incoming link. However, in order to establish the routing table, each intermediate node needs to know the next hop towards the destination. Thus, it is essential for the downstream node to inform the upstream node that it can route packets to the destination. Before a downstream node can inform its upstream node, it must first know if it can route packets to the destination, which requires its downstream node to inform it. By repeating this argument, the destination node must inform its upstream node first by the RREP. Thus the RREP must follow the reverse path through which the RREQ has passed.

Property 9.8: The basic AODV can only discover node-disjoint routes.

Proof: Since AODV requires the RREP to be forwarded through the reverse route of the RREQ, and each intermediate only remembers one upstream node and one downstream node, there can be no two routes to intersect with each other. Thus, every route discovered is node-disjoint with other routes.

Property 9.9: AODV can only work in bidirectional networks if no bidirectional link abstraction or unidirectional link elimination mechanisms are performed.

Proof: Since the feedback of RREP must be in the reverse path of RREQ, every link on the discovered route must be bidirectional, or else the RREP cannot reach the source and the route discovery can fail. Thus, AODV requires links need to be bidirectional.

9.2 Routing in Cognitive Radio Networks

After identifying 3 major challenges in CRN routing, we are trying to alleviate the third challenge, then deals with the first two to design the routing algorithm.

9.2.1 Trusted Cognitive Radio Networking

In order to tackle mathematically the challenge of *heterogeneous wireless networks* described in the previous section, we can introduce a *trust* mechanism in addition to typical network security schemes. Please note the interesting observation that the security in CRN lies on the ground of end-to-end nodes, and intermediate nodes in CRN (either CRs and/or nodes in primary systems) can simply forward the CR traffic packets (i.e., cooperative relay insideCRN). Such a cooperative relay of packets can be facilitated as *amplify-and-forward* (AF) and *decode-and-forward* (DF), while intermediate nodes in CRN have limited security treats by relaying packets. Compress-and-forward (CF) cooperative networking might jeopardise the security of the intermediate nodes due to mixing relay packets and own traffic together. In the following, the cooperative relay suggests either AF or DF, but not CF.

We can now classify a node in CRN and thus traffic/control packets from such a node into three categories during the operation of CRN:

- *Secure:* The node has executed a security check that is good throughout the entire heterogeneous wireless networks, such as through a public key infrastructure (PKI) check. The packets and messages from this node can go all the way in CRN as secure clearance. A node classified as 'secure' can be a CR and a node in a co-existing PS.
- *Trusted:* The level of security for 'trusted' is not as effective as 'secure'. As a CR is generally not able to complete a security check of several rounds of the handshaking protocol within the timing window of an available link (i.e., CR to CR, or CR to PS node), we create a security level of *trusted* that enables packet transmission over available opportunistic unidirectional links. In this case a CR source node generates packet(s) for opportunistic transmission, and the CR receiver node (either a CR or a node in the PS) recognises such a CR source node as 'trusted' and can relay packets towards a CR sink node via the appropriate routing mechanism. Please note that the CR source node and CR sink node will complete their end-to-end security check in advance by all means. A CR receiving node should always maintain a table of trusted/secure nodes, based on security checks and historical update. In other words, any node in the CR only allows reception of packets from its secure and trusted neighbouring node. Such a table is localised and is not large in the number of neighbouring nodes. The methodology of updated trusted-node tables is described in Chapter 10.
- *Lure:* A CR node is neither secure nor trusted by its target receiving node, and it is classified as 'lure'. The major reason to be rated as lure shall be from bad historical actions, such as spreading virus, wasting bandwidth in a wireless network, attacking wireless network, selfish behaviors, etc. Such a lure node actually loses its cognitive radio capability in practice of CRN operation.

The purpose for the introduction of a trust mechanism is clear: that is, to create a homogeneous networking functioning environment for heterogeneous wireless networks, and thus to allow the cooperative relay of packets in spite of the opportunistic and extremely dynamic link availability of CRN. In other words, we will encourage nodes from all kinds of wireless networks to act as nodes in the CRN by providing some incentive programmes (discussed in Chapter 11), so that these nodes can effectively relay packets from trusted CR source nodes, to form a large scale of *homogeneous multi-hop ad-hoc network* under the same *trust* level across different wireless networks.

We summarise some critical issues of the CRN network layer operation in the following list:

- The CRN consists of CRs and nodes from various co-existing primary systems, which may operate using different communication parameters, in different frequency bands and in different geographical locations. SDR inside a CR is capable of reconfigurable realisation for multiple systems operating at multiple frequency bands.
- The CR source node (initiation point of traffic) and CR destination (termination point of traffic) node should conduct their own end-to-end security beyond trust level by employing CRN nodes to complete bidirectional verification.
- CRN nodes are assumed to conduct only AF or DF cooperative relaying, under the trust domain of CRN.
- Nodes in the secure domain may reject relays from trusted nodes, which suggests that such links are not available in trusted multi-hop packet radio network routing. Similarly, nodes in trusted domain (i.e., typical nodes in CRN) may reject connection requests from lure nodes.
- Any packet from a CR source node, once getting into a primary system or infrastructure, follows the operation of the primary system or infrastructure, to enjoy the benefits from existing systems and networks. Assume, for example, that a CR source node wishes to relay its packets through *nearby* WiFi to access a web site of the Internet, where nearby means radio accessibility as a kind of localisation. As long as the packets from the CR source node are allowed to the access point of the WiFi, these packets transport as WiFi packets. A CR terminal device is therefore capable of conversion/re-configurability among multiple physical layer transmissions and multiple medium access control schemes.

The general CRN operation can therefore be summarised as in Figure 9.11. We have an infrastructure network as the core that might be just the Internet, and several radio access networks (RAN) that provide various ways to access the core infrastructure network. Mobile stations (MS) are associated with certain RAN technology. Each cognitive radio (CR) is capable of configuring itself into the appropriate radio system to transport packets for communication/networking purpose. RANs, MSs and the infrastructure can be any specific primary system, and there are a few possible primary systems co-existing in the figure. A CR may also be a MS of a PS. Bidirectional links have double arrows, and all links in primary systems will be bidirectional. Opportunistic links owing to CR's dynamic spectrum access and certain ad-hoc links have single arrows in the figure. From the CR source node, there are three different cooperative paths to transport the packets. There are also three cooperative paths to the CR sink as the final destination. Please note that outgoing path 3 and incoming path 3 generally represent cognitive radio relay networks (CRRN) described earlier.

As we can clearly observe from Figure 9.11, the CRN consists of CRs and PSs. CR dynamic spectrum access (DSA) at physical layer transmission and medium access control works between CR transmission and CR receiver in a CR (or DSA) link. CRN routing establishes on top of these CR links and bidirectional links in the primary systems. Let us summarise this again:

- The *CR transmitter* and *CR receiver* form a *CR link*, typically using dynamic spectrum access. The CR receiver may be a CR or a node in PS.

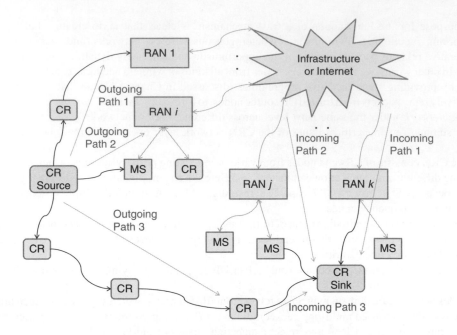

Figure 9.11   Routing packets in the CRN

■ The *CR source node* and *CR destination node* form a virtual link like a session in CRN. The CR destination node can be a CR or any node in PS. When the CR destination node is a CR, we may call it a *CR sink*.

9.2.2 *Routing of Dynamic and Unidirectional CR Links in CRN*

To conduct CRN routing over unidirectional CR links and usually bidirectional links in the PS (mobile stations in the PS can form ad-hoc with possible unidirectional links) as per earlier descriptions, we can extend on-demand routing protocols of MANET for CRN routing by the following:

• Each CR link is modelled by a 2-state Markov chain, independent of other CR links.
• Without knowing the specific PS, all links in the PS are assumed to be bidirectional and can support our routing protocol. As a matter of fact, the entire behaviour inside a PS can be treated as a 'link' by the queuing model of this PS if we just follow the PS operation.
• Typical MANET routing algorithms are trying to isolate unidirectional links, because they are likely to be very localised. However, unidirectional links are inevitable in CRN. Fortunately, we may assume the depth (i.e., number of hops) from CR to PS to be within Δ hops, due to their roles in wireless access to the infrastructure or purely *ad-hoc*.
• The fact of CR links being unidirectional is usually true at one instant. At the next instant, this CR link might be still unidirectional but reverse its direction depending on network situations. By introducing the trust mechanism, the CRN would function pretty much like an ad-hoc network with 'temporary' unidirectional links.

For routing in the CRN, we value one major purpose of the CRN: to reduce the latency of traffic due to more cooperative paths, especially as CR sources are not able to transport packets to the CR destination node without CRN technology. In the mean time, there are a few issues of which we want to make sure in CRN routing:

- Since CR does not interfere with the PS(s), we should avoid the global or periodic advertisement of any CR node, though such advertisement is common in ad-hoc network routing.
- For a CR link that is the link with the CR as transmitter, we will avoid acknowledgement over the link, because there might not be enough opportunistic time window to execute this acknowledgement.
- For the same reasons as the above two points, we will not use *hello* packets in common ad-hoc network routing.
- CRN routing must be able to detect and to minimise the possibility of any *loop* or any *dead-end*, where dead-end means 'no way to forward the packet further within a reasonable amount of time', and loop means 'the packet that was forwarded to another route will come back in a repeated way'.

We assume localised connectivity to be of concern in CRN routing, which is pretty much true for CRN operation and routing because the CR links are only opportunistic. Under a highly dynamic nature, trying large-scale or global optimisation is likely to be in vain. Our strategy is to forward the packet over an effective opportunistic CR link, towards the appropriate direction/trend. This exactly matches the philosophy of reactive (or on-demand) routing in ad-hoc networks.

Consequently, taking the spirit of AODV routing, we create CRN on-demand (CRNO) routing as follows. Each CR node executes routing only when there is a need (on-demand). The routing message includes the following routing overhead information:

- CR destination node IP;
- CR source node IP;
- Message ID (i.e., msg_id);
- CR relay node IP (cr_relay_ip);
- CR transmitter IP (cr_tx_ip) and its radio-type (cr_tx_type) for the received packet/frame;
- CR receiver IP (cr_rx_ip) and its radio-type (cr_rx_type) for the forwarding packet/frame;
- Sequence number seq_count associated with the path (cr_tx_ip, cr_relay_ip, cr_rx_ip), starting from 0, and adding 1 for each same path;
- Time counter at CR relay node, time_counter, starting from 0 and adding 1 for a new time slot duration.

When a new CR node or a new mobile station of the PS comes into the scenario, we may not be able immediately to acquire its IP address, and so we can therefore use an ID to serve the purpose of the table.

CRNO routing consists of three phases in operation: sensing phase, path discovery phase and table update phase (see Figure 9.12).

9.2.2.1 Sensing Phase

The CR node listens to the radio environments, that is, spectrum sensing of multiple co-existing systems (and even possibly different frequency bands), to update its *forward-path table*. The forward-path table records information regarding each potential CR receiver, history, estimate of its trust on the CR node and communication parameters to adjust the SDR. Each potential CR receiver is identified by an IP address that can be acquired from its past transmissions or by an ID designated by the CR node. History can be a simple flag to indicate whether the potential CR receiver is trustworthy or not, based on history and learning processes. Finally, communication system parameters can be obtained from spectrum sensing to adjust the SDR.

9.2.2.2 Path Discovery Phase

Once the CR node originates a packet/frame to the destination or receives a packet/frame for relay, it checks the backward-path table for any violation. In the case of no violation from the checking, the CR node selects another CR node from the forward-path table to relay the packet/frame. The selection is based on the availability of CR links and the forward-path table. Of course, links to the PS have the highest priority. On the other hand, when a violation happens, the CR relay node seeks an opportunity to 'negative-acknowledge' the CR transmitter based on the backward-path table. The CR transmitter node shall try to re-route the packet to another CR relay node if possible, or further back if no route available.

9.2.2.3 Table Update Phase

In addition to link selection to complete the routing, a *backward-path route* associated with this relay has to update as a part of the *backward-path table*. Each backward-path route consists of parameters msg_id, cr_rx_ip, cr_rx_type, cr_tx_ip, cr_tx_type and seq_count.

Both cr_rx_type and cr_tx_type are to specify the operation of co-existing multi-radio systems (or overlay wireless systems/networks) in the CRN.

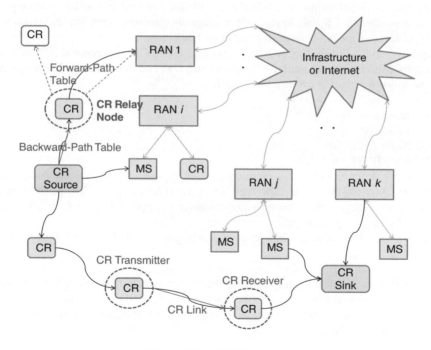

Figure 9.12 CRNO routing

It is obvious that the parameter *history* in the forward-path table plays a key role in routing.

The backward-path table prohibits routing disasters from loops and dead-ends. The violation is defined as the detection of either loop existence or dead-end existence. Seq_count plays its role in determining the existence of a loop. Time-out to indicate that it is not possible to relay a packet is issued to avoid dead-end, which is useful information to update the backward-path table.

When negative-acknowledgement (or positive-acknowledgement from destination) cannot trace back all the way to the CR source node, it is likely to be due to some permanent unidirectional links; end-to-end timeout can terminate the routing and re-start a new round of routing, which should avoid the earlier permanent unidirectional link that it can identify and isolate.

9.3 Control of CRN

9.3.1 Flow Control of CRN

Flow control can happen in two types in the CRN. End-to-end flow control between the CR source node and CR destination node is possible, in which a typical credit-based flow control such as a leaky-bucket does the work. However, for completely successful operation of CRN on-demand routing protocols, such as CRN-ODV or CRN-DSR, we need another function: flow control in the CRN network layer. In contrast to conventional first-type flow control in computer networks, flow control in the CRN is primarily for damage control purposes. Since it is not possible for us to ensure that neither dead-end nor loop happens in AODV, we have to detect these two cases and stop the CR link relaying packets under these scenarios, so that network bandwidth is not wasted. To achieve such a goal, *loop detection* and *dead-end detection* are needed and associated with routing.

Furthermore, we can observe that the entire CRN of CRs and PSs as shown in Figure 9.11 is actually formed by several segments as per Figure 9.13, and the packets are routed from the CR source node to the CR destination node (or CR sink), through the following segments:

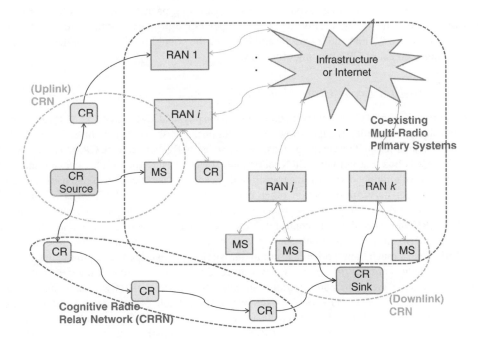

Figure 9.13 Segmentation or decomposition of CRN

- Uplink CRN;
- Co-existing Multi-Radio Primary Systems usually with infrastructure or core network (such as the Internet), which may be considered as a sort of *tunnelling* in CRN to transport the packets quickly.
- Downlink CRN;
- Cognitive Radio Relay Network (CRRN, described in Chapter 6).

CRRN can be considered as a special kind of CRN consisting of pure CRs, with the only purpose of relaying packets.

The traffic flow can be categorised as follows:

- CR source node → Uplink CRN → PS and infrastructure → downlink CRN → CR destination node;
- CR source node → CRRN → CR destination node.

Routing in CRN thus has another hidden agenda based on segmentation or decomposition. For the uplink CRN, the routing will try to reach the PS via opportunistic CR links. For example, in Figure 9.13, when the CR relay node is in the process of selecting a forwarding path, it has a tendency to select the node 'closer' to the PS, which is the node in RAN 1. The routing will, however, try to leave the PS via opportunistic CR links for the downlink CRN. When a CR node in the path discovery phase is based on the forward-path table, the parameter (or more precisely field) *history* plays a key role in providing such information in node selection. In other words, routing in the uplink CRN and downlink CRN is not totally stochastic, and there will be a drift along the direction inside a dynamic topology CRN. It reminds us of the movement of ants, and literature [7, 8] about ant routing provides more opportunities to develop effective update of the parameter/field history in the table.

We also note that CRN routing favours a way of forwarding packets that is effective for overlay/co-existing multi-radio systems. This suggests a longer range primary system is to be favoured in relaying packets for a CR relay node (as long as it is among the possible options) and is therefore another potential enhancement of CRN routing efficiency.

9.3.2 End-to-End Error Control in CRN

The conventional concept of packet error control lies in the physical layer and data link layer. However, error control will be useful in supporting CRN functions. Please recall that links in the CRN are dynamically available and it might not be feasible to conduct ARQ between the CR transmitter and CR receiver in a CRN link. Furthermore, CRN routing just tries its best to forward the packets and the CR sink might receive multiple copies of one transmitted packet, and these copies of one packet might not be correct because no error protection other than forward error control (FEC) is available. The conventional network layer requires an extremely low packet error rate, which is warranted by physical layer FEC, CRC check and data link control. For the CRN, data link control may or may not exist, and error control between the CR source node and sink node is needed, while re-transmissions will be minimised due to a much higher price than common (wireless) networks. We can immediately borrow the idea from *hybrid automatic request* (HARQ) to conduct the CRN network layer error control, to significantly reduce the error control traffic significantly between the CR source node and CR destination node. As Figure 9.11 shows, for the purpose of reliable packet transportation in wireless networks, the CR destination node may receive three (or more) copies of a packet from the CR source node, which suggests the application of HARQ to create more path diversity and to enhance error control capability.

The challenge for HARQ in the CRN lies in the uncertain number of copies of a packet to be received at the CR destination node and in the uncertain arrival times of these copies.

9.3.3 Numerical Examples

General CRN routing is an extremely complicated mechanism. However, we can design experiments to verify our proposed routing. The first experiment is to demonstrate the feasibility of the CRN, as the generalisation of a cooperative relay among CRs and nodes in the PS, being capable of forwarding packets from the CR-source node to the CR-destination node.

The objective of this simulation is to compute routing delay, when the routing path is establishing based on the available channel. The routing delay is defined as delay caused by routing through these

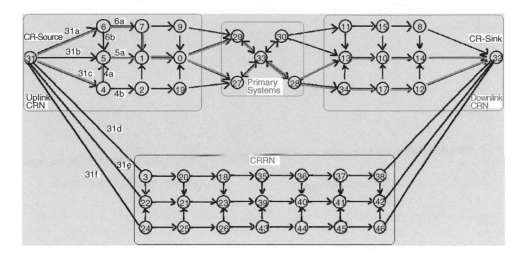

Figure 9.14   Topology of CRN in simulations

dynamically opportunistic links, without considering other factors such as transmission delay, processing delay, etc. Our simulation follows the topology shown in Figure 9.14, with the following assumptions:

- There is one CR source and one CR destination. Link direction is as shown in Figure 9.14. Arrows shows the direction of the link. Although there are unidirectional links in the scenario, in this stage of the simulation, they do not have a special effect or function.
- Every node has a routing table (forward path table) to summarise potentially available links with receiving nodes. We assume spectrum sensing capability.
- So far there is no backward path table, because no acknowledgement is sent by the receiver.
- When data starts to transmit, the node checks the available channels in order. At this stage of the work, none of the channels has priorities. For example, in Figure 9.14, source node 31 has 6 channels and in every iteration it always starts to scan with 31a and then 31b, 31c and so on.
- Every channel has a Markov based availability function. We properly select the seed to generate random numbers to ensure statistical meaning.
- Channel propagation delay or delay in the primary systems is neglected; computing delay is the only delay caused by routing. In one slot a node can scan only one channel if it is available, and the delay counter is either not increased or is increased.
- During the packet transmission, zero delay means that all channels which are checked first are available, and so nodes do not have to check the second channel to forward the packet.
- Simulations repeat 10^5 times; that is, 10^5 packets are sent.

Figure 9.15 summarises the simulation of the CRN example shown in Figure 9.14. Connectivity means the probability of a link available to CR transmission. According to a wide range of study, the spectrum of primary systems may be used with a 10% to 20% duty cycle, and thus 90% and 80% connectivity may have more reference value. So we can clearly see that our proposed routing can work under the dynamically available uni-directional links, with tolerable routing delay, in a well-behaved but general network topology case. The variable, *maximum resend attempt* or *time-to-live* (TTL), indicates the life time for a packet at a (CR) relay node. Beyond this maximum resend attempt or TTL, the relay node will drop the packet, which the source node will learn later by backward table or end-to-end control function.

We now consider a more dynamic network topology to verify our idea in the proposed CRN routing. Recalling the decomposition of CRN in Figure 9.13, the most general path can be treated as CR-Source

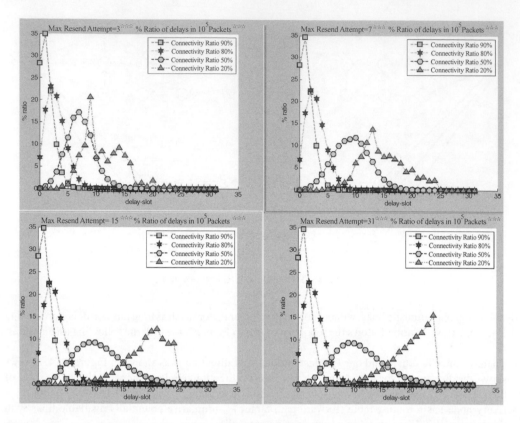

Figure 9.15   Distribution of routing delay caused by channel connectivity (available to CR opportunistic transmission in terms of probability), where x-axis is routing delay and y-axis is percentage of packets

to CR(s) to PS-tunnel to CR(s) to CR-Destination. The primary system trunk here plays a role like tunnelling with just propagation delay (assuming a unit-time slot for the time being). For CRRN, we can simply take out the PS-tunnelling. For each hop, we also assume the packet transmission delay to be a unit-time slot. Now, the problem is to calculate the accumulated delay (latency) from the CR source to CR destination under the unidirectional link (the link might be unavailable and the latency increases).

First of all, we will study the 1-D case (linear case) (see Figure 9.16) where the state transition of the CR node can be modelled as a Markov chain as shown in Figure 9.1(b), with $P_{AA} + P_{AN} = 1$ and $P_{NA} + P_{NN} = 1$. Since there is no guaranteed end-to-end route between the CR source and CR destination under the unidirectional link, where the network topology might change very quickly, each packet shall be sent directly from one node to another.

Each node has two states, namely available (state A) and unavailable (State N). If a packet arrives at certain node, it will wait for one or more time slot(s) until the node state value turns to be 1. For example, if the previous state value is 0, which means the current link is not available, the packet will wait for one more time slot. After that, if the state value is changed to 1 with probability P_{NA}, the packet will be transmitted to the next hop according to the current node's route table. If the state value is still 0 with probability P_{NN}, the packet has to wait for the next time slot. Finally, the packet will be discarded if the maximum time slot (TTL) exceeds.

Now, let us suppose there are N nodes on a route path from source to destination. Each of the nodes is unidirectional with the Markov chain state transmission. If the link is available, each hop is equivalent to one time slot, or else the one hop latency will be larger than 1, which depends on Markov chain

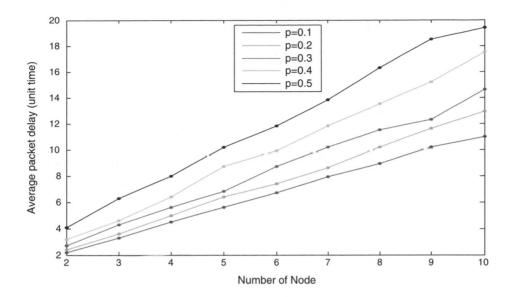

Figure 9.16   Delay performance of linear topology

probability. Intuitively, we can see that the latency from source to destination is determined by the number of nodes on the route, the Markov chain probability as well as the TTL. Let the initial states of N nodes on the route be: $\{1\,0\,0\,0\,1\dots0\}$ and $P_{AA}=P_{NA}\in[0,1]$. Taking $N=4$, the initial state is $\{0\,1\,0\,1\}$, and $P_{AA}=P_{NA}=0.1$ as in the example shown below: CR-Source to CR to CR to CR-Destination. The end-to-end source to destination packet delay is 4.5, according to the simulation. We can observe as follows:

- the average packet delay increases with P_{AN} (and/or P_{NN});
- the growth of delay is faster than the growth of the number of nodes (or P_{AN}/P_{NN});
- for a given N and P_{AN}/P_{NN} as well as TTL, we can estimate the average packet delay (whether the packet arrives or will be discarded).

Next, we look into the general network topology in a 2-D scenario as in Figure 9.17. Our simulations assume 50 randomly deployed CR nodes in a 200 (unit length) by 200 (unit length) rectangular area. Each CR has a communication range of 50. That is, $N=50$, $[X, Y]=200*200$ (unit length)^2, $R=50$ (unit length). Each of the 50 CRs wishes to transmit a data packet to the CR Destination (CR Sink), which is located at (100, 225) outside. We temporarily do not consider MS/PS tunnelling and it is a pure CRN. Later on, the network performance can be improved with the introduction of such tunnelling.

Based on our routing algorithm, we can build the corresponding route table finally to the CR Destination via a greedy algorithm. Suppose there is a source CR node (4) with data to send to the CR destination node. It takes the route $\{4 \rightarrow 25 \rightarrow 8 \rightarrow 14 \rightarrow 38 \rightarrow 17\}$ (see Figure 9.18). Considering the unidirectional link with a Markov chain property, we calculate the accumulated end-to-end packet delay based on the 1-dimensional experience we studied above. If we let $P_{AN}=P_{NN}=0.1$ and TTL $=10$, we can get an averaged packet delay of 6.7. If $P_{AN}=P_{NN}=0.4$, the final end-to-end delay is 9.9, which is less than TTL. If $P_{AN}=P_{NN}=0.5$, the packet from node 4 will be discarded since the final delay is larger than TTL. In this case, the packet cannot be successfully transmitted to CR sink.

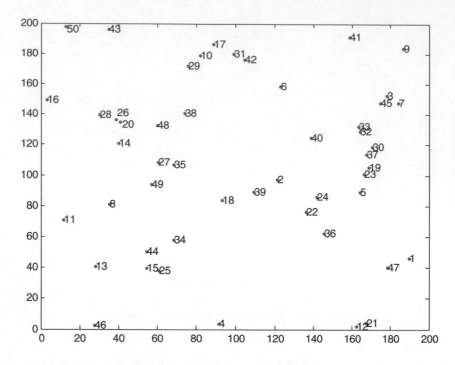

Figure 9.17   Random network topology in 2-D

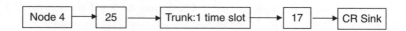

Figure 9.18   Segmentation of randomly generated topology CRN

It is interesting to observe that the network topology (density) plays a very important role in the packet transmission delay. If the CR increases its transmission radius, the network density will be higher and it will take less latency to reach CR sink. On the other hand, the interference as well as energy consumption would also increase. It is a kind of tradeoff between latency and energy consumption. For the same network topology, if $R = 80$, the corresponding route of node 4 is $\{4 \rightarrow 18 \rightarrow 6\}$ and the corresponding end-to-end packet delay becomes smaller. Finally, if we replace some of the CR nodes with MS/PS, the end-to-end packet delay will be much shorter since there is a backbone (trunk) network between the CR source and CR destination (or sink). Taking the same network topology as an example, when $R = 50$, if we replace the route from node 8 to node 38 with a trunk network, the route from node 4 to CR sink is as follows, with a final average end-to-end packet delay $3.3 + 1 = 4.3$ when $P_{AN} = P_{NN} = 0.1$.

9.4 Network Tomography

As we described earlier for routing and related network layer functions for the CRN, we note the need for information about potentially available neighbouring links (potentially up to several links in depth). Furthermore, since the CRN is pretty dynamic and can be inter-connected with heterogeneous networks among co-existing multi-radio networks, end-to-end routing is unlikely and thus network-level information about localised, potentially-available links is essential. Information such as traffic activity to suggest link availability for the CR can be very useful in routing and other network-layer functions.

Although we have already introduced techniques to detect co-existing multi-radio systems with some network information in Chapter 7, the acquisition of such network-level information from potential links in the CRN requires different methods. The immediate idea is to leverage existing technology in *network tomography* [10,11]. Network tomography originally intends the acquisition of information for network monitoring and inference in large scale networks such as the Internet, without extra cooperation among nodes. The terminology, network tomography, is a mixture of ideas behind network inference and medical tomography. Two types of network tomography are usually considered:

- *Link-level parameter estimation based on end-to-end and path-level traffic measurement:* The traffic measurements typically consist of counts of packets transmitted and/or received between two nodes, or delays between packet transmission and reception. The delay results from propagation, queuing and routing. The measured path delay is the sum of delays in all links. A packet may be dropped and the delay is generally random.
- *Sender-receiver (i.e., CR course node and CR destination node in our current case of concern) path-level traffic estimation based on link-level traffic measurement:* Based on packet counts, the goal is to estimate the amount of traffic originated from a specific node and destined to a specific node. The combination of the traffic of all origination-destination pairs for a traffic matrix is random in nature (for both link level and path level).

We usually assume that node cooperation in network tomography sends some probing packets. Most network tomography problems then start from the linear model

$$y = R\theta + \varepsilon \tag{9.2}$$

where y is the observation (or measurement) vector, R is the routing matrix, θ is a vector of packet parameters and ε is the noise vector. For Internet tomography, R is a binary matrix (i.e., every element is either 1 or 0). As a matter of fact, to reflect the time varying property in networks, we may introduce the time index to rewrite the equation as

$$y_t = R_t\theta_t + \varepsilon_t$$

This is more reasonable for the CRN due to its time-varying nature. Another approach is to introduce a new dimension of problems, and we shall treat each element in R as a probability within [0, 1], as a random matrix for the CRN.

We can now outline how network tomography works. Packets from a source are sent to a number of destination nodes, as shown in Figure 9.19(a). The paths from the source to destinations may go through some regions of which we have no or limited information, the cloud region in the figure where direct probing does not work. The logical link topology can be represented as in Figure 9.19(b).

From the example, we consider the simplest tree-structure topology as in Figure 9.19(c). Generally speaking, we can label the nodes $j = 0, 1, \ldots, m$, and there are n distinct measurement paths through the network, by considering one path from the source to one destination. We can define r_{ij} the as probability that the ith measurement path contains the jth link. For Figure 9.19(c), we have

$$R = \begin{pmatrix} 1 & 1 & 0 \\ 1 & 0 & 1 \end{pmatrix}$$

As in this example, the matrix may not be full-rank. By R, we can establish link estimation or statistical inference to either unicast networks or multicast networks, and thus various end-to-end network properties. Please recall our discussion of radio resource exploration for CRs in Chapter 7 by active sensing or probing, which can be considered to be a sort of *cognitive radio network tomography* to open a new dimension of research via statistical inference or learning algorithms [15]. There is more that we need to know in this frontier to ensure successful and smooth operation of the CRN.

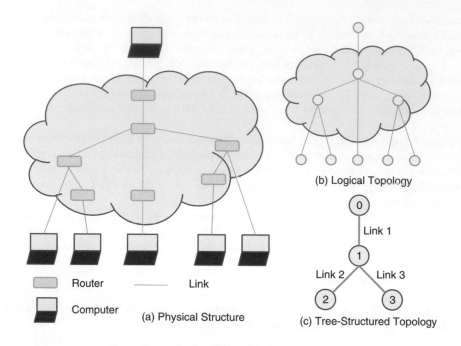

(b) Logical Topology

Router ———— Link

Computer   (a) Physical Structure

(c) Tree-Structured Topology

Figure 9.19 With cloud region as inaccessible by direct probing/measurement (a) Physical structure for single-source multiple-receiver network (b) Logical Topology (c) Simple Tree-Structured Topology

9.5 Self-organisation in Mobile Communication Networks

In order to serve the tremendous amount of wireless devices in the future, cellular type systems tend to use micro base stations to serve small cells, and more relay stations to extend radio coverage with limited complexity of network management. With large numbers of base stations and *relay stations*, along with the trend for CRN, self-organisation is an essential feature for future mobile communication networks.

9.5.1 Self-organised Networks

Self-organised network (SON) systems ideally mean that the systems are organised without any external or central dedicated control entity. It is a general concept applied to wired and wireless networks. With migration of the Internet and its role in the backbone of future wireless networks, SON definitely plays a key technology role in the CRN. The Wireless World Research Forum (WWRF) SIG-3 is dedicated to the technology study of SON [9]. The basic framework of self-organisation network operation is shown in Figure 9.20. The self-organisation network operation is composed of

- *Self-configuration:* The terminal connectivity towards the network is achieved by configuring the radio/network parameters to this newly associated network/system. In other words, as described in Chapter 5, the terminal will have the capability to reconfigure in co-existing multi-radio environments.
- *Self-optimisation:* Based on various direct or cooperative measurements, optimisation of networking parameters can be achieved for appropriate QoS and network/spectral efficiency.

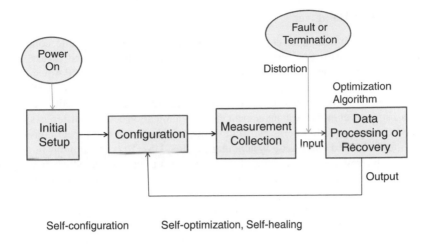

Self-configuration   Self-optimization, Self-healing

Figure 9.20   Basic framework of self-organisation

- *Self-healing:* The self-healing process can automatically detect and recover network failure by reconfiguring the radio and/or network operating parameters, so that system performance loss due to failures can be compensated for appropriately.

Some typical network operating situations can further explain how SON works as follows:

- *Association:* A newly coming node to join a wireless network should perform an initial setup and configuration process, for IP address, detection/association with network management component, authentication, downloading node software for radio and networking operation, updating cell information and information regarding neighbouring cells and stations, adjusting reconfigurable radio networking and communication parameters, etc.
- *Node failure and termination:* The performance measurement will indicate problems. Then, fault detection and recovery methods of SON functionality, along with self-healing, come into effect.
- *Operating environmental changes:* Dynamic changes due to mobility of nodes and traffic (likely in any moment of the CRN), self-optimisation and self-configuration may work to adjust matching to any change.
- *Heterogeneous networking environment:* Co-existing multi-radio systems create the need to tackle this challenge and to leverage this advantage of opportunistic networking, especially in CRN.

It is clear that self-configuration is pretty much based on the reconfigurable radio transceiver and networking functions described in Chapter 2. Self-healing lies in the detection of system/node failures or malfunctions, which is more like system sensing except that it focuses more on identification of problems. The major intellectual challenge is how to achieve self-optimisation in SON and also in CRNs, especially for coverage, handover and radio resource management of co-existing multi-radio networks.

9.5.2 *Self-organised Cooperative and Cognitive Networks*

To extend coverage and to enhance radio resource utilisation, we can combine the concept of cooperation with SON, to develop more powerful CR networking. We can realise this aim from either inter-system handover [13] or even joint radio resource allocation among different systems.

9.5.2.1 Inter-system Handover

Since future wireless communication networks are supposed to be flexible, scalable, context-aware and resilient in network infrastructure, inter-system handover can realise terminal access selection and mobility management among candidate systems and personal networks. IMT-Advanced candidate systems are characterised by a simplified architecture where the number of physical entities is reduced in favour of enhancement of the logical functionalities (e.g., moving RRM functions closer to the air interface). As a support of mobility management the communication between different systems should be established through generic interfaces. Multi-mode terminals are one aspect considered for 4G systems. It should be noted that legacy system (e.g., GPRS, UMTS) configurations might impact the operability of such terminals, and therefore this is one aspect to be considered when designing inter-system handover mechanisms. A 4G multi-mode terminal, connecting to a 4G system, should be able to decode the encapsulated information coming from a determined legacy system for inter-system handover and other functional purposes. To enable fast frequency scanning and cell registrations to the proper RATs for the multimode terminals, the 4G system should broadcast the selected system information from the legacy systems according to the operators' policy. Such system information can be the frequency allocation to the specific legacy RATs.

A step further is the interworking between the 4G system and a global spectrum coordination system when applicable. The global spectrum coordination system operates on a set of out-band signalling carrying the spectrum coordination with respect to the operators. Such interworking enables advanced spectrum management of the 4G system. An architecture supporting inter-system mobility management can be realised by implementing a number of functionalities in the physical entities part of the 4G RAN architecture. An example scenario for inter-system mobility management including a 4G system, a WLAN system and an UMTS system is shown in Figure 9.21. A gateway (GW) entity can be introduced as an anchor point for inter-system communication and that provides the interface (I_G) to the Internet and communicates with external routing functional entities. It can be anticipated that in the future, GWs will connect different RANs and the RANs will be considered as elements of the whole

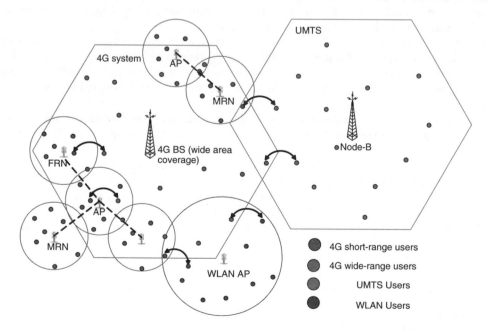

Figure 9.21   Scenario for inter-system mobility management including a 4G system

communication networks. A base station (BS) would perform almost all radio-related functions for the active terminals (i.e., terminals sending data) and would be responsible for governing radio transmission to and reception from the terminal and relay nodes (RNs) in one or more cells. The BS is in control of relays (if used) and determines routes, forwards packets to the respective relay and takes care of flow control for the relays to ensure that they can forward the data to their associated terminals.

The RN is equipped with relaying capabilities that are wirelessly connected to a BS, mobile terminal and/or another RN. As such it contains forwarding functions and schedules packets on the radio interface. Furthermore, system information broadcast, provided by the BS, is relayed by the RN for an extension of the system coverage. One BS may communicate with terminals through multiple RNs, i.e., multi-hop communication.

The I_G interface will ensure interworking between the different systems. If we assume a split of the control and user plane functions as proposed by 3GPP, the GW control plane has to support the necessary requests, measurement reports and confirmations through the I_G interface. An entity located outside of the RAN architectures can be implemented for coordination between interworking functions related to inter-system mobility management that would also communicate through the I_G interface. This is similar to the known common RRM (CRRM as in TR 25.881) approach to interworking as adopted by 3GPP. The generic RRM functions should be allocated to the GW in order to facilitate inter-system interworking without causing too much signalling load.

It is proposed that inter-system interworking for a 4G scenario is based on the tight coupling principle (i.e., the external entity in charge of inter-system interworking will be involved in each RRM decision). An internetworking architecture should be generic enough to include any type of system into the interworking process without requiring changes in the specific RRM mechanisms developed for each system. It is proposed that cooperation is realised based on the reference architecture shown in Figure 9.22. A specific RRM (SRRM) is the entity in charge of adapting the RANs to the cooperation implemented in the external entity in charge of inter-system RRM. The SRRM has two types of functionalities and interfaces with the RANs, one for traffic monitoring and reporting of status of physical nodes and the other devoted to the direct actuation in the RAN nodes; basically it translates RRM commands of the external entity to the RANs. It is proposed that the SRRM in the 4G RAN

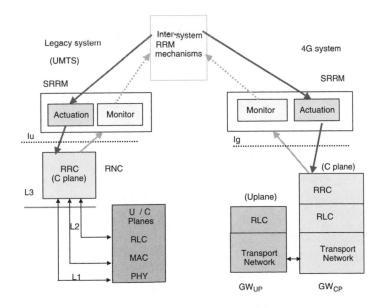

Figure 9.22   Realisation of inter-system interworking in a 4G scenario

implements in addition internal RRM functionalities distributed in the RAN nodes (i.e., GW and BS, respectively).

In another implementation that supports roaming, the different interworking systems can be coupled by an entity similar to the home subscriber server (HSS) or the home location register (HLR). Open coupling is another approach to interworking, where there is no mechanism for interchange of information between the networks. All actions are coordinated by the mobile terminal, which acts as the only relay for the inter-system handover. This option is possible for all types of systems; however, information about the new system and the possibility of performing a seamless and reliable handover is very restricted, unless a 'make-before-break' strategy is employed for handover, which usually requires two radio transceivers.

The concept of a pool of GWs [13] (see Figure 9.23) decouples the physical relation between a unique GW and a number of BSs associated with it. Instead, each GW may associate to each BS in the pool area. Therefore, it is by default not necessary to employ inter-GW handover (HO) within a pool area when all GWs are equally balanced. Following that, the pool area is defined as an area in which the UT may roam without the need to change the GW. The GW capacity of a pool area can be scaled simply by adding more GWs. In contrast, in the classic hierarchical structure, each GW is associated with a set of BSs serving their own location area, providing a direct mapping between a GW and the area covered by associated BS.

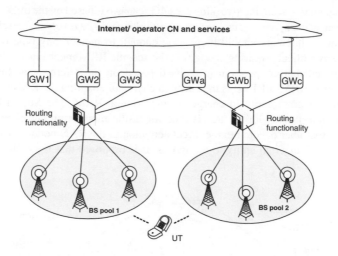

Figure 9.23   GW pools for support of mobility management in a 4G scenario

Operators can deploy multiple 4G systems (e.g., in a large country where a single network is insufficient to cover it all) and use regular macro mobility mechanisms using the Mobile IP (MIP) protocol for handover between both systems. The MIP protocol can be used also for load balancing between GWs. An inter-GW load balancing mechanism offers at least two advantages. On one hand, the average flow processing time will be considerably reduced when the GWs are not equally loaded. On the other hand, the network can be designed in a more cost effective manner, which does not demand an over-engineered network using the conventional peak-hour based traffic analysis.

The network architecture jointly designed with handover offers an overall optimised performance. It enables the introduction of a hybrid handover mechanism, which perfectly matches the pool area deployment concept. The key reasons of using the pool area of GWs are load balancing and avoiding unnecessary IP handover. This would lead to maximum use of system resources and improvement of the system performance. Network architecture can be realised through a pool of gateways and also a pool of spectrum bands. As mobile WiMAX migration of systems is already part of 3GPP, under an all-IP networking structure, inter-system handover may facilitate a realistic state-of-the-art and scalable gigabit wireless access.

9.5.2.2 Cooperative Joint Radio Resource Allocation

An even more aggressive scenario is to have a fully 'integrated' system network architecture to allow joint radio resource allocation among different systems, to approach the ultimate radio resource (in terms of frequency/subcarrier, time, code, spatial/antenna in MIMO) utilisation. It is worth considering because there are more and more wireless communication systems appearing due to different applications and demands. For example, the GSM/GPRS has been providing mobile phone services since 1990, and the IEEE 802.11 series as local area networks provides medium distance wireless access inside buildings. Bluetooth and UWB are often used for short distance ($\sim$1 m) transmission and Zigbee as a sensor network is used for some distributed applications. In addition to all the above existing systems, there are emerging wireless systems such as 3G (CDMA series), HSPA, 3G LTE and IEEE 802.16 series. As we know, they all eat bandwidth and make the available spectrum scarcer and scarcer. Allocating a fixed spectrum for one specific system causes spectrum inefficiency and will become more difficult in the future.

Co-existence/co-operative communication is one way to solve this difficulty. In Reference 14, the authors built a Joint Radio Resource Management (JRRM) for WiMAX and HSDPA co-existence. Traffic from different users was conducted to each of the two systems according to the real-time condition of each system. A more insightful scenario was considered for the European WINNER II project, which integrated IP handover and radio handover in mobility management into co-existence radio resource management for new and legacy systems. However, considering the co-existence of multiple systems by network selection or radio resource management only may be too conservative. That is why we introduce the cross-three-layer radio resource allocation as follows. To maximise spectrum efficiency, we try to integrate factors involving radio resource utilisation from all layers into the system design. The spectrum re-farming among cells, bandwidth allocation among systems and users inside cells, antenna usage and power allocation for each system are all considered simultaneously. This is undoubtedly a top-down cross-layer design from radio resource all the way to the physical layer, and of course self-organised network coordination among systems is needed to realise cooperative radio access and mobility management. In the following we clarify each of those ideas one by one.

We start with a network including many cells and where there are several co-existence systems in all cells. Since frequency reuse is adopted in most wireless systems, we can reduce the complexity of spectrum re-farming to only consider N cells in one frequency reuse pattern. We can adopt a frequency-reuse pattern $N = 4$ as an example. Each hexagon represents a cell and different colours represent non-overlapping frequency bands used by different cells. Apparently, neighbouring cells should have different frequency bands to avoid co-channel interference.

Spectrum re-farming can be considered to be allocating a pre-determined frequency band B among those N cells to maximise total network capacity according to the demands of each cell. If there are X_i systems in ith cell and the capacity of the εth system is C_i^ε, the re-farming problem can be formulated as Equation (9.3), where B_i is the frequency band assigned to cell i and $B_1 \cup B_2 \cup B_3 \ldots B_N = B$.

$$C_T = \max_{B_i} \sum_{i=1}^{N} \sum_{\varepsilon=1}^{X_i} C_i^\varepsilon \tag{9.3}$$

Here we do not specify the definition of each C_i^ε. It is directly related to the characteristic of each co-existence system.

We next consider spectrum allocation inside each cell. In one cell i, all systems and their users share the same bandwidth B_i, whether data link, control link or relay link. Here we focus on centralised architecture; there is a BS for each system and users are distributed in the whole cell. In this scenario we can further divide the radio resource into time slots and frequency bands. Many studies report on how to allocate such a time-frequency resource jointly with the power among users within one system in order to maximise the system capacity such as [14]. Here we propose to extend these ideas to multiple co-existence systems and to combine the allocation with upper layer consideration. Let us take OFDMA and MIMO-OFDMA co-existence as an example; these two systems are already adopted in many wireless systems due to their high spectral efficiency and flexibility of spectrum utilisation. Bandwidth in the two systems is divided into orthogonal subcarriers and power can be distributed adaptively on them. If MIMO is used, antennas provide another dimension to increase system capacity. The maximisation of cell capacity can be formulated as in Equation (9.4):

$$C_i = \max_{\omega^{i,\varepsilon,u}_{t,\alpha,k}, p^{i,\varepsilon}_{t,\alpha,k}} \sum_{\varepsilon=1}^{X_i} \sum_{u=1}^{U^\varepsilon_i} \sum_{t=1}^{T} \sum_{\alpha=1}^{N^\varepsilon_T} \sum_{k=1}^{K_i} \frac{\omega^{i,\varepsilon,u}_{t,\alpha,k} \log(1 + SNR^{i,\varepsilon,u}_{t,\alpha,k}(p^{i,\varepsilon}_{t,\alpha,k}))}{TN^\varepsilon_T K_i} \tag{9.4}$$

In Equation (9.4) we assume that the bandwidth B_i of this cell was divided into K_i subcarriers and we consider T time slots once at a time. For systems using MIMO, the transmit antenna number is represented as N^ε_T: those systems with one more diversity than just time and frequency. The u, t, α and k are user, time slot, antenna and subcarrier index, respectively. For this cell i, we assume there are U^ε_i users belong to system ε. The variables ω and p in Equation (9.4) are subcarrier and power allocation, $\omega^{i,\varepsilon,u}_{t,\alpha,k} = 1$ means that the subcarrier k in antenna α of time slot t was assigned to user u in system ε, or $\omega^{i,\varepsilon,u}_{t,\alpha,k} = 0$ and $p^{i,\varepsilon}_{t,\alpha,k}$ are the power allocated to that subcarrier.

We also do not specify the definition of signal to noise ratio (SNR) in Equation (9.4), because it is also directly related to the system architecture and channel condition of each system. But for systems that serve non-mobile users, or systems with excellent equalisation, the SNR can be formulated as in Equation (9.5), where X is the transmit data and H is the channel gain on each subcarrier:

$$SNR^{i,\varepsilon,u}_{t,\alpha,k} = \left(p^{i,\varepsilon}_{t,\alpha,k} |H^{i,\varepsilon,u}_{t,\alpha,k}|^2 |X^{i,\varepsilon}_{t,\alpha,k}|^2 \right) \Big/ \left(N_0 \frac{B_i}{K_i} \right) \tag{9.5}$$

From these examples we have some idea about the cross-three-layer resource allocation for co-existence systems. For our example of MIMO-OFDMA/OFDMA co-existence, substituting Equations (9.4) and (9.5) into Equation (9.3) results in one such optimisation. If we want to consider a more general case such as single carrier systems, we just modify Equation (9.4) and then we can obtain a different optimisation.

References

[1] J. Mitola, III, *Cognitive Radio Architecture*, John Wiley & Sons, Inc., New Jersey, 2000.

[2] K.-C. Chen, 'Medium Access Control of Wireless LANs for Mobile Computing', *IEEE Network*, September–October, 1994.

[3] D.B. Johnson, D.A. Maltz, 'Dynamic Source Routing in Ad-hoc Wireless Networks', in *Mobile Computing*, T. Imielinski and H. Korth (eds), chapter 5, pp. 153–181, Kluwer Academic Publishers, Boston, MA, 1996.

[4] C.E. Perkins, E.M. Royer, 'Ad-hoc On-demand Distance Vector Routing', *Proceedings of the second IEEE Workshop on Mobile Computing Systems and Applications*, 1999.

[5] A. Boukerche, 'Performance Evaluation of Routing Protocols for Ad-hoc Networks', *Mobile Networks and Applications*, **9**, 2004, 333–342.

[6] R. Prakash, 'A Routing Algorithm for Wireless Ad-hoc Networks with Unidirectional Links', *Wireless Networks*, 7, 2001, 617–625.

[7] R. Beckers, J.L. Deneubourg, S. Goss, 'Trail and U-turns in the Selection of the Shortest Path by the Ant Lasius Niger', *Journal of Theoretical Biology*, **159**, 1992, 397–415.

[8] L. Rosati, M. Berioli, G. Reali, 'On Ant Routing Algorithms in Ad-hoc Networks with Critical Connectivity', *Ad-hoc Networks*, **6**,2008, 827–859.

[9] H.J. Byun, M.S. Do, Y. Liu, J. So, R. Taori,'Self-Organization in Future Mobile Communication Networks', White Paper, Wireless World Research Forum (WWRF) Special Interest Group (SIG) 3, Version 1.0, 22 April, 2008.

[10] M. Coates, A.O. Hero, III, R. Nowak, B. Yu, 'Internet Tomography', *IEEE Signal Processing Magazine*, May 2002, 47–65.

[11] M.G. Rabbat, M.A.T. Figueiredo, R.D. Nowak, 'Network Inference from Co-Occurrences', *IEEE Transactions on Information Theory*, **54**(9), 2008, 4053–4068.

[12] Y.C. Peng, *Routing Challenges in Cognitive Radio Networks*, MS Thesis, National Taiwan University, 2008.

[13] E. Tragos, A. Mihovska, E. Mino-Diaz, P. Karamolegkos, 'Access Selection and Mobility Management in a Beyond 3G RAN: The WINNER Approach', *ACM MobiWac*, 2007.

[14] F.S. Chu, K.-C. Chen, 'Radio Resource Allocation for Mobile MIMO-OFDMA', *IEEE Vehicular Technology Conference, Spring*, Singapore, 2008.

[15] C.K. Yu, K.C. Chen, 'Radio Resource Tomography of Cognitive Radio Networks', *IEEE Vehicular Technology Conference, Spring*, 2009.

10

Trusted Cognitive Radio Networks

It is clear that CRN is a temporarily organised network with both ad-hoc networking and heterogeneous networking features, except that the larger dynamics in the CRN topology and link availability make static security and traditional networking functions infeasible. CRN, however, can be considered through the seven-layer OSI model like all wireless networks and Internet applications. From the earlier chapters regarding network layer functions, we know that trust is one of the key parameters in the facilitation of CRN, due to its role of gluing together the nature of the various heterogeneous (wireless) networks in CRN.

There are two important steps that need trust in CRN operations: association and routing, and we focus on these major issues in this chapter. When a CR initially tries to connect a node to join an existing (cognitive radio) network or to form a CRN, it is practically impossible to execute conventional security functions, because

- such an action may create security holes;
- it is unwise to consume huge amounts of computation power for security without making sure it is a valid request to form a CRN;
- there is not enough available time for an opportunistic CRN link to exist for a complicated hand-shaking security protocol.

A trusted mechanism is therefore needed, and authentication is a part of trust along with other technical and nontechnical factors. The next challenge occurs when a node in the CRN tries to route the traffic through another node or some part of another network. Typical ad-hoc networks using the *public key infrastructure* (PKI) scheme are used to achieve secure routing and other purposes in the literature. Such a CRN node under the request of a routing packet or packets may not be able to execute security such as the typical PKI scheme in ad-hoc networks, due to it not being practical to perform checking under limited communication and computation resources and due to the increasing possibility of being attacked. Consequently, trusted networking and updating of trust measures can be very useful in reaching a compromise in the face of these technical challenges. Once a node passes the trust evaluation, which must be quick and fairly reliable, it can function as a node in the CRN to transport/relay packets; other security and service mechanisms are still executed at later, appropriate moments, however, which is the fundamental spirit behind routing in (heterogeneous) CRN as discussed in Chapter 9.

Cognitive Radio Networks   Kwang-Cheng Chen and Ramjee Prasad
© 2009 John Wiley & Sons, Ltd

10.1 Framework of Trust in CRN

Trust has been long and widely studied in the research literature, from social sciences to computer sciences. The immediate challenge in studying trust in CRN is the mathematical definition of trust, and we want to keep in mind that *distrust* may be just as important as *trust*. Following the efforts in [5,6], we may observe that CRN is a sort of cooperative networking and heterogeneous networking. Although CRN can be generally considered as a multi-hop (ad-hoc) network, it is fundamentally different from a homogeneous wireless ad-hoc network. Consequently, to ensure smooth operation of CRN in supporting ubiquitous computing, trust forms the foundation of CRN. Trust has been widely mentioned in the research literature regarding trusted computing and Internet/web computing, ad-hoc networks and even social science. However, trust for CRN is quite different from these application scenarios. Trusted computing deals with components inside a set or territory. Internet/web computing treats trust as a kind of reputation or credit given by a mechanism (such as other's scoring). We therefore need to develop a mathematical framework to model trust in CRN to quantitatively apply trust in CRN design and operation.

Trust is critical in CRN operation above and beyond security design, because security usually needs communication overheads in advance. We can use the following examples to explain the need for trust other than security:

- A cognitive radio that senses a spectrum hole or opportunity and dynamically accesses the spectrum for transmission requires 'trust' from the originally existing system (i.e., primary system) and regulator, even without creating interference to PS.
- A cognitive radio may want to leverage another existing cognitive radio to route its packets, even though another CR is not the targeted recipient terminal. It therefore requires 'trust' from another CR.
- A cognitive radio can even leverage the PS to forward its packets to realise the goal of packet switching networks. It needs 'trust' from the PS, not only at the network level but also in the service provider (or network operator).

10.1.1 Mathematical Structure of Trust

If we temporarily ignore the operator side, we can just consider the trust in the CR network. Trust must be measurable so that networks can operate based on it, as per routing discussed in Chapter 9. Intuitive measures of trust may be as follows.

Definition 10.1: Trust is a measure in [– -1, 1].

Remark: $\tau(i, j)$ denotes the trust measure for node j to handle (receive and/or forward) a packet from node i. It is actually measured over $(-\infty, \infty)$ and can be normalised as $\tau(i, j) \in (-1, 1)$. 1 as the normalised value means trust in full (confidence); 0 means no information regarding trust or not; −1 means no trust at all. For a decision or a policy in CRN, it is obvious that negative trust and zero trust come under the same action (rejecting the packets). We can modify Definition 10.1 as follows.

Corollary 10.1: Trust in CRN is a measure in [0,1].

Remark: This is just like the probability measure, and enables our mathematical framework using probabilistic development and statistical decision theory. Please note that this degeneration from Definition 10.1 may result in an equivalent probabilistic mass for the trust measure at zero (i.e., $\int_{-1}^{0} 1_{\tau(i, j) \in [-1, 0]} d\tau$).

Remark: The trust measure at zero means *distrust*. Any node in the CRN being identified with zero trust shall be rejected by any action of the CRN. That is, any possible link to such a node should be removed from the CRN. It does not make any sense to be measured as zero or negative.

Lemma 10.1: Trust in CRN is generally *irreversible*. That is,

$$\tau(i,j) \neq \tau(j,i) \tag{10.1}$$

Remark: It is generally true for all consequences to adopt the concept of trust. The degree of Alice trusting Bob is not equal to the degree of Bob trusting Alice.

Definition 10.2: (Metric Space [7]) Every normed space can be regarded as a metric space, with distance $d(x,y)$ between x and y, with the following properties:

(i) $0 \leq d(x,y) < \infty \forall x, y$

(ii) $d(x,y) = 0$ if and only if $x = y$

(iii) $d(x,y) = d(y,x) \forall x, y$

(iv) $d(x,z) \leq d(x,y) + d(y,z) \forall x, y, z$

Lemma 10.2: Trust in CRN is not a metric.

Proof: To form a metric space, $\tau(i,j) \geq 0$, which can be resolved by introducing a bias. However, $\tau(i, j) + \tau(j, k) \leq \tau(i, k)$ violates the requirement of metric space. This equation means that the trust through an intermediate node is not higher than the trust directly from the originating node. Furthermore, trust is *irreversible*, that is, $\tau(i, j) \neq \tau(j, i)$. This can be explained by the case when node i is a mobile CR and node j is the base station of a cellular network.

Remark: This lemma is usually ignored in most literature modelling trust in research areas. However, it is critical here as many trust measuring systems are constructed based on the assumption of trust measure being a metric, and such a fundamental assumption might be in jeopardy during further development.

Remark: However, if we define 'distrust' instead of trust (i.e., D as distrust measure), the triangular inequality actually holds:

$$D(i,j) + D(j,k) \geq D(i,k) \tag{10.2}$$

We would like to point out, as many authors do, that modelling *distrust* might not be less important than modelling trust in networking research (either Internet or CRN).

Definition 10.3: Trust in CRN is contributed to by reputation (trust measured by other nodes) and collaboration (behaviours observed by the targeting node and possibly other nodes). Any zero-trust implies distrust.

Remark: Reputation is a terminology borrowed from trust in e-commerce. A CR node can increase its reputation by executing more actions under the operation rules. Both reputation and collaboration follow Definition 10.1. However, any part of reputation or collaboration being equal to zero results in zero/no trust. Definition 10.3 is different from the common additive definitions in the research literature, and suggests that 'multiplication' (or semi-group) is more suitable.

Proposition 10.1: (Trust-Path Theorem) Trust in CRN is a function of routing path. The overall trust is the multiplication of trust in each segment. That is,

$$\tau(n_0, n_1, \cdots, n_L) = \prod_{l=0}^{L-1} \tau(n_l, n_{l+1}) \tag{10.3}$$

and

$$\tau(n_0, n_1, \cdots, n_L) \neq \tau(n_0, n_L) \tag{10.4}$$

Remark: For 3-node (2-segment) packet forwarding

$$\tau(i, j, k) = \tau(i,j)\tau(j,k) \neq \tau(i,k)$$

Remark: In the literature, many researchers note that trust relies on its past history, which coincides with this proposition. However, this proposition suggests that it is more than just history; the order of the historical path makes sense.

Proposition 10.2: (Trust Processing Theorem) More processing (i.e., packet/traffic transportation) in CRN cannot increase trust.

Remark: This is a direct result from Definition 10.2 and Proposition 10.2. It is easy to understand in CRN: for any packet from the originating node, after more segments of transportation, we have no more trust.

Definition 10.4: (Semi-group [7]) Let X be a Banach space, and suppose that to every $t \in [0, \infty)$ is associated with an operator $Q(t) \in B(X)$ in such a way that

$$\text{(i)} \quad Q(0) = I$$

$$\text{(ii)} \quad Q(s+t) = Q(s)Q(t) \forall s, t \geq 0$$

Proposition 10.3: Trust in a homogeneous ad-hoc network is degenerated into a semi-group as in [8].

Remark: Please note that we require an homogeneous ad-hoc network condition such that the ad-hoc network implies that the reversible (or exchange) property of trust holds as in Equation (10.1). CRN, in general, however, is neither homogeneous nor ad hoc (CRN is likely to have an infrastructure in some parts). From this mathematical property, we can also easily tell the difference of trust in a CRN and trust in an ad-hoc network.

It might be disputable to identify a mathematical measure for trust in CRN. However, it is widely agreed that trust measures have the property of a probability measure, as Corollary 10.1 suggests. Let us start from a branch of artificial intelligence: mathematical reasoning.

Lemma 10.3: (**Cox Axiom**) We want to measure any 'certainty' that is consistent with the following conditions:

- degree of certainty can be ordered;
- there exists a function to map certainty of a statement to its negation/complement;
- degree of reasoning R(A∧B) is related to the conditional reasoning R(A|B) and R(B) by some function g

$$R(A \wedge B) = g(R(A|B), R(B))$$

Therefore, this reasoning system must be equivalent to the measure of probability.

Remark: The trust model in CRN indeed applies. Therefore, we can subsequently measure trust just like probability.

10.1.2 Trust Model

The primary objective of the trust model in CRN is to provide us with a kind of mathematical framework to sense, measure, analyse and learn the topology variation and behaviour of neighbours in such heterogeneous environments. In these networks, the trust model plays an important role among entities belonging to different systems and it will provide a mechanism for nodes to establish trust association. After trust association, cooperation between entities is possible and, therefore, they can communicate with each other to work together such as in relay traffic, partner selection and trusted routing. In this chapter, we try to build up the trust model for CRN and this should provide two major components for nodes:

- *Trusted association:* The initial decision of a node is to accept or reject the trusted association from a neighbouring CR node. The central concept for this initial decision is to conduct networking of minimum risk.
- *Learning algorithms:* Each node in a CRN should keep records of the others and employ a proper learning algorithm to adapt the long-term trend of probability/trust measure of packet forwarding, so that timely judgment to hold on the trusted route or to retract it could be made.

We depict the basic components of our proposed trust model in Figure 10.1, which illustrates the interaction between these two functions and provides the whole scenario of the trust model. In wireless networks, channel condition is always changing and trust is thus dynamic in CRN. We need to develop a proper learning algorithm to adapt the tendency of cooperation from neighbours in all kinds of environments. Each node in a CRN could determine its trust policy about its neighbours according to the existing environment. The algorithm must catch bad behaviour and punish the nodes for this attempt to deteriorate the throughput in wireless networks. On the other hand, this mechanism needs to give the nodes an opportunity to recover these terrible isolation situations.

10.2 Trusted Association and Routing

One of the most challenging tasks for CRN is when a CR (transmitting node) initially requests association with a cooperative PS node or another CR (receiving node) to conduct cognitive radio networking functions. Similar to cooperative networks described in Chapter 4, the receiving node in the CRN can carry out one of the following: (i) reject association; (ii) amplify-and-forward (AF) mode; or (iii) compress-and-forward (CF) mode. The difference between AF and CF in routing is that AF just executes the physical layer function, and we do not need to worry about attacks, and CF actually

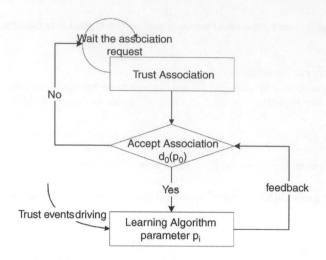

Figure 10.1 Flowchart of the trust model in CRN

decodes the packets to the upper layers under the threat of security threats. In other words, the cost for AF is simply communication bandwidth (and possible battery energy) waste even for a wrong decision, but the cost for CF of a wrong decision might be to jeopardise the entire network, which suggests a security check such as PKI before association of CF.

10.2.1 Trusted Association

We illustrate the association of CRN in Figure 10.2 and adopt the Neyman-Pearson criterion (since no *a priori* probability or cost function is available in such a decision) as follows.

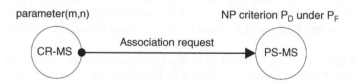

Figure 10.2 Neyman-Pearson theorem in trust decision

Proposition 10.4: There are only two possible decisions for association in a CRN based on trust measure, that is, to accept association and to reject association. Let X denote the trust measure with distribution $F_\theta(x)$ representing the information inside the association request from CR-MS to PS-MS. Let $\Theta = \Theta_0 \cup \Theta_1$ be a disjoint covering of the trust space, and H_i denote the hypothesis that the parameter θ belongs to the trust space $\theta \in \Theta_i$. Then, the decision problem is now to distinguish between the two hypotheses by considering a CRN architecture as in Section 6.2:

(i) H_0: $\theta \in \Theta_0$: This means that CR-MS would not be worth being trusted. The probability density function for CR-MS is $f_{\tau|0} = f_\tau(x|0)$, where x means the trust measure of CR-MS.

(ii) H_1: $\theta \in \Theta_1$: This means that the CR-MS would be worth being trusted. The probability density function for CR-MS is $f_{\tau|1} = f_\tau(x|1)$, where x means the trust measure of CR-MS.

The PS-MS decides to trust CR-MS if the probability density function of trust measure, x, from CR-MS under trust space, $\theta \in \Theta_1$, is larger than that under trust space, $\theta \in \Theta_0$.

Remark: For AF, once accepting association, the packet(s) is relayed. For CF, once accepting association, the transmitting node has to go through a security check, and then the receiving node compresses and relays the packet(s).

Proposition 10.5: When a node in a CRN (primary system or secondary systems) receives a request of association from a new node (i.e., to join this CRN), it forms a statistical decision as follows: Based on the trust measure τ associated with this node, decision $\delta(x) = a_i$ can be formed, where a_1 means 'accept' the association and a_0 means 'reject' the association.

The PS-MS will decide to trust CR-MS if the probability density function for trust measure, x, from CR-MS under trust space $\theta \in \Theta_1$ is larger than that under trust space $\theta \in \Theta_0$. Therefore, with these two hypotheses, we can form the decision rule for the trust measure:

$$\delta(x) = \begin{cases} 1 & , f_{\tau|1}(x|1) > \gamma \cdot f_{\tau|0}(x|0) \\ k & , f_{\tau|1}(x|1) = \gamma \cdot f_{\tau|0}(x|0) \text{, for some } \gamma \geq 0,\ 0 \leq k \leq 1 \\ 0 & , f_{\tau|1}(x|1) < \gamma \cdot f_{\tau|0}(x|0) \end{cases} \quad (10.5)$$

The decision that maximises the probability of detection for a given probability of false alarm is the likelihood ratio test as specified by the Neyman-Pearson theorem. In order to maximise P_D for a given $P_F \leq \alpha$, PS-MS decides to trust CR-MS if

$$l(x) = \frac{f_{\tau|1}(x|1)}{f_{\tau|0}(x|0)} > \gamma \quad (10.6)$$

In the same way, PS-MS decides not to trust CR-MS if

$$l(x) = \frac{f_{\tau|1}(x|1)}{f_{\tau|0}(x|0)} < \gamma \quad (10.7)$$

The likelihood ratio

$$l(x) = \frac{f_{\tau|1}(x|1)}{f_{\tau|0}(x|0)},$$

indicates the likelihood of H_1 versus the likelihood of H_0 under each trust measure. We can transform the decision into:

$$\delta(x) = \begin{cases} 1 & , l(x) > \gamma \\ k & , l(x) = \gamma \text{ for some } \gamma \geq 0,\quad 0 \leq k \leq 1 \\ 0 & , l(x) < \gamma \end{cases} \quad (10.8)$$

Remark: We can model the association process successfully as a binary statistical decision problem. It can be applied to a more realistic study of dynamic spectrum access to CRNs. Please also note that the *randomised decision rule* is still possible here. However, to define the binary decision problem completely, we still need to explore definitions of cost and *a priori* probability (i.e., trust measure) distribution. Since it is difficult to assign the cost function or *a priori* probabilities before establishing trust association in realistic situations, we consequently adopt the *Neyman-Pearson* theorem to solve the problem.

Definition 10.5: Following the earlier Propositions, we can define the probability of false alarm and probability of detection with the meaningful interpretation for the trust association. We first define *the probability of PS-MS trusting CR-MS, given that CR-MS does not trust PS-MS as a probability of false alarm*:

$$P_F = \int_{x \in Z_1} f_\tau(x|0)dx \qquad (10.9)$$

Then, we define *the probability of PS-MS trusting CR-MS, given that CR-MS trusts PS-MS as a probability of detection*:

$$P_D = \int_{x \in Z_1} f_\tau(x|1)dx \qquad (10.10)$$

Remark: We can easily tell the importance of the trust measure. The trust measure is composed of true trust and observation deviation (i.e., like observation noise in common communication theory). Such an observation deviation has the tendency to be negatively distributed because more observation cannot increase trust as in Proposition 10.2(trust processing theorem). However, for malicious nodes, such a conclusion may not apply and observation deviation could be two-sided in distributions.

What we plan to do with the *Neyman-Pearson* criterion in this decision problem is to minimise the risk of PS-MS trusting CR-MS given the constraint that CR-MS is not worth being trusted. We do not make any assumptions on the probability density function of trust measure here. It could be discrete or continuous and we only want to derive a general trust decision rule for our trust model.

Proposition 10.6: When *a priori* information of trust is unknown and the cost function cannot be well defined, the decision $\delta(x) = a_i$ can be based on the *Neyman-Pearson* criterion. That is, given $P_F \leq \alpha(0 \leq \alpha \leq 1)$, optimise P_D.

Remark: The approach that maximises the probability of detection for a given probability of false alarm is the likelihood ratio test. For a Neyman-Pearson test, the values of P_F and P_D completely specify the test performance. The problem in the trust decision is changed to

$$\text{maximise } P_D \text{ for a given } P_F = \alpha \qquad (10.11)$$

$$\text{decide } \delta(x) = \begin{cases} 1 & , l(x) > \gamma \\ k & , l(x) = \gamma \text{ for some } 0 \leq k \leq 1 \\ 0 & , l(x) < \gamma \end{cases} \qquad (10.12)$$

where the threshold, γ, is specified from the system constraint:

$$P_F = \int_{\{x:l(x)>\gamma\}} f_{\tau|0}(x|0)dx = \alpha \qquad (10.13)$$

If trust is based on individual perception, it is likely that different observers observe different situations; therefore, each node in CRN would adopt a different significance level upon its acceptable level of malicious behaviour. Increasing γ makes the test less sensitive for the disturbance and we accept a higher probability of detection in return for a lower probability of false alarm.

We now give an example of how nodes in CRN can make decisions using the Neyman-Pearson criterion we have described above. We take the normal distribution for trust measure to realistically solve the trust problem here. Consider the case in Figure 10.3 where CR-MS-A independently relays/ forwards the packet with probability p and drops/ignores it with probability $1 - p$ as receiving packets from CR-MS-B. According to this information, PS-MS has to accept or reject the association from CR-MS-A by maximising P_D under the system constraint, P_F.

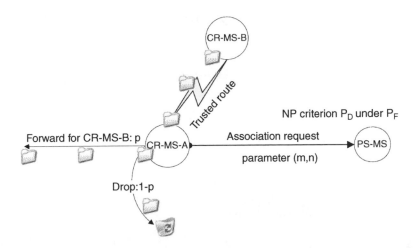

Figure 10.3   Execution of association based on the Neyman-Pearson criterion (using normal distribution)

Let X be the number of packets forwarded by CR-MS-A and we assume that every packet is forwarded or dropped independently. This assumption is reasonable since the amount of the data transferred in the network is large enough. Then, in this situation, X is a binomial distribution with parameters $(m + n, p)$ in which p is the probability of success. As $m + n$ is large and by the *De Moivre-Laplace* theorem, we know that for any numbers a and b, $a < b$,

$$\lim_{n \to \infty} P\left(a < \frac{X - (m+n)p}{\sqrt{(m+n)p(1 - p)}} < b\right) = \frac{1}{\sqrt{2\pi}} \int_a^b e^{-t^2/2} dt \qquad (10.14)$$

where the expected value is

$$E(X) = (m+n)p \qquad (10.15)$$

and the variance of probability of success is

$$\sigma_X = \sqrt{(m+n)p(1 - p)} \qquad (10.16)$$

Since the probability of packet transmission being successful in the next stage may be impossible to compute analytically, we could approximate this probability according to the experience rating:

$$\Pr(cooperation) \approx \frac{\text{number of packets forwarding}}{\text{number of packets received}} \qquad (10.17)$$

With such an approximation, the probability of packets forwarding successfully by CR-MS-A is $\frac{m}{m+n}$, which we define as *probability of trusted cooperation*. We can model the packet forwarding behaviours of CR-MS-A with normal distribution in this problem. Now, the problem is turned into the hypothesis problem:

- H_0: *CR-MS-A would not be worth being trusted.*

$$X \sim N(\mu_0, \sigma_0^2)$$

where

$$\mu_0 = (m+n) \cdot (1-p) \text{ and } \sigma_0^2 = (m+n) \cdot p(1-p)$$

The probability density function of x is

$$f_{\tau|0}(x|0) = \frac{1}{\sqrt{2\pi}\sigma_0} e^{\left\{-\frac{1}{2}\cdot\left(\frac{x-\mu_0}{\sigma_0}\right)^2\right\}} \tag{10.18}$$

- H_1: *CR-MS-A would be worth being trusted.*

$$X \sim N(\mu_1, \sigma_1^2)$$

where $\mu_1 = (m+n)p$ and $\sigma_1^2 = (m+n)p(1-p)$
The probability density function of x is

$$f_{\tau|1}(x|1) = \frac{1}{\sqrt{2\pi}\sigma_1} e^{\left\{-\frac{1}{2}\cdot\left(\frac{x-\mu_1}{\sigma_1}\right)^2\right\}} \tag{10.19}$$

We can derive the likelihood ratio by the hypothesis if we decide to choose H_1:

$$l(x) = \frac{f_{\tau|1}(x|1)}{f_{\tau|0}(x|0)} = \frac{\frac{1}{\sqrt{2\pi}\sigma_1} e^{\left\{-\frac{1}{2}\cdot\left(\frac{x-\mu_1}{\sigma_1}\right)^2\right\}}}{\frac{1}{\sqrt{2\pi}\sigma_0} e^{\left\{-\frac{1}{2}\cdot\left(\frac{x-\mu_0}{\sigma_0}\right)^2\right\}}} > \gamma \tag{10.20}$$

Finally, it is equivalent to

$$x > \frac{\sigma_1^2}{\mu_1 - \mu_0} \ln \gamma + \frac{\mu_1 + \mu_0}{2} = \gamma' \tag{10.21}$$

We could determine the threshold γ from the false alarm constraint

$$P_F = \Pr\{x > \gamma'|H_0\} \leq \alpha \tag{10.22}$$

Then, PS-MS can make the decision by maximising the probability of detection

$$
\begin{aligned}
P_D &= \Pr\{\text{decide } H_1 \text{ given } H_1\} \\
&= \Pr\{x > \gamma'|H_1\} \\
&= \int_\gamma^\infty f_{\tau|1}(x|1)dx \\
&= Q\left(\frac{\gamma' - \mu_1}{\sqrt{\sigma_1^2}}\right)
\end{aligned}
\tag{10.23}
$$

under the system constraint, probability of false alarm, which also can derive the threshold by

$$
\begin{aligned}
P_F &= \text{Pr}\{\text{decide } H_1 \text{ given } H_0\} \\
&= \text{Pr}\{x > \gamma' | H_0\} \\
&= \int_{\gamma'}^{\infty} f_{\tau|0}(x|0) dx \\
&= Q\left(\frac{\gamma' - \mu_0}{\sqrt{\sigma_0^2}}\right)
\end{aligned}
\tag{10.24}
$$

where $Q(x)$ is the right-tail probability of a Gaussian random variable with zero mean and unit variance

$$
Q(x) = \int_{x}^{\infty} \frac{1}{\sqrt{2\pi}} e^{-t^2/2} dt
\tag{10.25}
$$

Using the constraint, $P_F = \alpha$, we can derive the trust threshold as

$$
\gamma' = \sigma_0 \cdot Q^{-1}(\alpha) + \mu_0
\tag{10.26}
$$

and we can obtain the probability of detection as

$$
\begin{aligned}
P_D &= Q\left(\frac{\sigma_0 \cdot Q^{-1}(\alpha) - (\mu_1 - \mu_0)}{\sqrt{\sigma_1^2}}\right) \\
&= Q\left(Q^{-1}(\alpha) - \frac{(\mu_1 - \mu_0)}{\sqrt{\sigma_1^2}}\right)
\end{aligned}
\tag{10.27}
$$

If we define the coefficient c^2

$$
c^2 = \frac{(\mu_1 - \mu_0)^2}{\sigma_1^2} = \frac{(m+n) \cdot (2p)^2}{p(1-p)}
\tag{10.28}
$$

as the trust confidence equalling to the definition of deflection coefficient often used to approximate detection performance evaluation for most detection problems, we can derive the maximum probability of trusted cooperation from this system constraint:

$$
P_D = Q\left(Q^{-1}(\alpha) - \sqrt{c^2}\right)
\tag{10.29}
$$

Given the constraint on the probability of false alarm, and therefore, the trust threshold, the optimal trust decision is to:

■ Decide to trust if the probability of trusted cooperation from neighbours maximises P_D.
■ Decide not to trust if the probability of trusted cooperation from neighbours does not maximise P_D.

10.2.2 Trusted Routing

Once a node is accepted into CRN after association, as in typical multi-hop networks, CRN shall update its network topology and routing table. If we treat CRN as a homogeneous amplify-and-forward network, routing is the same as any multi-hop ad-hoc network. However, it is obvious that CRN is not homogeneous. As a particularly severe risk of security applies to CRN, it is important to develop a trusted network layer function, especially with regards to topology and routing. Based on the developed mathematical framework of trust in CRN, we are ready to derive the fundamental operation of CRN at the network layer, network topology establishment and routing in CRN.

Typical routing algorithms in a homogeneous network proceed on the distance measure that accounts for available bandwidth/capacity and transmission cost. For routing of CRN, we have to consider trust in heterogeneous networking environments (e.g., PS is a cellular and CR is ad-hoc WiFi station).

Proposition 10.7: (Trusted Routing) A routing metric between node i and node j in CRN is defined through a state-machine of state $(\tau(i,j), d(i,j))$, which represents trust measure and distance measure.

To deploy the routing algorithm, such as say the Dijkstra algorithm, we may simply define a new trusted distance $D(i,j) = d(i,j)/\tau(i,j)$ to iterate the algorithm to find the route, under the assumption of a reversible trust relationship between any two nodes. For the case of no trust (i.e., $\tau(i,j) = 0$), the link is practically eliminated due to the infinite distance. Figure 10.4 depicts both Bellman-Ford and Dijkstra routing algorithms using this new trusted distance measure, with a reversible trust relationship.

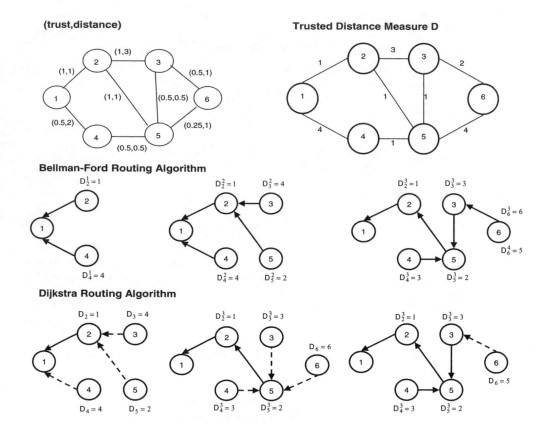

Figure 10.4   Trusted routing examples

There exists an open problem to route under a general irreversible trust relationship that invokes an unequal distance measure for two directions between two nodes, which is the asymmetric nature of uni-directional links in CRN as discussed in Chapter 9.

Definition 10.6: A CRN node (which can be an access point of the PS), deciding the trust and taking action to a packet from another CRN node, can form a Markov decision process (with a decision/policy associated with a state-space) [5].

Proposition 10.8: (Markovian Trust Process) For packet transportation from node i to node j, the recipient node (i.e., node j in this case) can form a binary hypothesis testing (trust as H_1 and non-trust as H_0) based on a certain decision policy. Trusted routing in Proposition 10.86 becomes a Markov process.

Remark: Trusted routing therefore becomes a kind of Markov decision process. As a matter of fact, randomised decision of trust is possible and meaningful in CRN. For example, $\tau(i,j) = \tau_r, 0 < \tau_r < 1$ may suggest a discount factor for trust status, due to various reasons such as roaming user/node, robustness against attacks or simply insufficient credits in account. The number of states can be finite or infinite.

10.3 Trust with Learning

The real challenge for trust in CRN is not only to construct the measure for trust to take the proper actions, but also subsequently to update the trust 'distribution function' for each network node to possibly conduct communication/networking functions such as requesting relays of packet(s). Each node in PSs supporting CRN or each CR shall be able to update and/or to maintain a trust table of neighbouring nodes for applications described in later sections. Although identifying mathematical measures for trust in CRN may be a matter of dispute, it is suggested that trust measure has the property of a probability measure. Based on such measures, we can precede decisions based on trust measures in different application scenarios. Hereafter in this section, we focus on ways to update the distribution function of trust measure.

10.3.1 Modified Bayesian Learning

Suppose the probability distribution of trusted cooperation is $f_P(p|\alpha_t, \beta_t)$, $0 \le p \le 1$, where the superscript t means the discrete-time index with $t=0$ as the initial distribution. $f_P(p|\alpha_t, \beta_t)$ can be updated recursively by $f_P(p|\alpha_{t-1}, \beta_{t-1})$ based on current trust evidence, observation and any further information. To resolve this challenge, we may adopt a learning algorithm for nodes in CRN to adapt the probability distribution of packets forwarding from neighbours at each (proper) time instance.

Lemma 10.4: (Modified Bayesian Learning Algorithm) Suppose the beta density $f_P(p|\alpha_t, \beta_t)$ denotes the probability distribution of trusted cooperation at time t in our trust model. The learning algorithm consists of three parts: the prediction of probability, decaying correction and measurement modification:

$$\hat{p}_t = \int p(x_t|P = p) \cdot f(p|\alpha_t, \beta_t) dp \tag{10.30}$$

$$f_{t-1}(\hat{p}) = f(p|k_1\alpha_{t-1}, k_2\beta_{t-1}) \tag{10.31}$$

and

$$f(p|\alpha_t, \beta_t) = c \cdot f((m_t, n_t)|p) \cdot f_{t-1}(\hat{p}) \tag{10.32}$$

where t is the discrete-time index, k_1 and k_2 are the decaying factors, (m_t, n_t) is the new trust evidence, c is the modification factor representing the constant of the integral and $\hat{p}_t$ is the prediction probability of trusted cooperation.

Remark: The prediction is used to predict the (probability) measure of trusted cooperation in the next stage and it is primarily designed for the decision criterion. Lemma 10.4 consists of the update probability distribution and the update rule. The probability density function, $f(p|\alpha_t, \beta_t)$, for the prediction function incorporates into new trust evidence and an *a priori* (probability) density function. It includes all the information prior to $T = t - 1$ where all trust evidence including the initial value is decayed upon receiving a new one:

$$
\begin{aligned}
\alpha_t &= k_1\alpha_{t-1} + m_t = k_1^t m_0 + k_1^{t-1}m_1 + k_1^{t-2}m_2 + \ldots + k_1 m_{t-1} + m_t \\
\beta_t &= k_2\beta_{t-1} + n_t = k_2^t n_0 + k_2^{t-1}n_1 + k_2^{t-2}n_2 + \ldots + k_2 n_{t-1} + n_t
\end{aligned}
\tag{10.33}
$$

Then, we carry on the decaying correction of past information prior to $T = t$ by Equation (10.31) and incorporate a new trust measurement after $T = t$ in Equation (10.32). We use the decaying correction to 'forget' the past trust evidence to explain the limitation of the period of validity because the trust model should gradually ignore the oldest record in order to catch the newest one. We use two constants k_1 and k_2 to represent the decaying factor as time goes on and usually k_1 is smaller than k_2 in order to catch the bad behaviour of nodes such as deception or throwing packets away with bad intentions. As we receive the new trust evidence at time $T = t$, it would be appropriate to give the latest record more weight in order to support the dynamic and fast operation in CRN. Therefore, we can maintain the parameter update for trusted cognitive radio networking through this learning mechanism.

Remark: Before deriving the prediction probability, $\hat{p}_t$, we need to describe the details in the correction equation. At the end of time $T = t$, we have the new trust evidence (m_t, n_t) and the prior probability distribution and we can calculate probability density function as

$$
\begin{aligned}
f(p|\alpha_t, \beta_t) &= \frac{f_\tau((m_t, n_t)|p) \cdot \hat{f}_{t-1}(p)}{\int f_\tau((m_t, n_t)|p) \cdot \hat{f}_{t-1}(p)\, dp} \\
&= \frac{p((m_t, n_t)|P = p) \cdot f(p|k_1\alpha_{t-1}, k_2\beta_{t-1})}{\int p((m_t, n_t)|P = p) \cdot f(p|k_1\alpha_{t-1}, k_2\beta_{t-1})\, dp} \\
&= \begin{cases} \dfrac{\Gamma(\alpha_t + \beta_t)}{\Gamma(\alpha_t)\Gamma(\beta_t)} p^{\alpha_t - 1}(1 - p)^{\beta_t - 1}, & 0 \le p \le 1 \\ 0, & \text{elsewhere} \end{cases}
\end{aligned}
\tag{10.34}
$$

We provide a flowchart for the technological processes in the learning mechanism in Figure 10.5. This provides two important concepts in our trust model. The first is the measurement correction on a time scale. The past trust evidence must decay to support the immediate interaction in the CRN and provide a way to recover from the bad channel condition. The second is the prediction. The trust decision is made at every stage mainly based on this prediction value; whether to hold or retract it. The prediction function given in our approach is not the only solution for the learning algorithm and it depends on the system model that one chooses. After receiving new trust evidence (m_t, n_t), each node in the CRN should predict the probability of trusted cooperation in the next stage according to the past record and latest trust evidence. What we try to do in this learning model is to make inference about the probability of $\hat{p}_t$ given the probability distribution of p, $f_P(p)$, and new trust evidence (m_t, n_t).

The probability that the next transmission is successful depends on the prediction probability of trusted cooperation, $P = \hat{p}_t$, and then we can compute the probability:

$$
P(x_t|P = \hat{p}_t) = \begin{cases} \hat{p}_t & \text{if } x_t = 1 \\ 1 - \hat{p}_t & \text{if } x_t = 0 \\ 0 & \text{elsewhere} \end{cases}
\tag{10.35}
$$

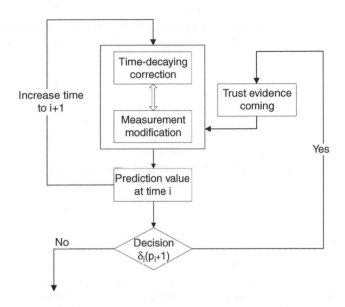

Figure 10.5   The flowchart of the learning mechanism.

As we receive the new trust evidence at time t, it would be appropriate to give the last record more weight in order to support the dynamic and fast operation in CRN. Finally, we can directly calculate the prediction probability $P(x_t = 1)$:

$$
\begin{aligned}
P(x_t = 1) &= \int P(x_t = 1 | P = \hat{p}_t, (m_t, n_t)) f(p | \alpha_t - 1, \beta_t - 1) dp \\
&= \int P(x_{t+1} = 1 | P = p) f(p | \alpha_t, \beta_t) dp \\
&= \int p \cdot f(p | \alpha_t, \beta_t) dp \\
&= \frac{\Gamma((k_1 \alpha_t - 1 + m_t) + (k_2 \beta_t - 1 + n_t))}{\Gamma(k_1 \alpha_t - 1 + m_t) \Gamma(k_2 \beta_t - 1 + n_t)} \int_0^1 p^{k_1 \alpha_t - 1 + m_t} (1 - p)^{k_2 \beta_t - 1 + n_t - 1} dp \\
&= \frac{k_1 \alpha_t - 1 + m_t}{(k_1 \alpha_t - 1 + m_t) + (k_2 \beta_t - 1 + n_t)}
\end{aligned}
$$
(10.36)

We rewrite the equation in another form to denote that it is the weighted average of the maximum estimate of $P = p$, given (m_t, n_t) and the mean of the prior information:

$$
\begin{aligned}
P(x_{t+1} = 1) = & \left(\frac{m_t + n_t}{k_1 \alpha_t - 1 + k_2 \beta_t - 1 + m_t + n_t} \right) \frac{m_t}{m_t + n_t} \\
& + \left(\frac{k_1 \alpha_t - 1 + k_2 \beta_t - 1}{k_1 \alpha_t - 1 + k_2 \beta_t - 1 + m_t + n_t} \right) \frac{k_1 \alpha_t - 1}{k_1 \alpha_t - 1 + k_2 \beta_t - 1}
\end{aligned}
$$
(10.37)

When we obtain the prediction at each stage, we can make the decision by the decision criterion:

$$
\delta_t(m_t, n_t) = \begin{cases} 0, & \text{means ``reject'' if } \hat{p}_{t+1} \leq \gamma_i \\ 1, & \text{means ``accept'' if } \hat{p}_{t+1} > \gamma_i \end{cases}
$$
(10.38)

If the probability measure of trusted cooperation is larger than the threshold γ_i, it means more probability that the packet would be delivered successfully at the next stage than dropped, and *vice versa*.

10.3.2 Learning Experiments for CRN

We can describe several scenarios to demonstrate the suitable properties of the learning algorithm applied in CRN, and we can further conclude some general rules when we build up the trust model for CRN. Since the CRN is a highly dynamic heterogeneous network, the nodes leave or join the network dynamically and promptly, and the CRN topology may change very frequently. The learning algorithm should follow up the channel variations and user behaviour instantly, to learn the update (favourable or unfavourable) changes in the behaviours of packet forwarding.

10.3.2.1 Nodes Disconnect and the Effect of Initial Value

Disconnection from nodes is frequently encountered in CRN and the learning algorithm should be able to catch this extreme case as soon as possible in order not to deteriorate the trust topology in the network. As we show in Figure 10.6, we accept the trust association from the neighbour with probability of trusted cooperation $p = 0.8$ at time $t = 0$ under $k_1 = 0.2$, $k_2 = 0.5$. At time $t = 21$, the neighbour disconnects from the network and the learning algorithm catches the prediction of probability immediately because the probability of trusted cooperation drops quickly. We retract the trust

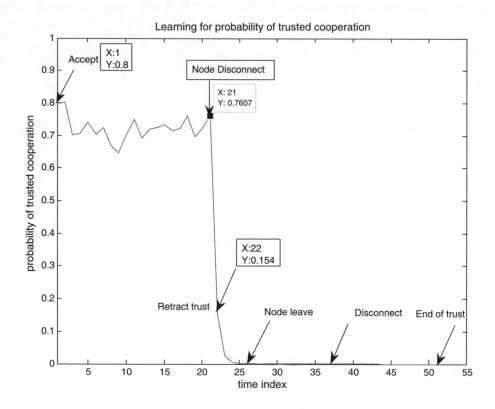

Figure 10.6   Nodes disconnect from the network under $m_0 + n_0 = 100$

association at time $t = 22$ and determine the node has left the network at time $t = 27$. Since nodes in CRB are dynamic, they may leave and join the network for a while and we have to observe more time before we can ascertain that it has disconnected from network. In Figure 10.7, we show the same scenario except that information inside the association request is different. The sum of the packets in Figure 10.6 is 100, and that of Figure 10.7 is 1300. We note that the initial value affects the subsequent trust decisions at times $t = 3$ and $t = 4$. Although we learn the behaviour of the node immediately, we should make assumptions on the number of initial value to avoid such situations in the learning model. The simulations later should adopt an initial value of $m_0 + n_0 = 100$, in order to fully catch up the latest trust evidence of the nodes. Now, we propose the first rule of thumb from this example.

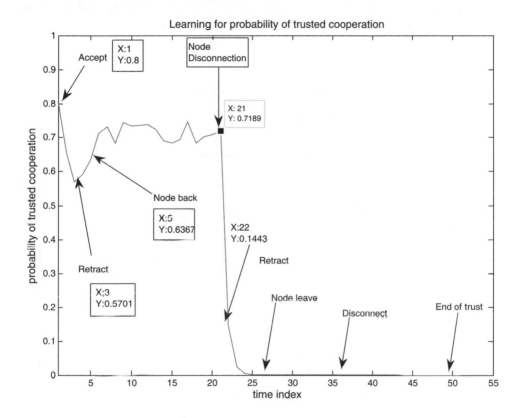

Figure 10.7   Nodes disconnect from the network under $m_0 + n_0 = 1300$

Proposition 10.9: (Rule of Thumb in the Different Traffic model) The learning algorithm in Lemma 10.4 can quickly learn the new trust evidence under heavy traffic density. Even when the initial value is large or the environment is varied frequently, we can still learn quickly to catch the latest trust evidence if the traffic is heavy and the decaying factors are small enough.

10.3.2.2 Nodes Leave and Join the Network Suddenly

The neighbour sends an association request including the past 100 records with probability of trusted cooperation $p = 0.8$ at time $t = 1$ under $k_1 = 0.2$, $k_2 = 0.5$. The node leaves the network at time $t = 21$

and comes back at time $t = 32$. Owing to the considerable amount of packets dropped at time $t = 22$, we retract the trusted route right away and declare that the node has left the network. However, when the node comes back, we notice the considerable probability of trusted cooperation and we do not reestablish the trusted route immediately. We observe for more time to build up the trusted route again. It is used to punish for the network performance drop and make sure that the node does not come back and leave again. We denote that the node may be back at time $t = 35$ and reestablish the trusted route at time $t = 36$. In Figure 10.8, we show the same scenario as in Figures 10.6 and 10.7 except that the traffic density in each time slot is different. The traffic density of Figure 10.8 is 100 and that of Figure 10.9 is 30, which represents the heavy and light traffic network, respectively. From Figure 10.8, we observe that the initial value significantly affects the learning algorithm because the sum of the initial value is 100 and it is larger compared to the traffic. Therefore, at times $t = 6$ and $t = 7$, the trust decisions are retracted and reestablished in successive order. We are now ready to propose other rules of thumb by the observation of the effect of the initial value problems.

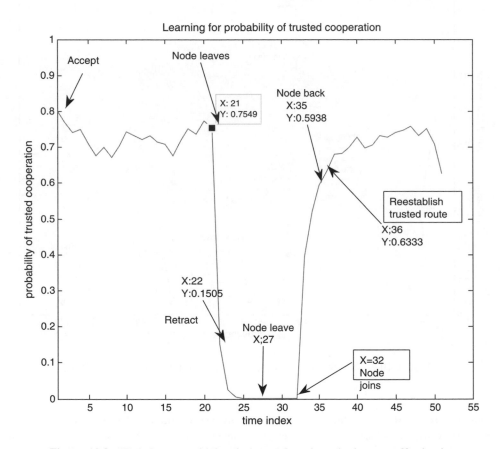

Figure 10.8 Node leaves and joins the network again under heavy traffic density

Proposition 10.10: (Rule of Thumb in the Importance of the Decaying Factor) The learning algorithm of Lemma 10.4 responds to the new trust evidence slowly under light traffic density. The only way we can adapt to this situation is to keep the decaying factor as small as possible and making subsequent decisions between longer time intervals.

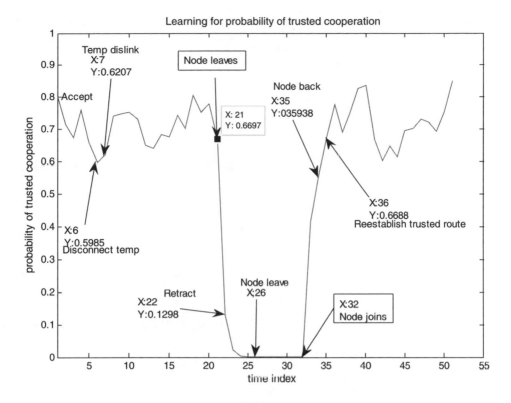

Figure 10.9 Node leaves and joins the network again under light traffic density

Proposition 10.11: (Rule of Thumb in the Initial Value Problem) The initial value should be decayed as soon as possible since it reflects the past information. It contradicts the basic concept of the learning algorithm, which trying to respond to the latest trust evidence. We can use adaptive decaying factors to 'forget' the initial value properly and quickly under the realistic network condition.

10.3.2.3 Variation on the Behaviour of Nodes

In this experiment, we consider the probability of trusted cooperation tending to perform better or worse. The nodes in CRN could incur a bad channel condition and, therefore, alter their behaviour on the trusted cooperation. The learning algorithm should be able to analyse the possible temporal disconnect and make further decisions. In Figure 10.10, we show that the node changes its behaviour at time $t = 16$ from a probability of trusted cooperation from $p = 0.9$ to $p = 0.7$ and to $p = 0.9$ at time $t = 36$ under $k_1 = 0.2, k_2 = 0.5$. The learning algorithm detects the variation at $t = 20$ although the probability of trusted cooperation is still larger than the trust threshold, $p = 0.7 > 0.6 = \gamma_i$. The reason is that the decaying factor punishes the bad behaviour more than we reward good behaviour. Since $k_1 = 0.2$ $< k_2 = 0.5$, at time $t = 16 \sim 20$, the learning algorithm detects the decline in the prediction of probability and retracts the trusted route at time $t = 20$. From time $t = 16$ to time $t = 35$, the learning algorithm also catches the stable packets forwarding from the node, although the prediction of probability has a notable decline compared to $t = 1 \sim 15$. This may come from many causes and the variation between $t = 0.9$ and $t = 0.7$ can possibly result from the channel condition. The learning

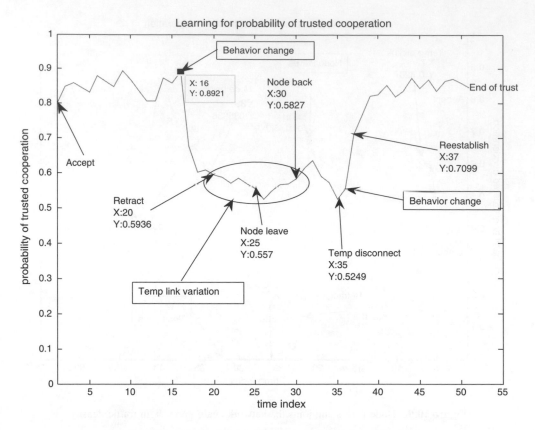

Figure 10.10   Node changes behaviour under light traffic density

algorithm will catch the probability from $p = 0.9$ to $p = 0.7$ immediately at time $t = 37$ and reestablish the trusted route when the channel condition turns better. In Figure 10.11, we repeat scenario again except that the traffic density in each time slot is 30. We note that the prediction curve in a light traffic network triggers the disturbance in the variation of behaviour and we denote this phenomenon as 'temp disconnect' in Figure 10.11. This does not indicate a real retraction of the trusted route but represents the unstable oscillating across the trust threshold. However, the learning algorithm still works well in capturing the behavioural changes, such as those at times $t = 16$ and $t = 36$, and finally detects the reestablishment of the trust association.

10.3.2.4 Intentionally Drop the Traffic

The learning algorithm should not only solve the frequent dynamic disconnections from the network, but also some special cases such as dropping the traffic intentionally. In such situation, the nodes do not drop all the traffic. Instead, they drop some fixed portion of the traffic. This could result in serious damage to the entire CRN if the dropped portion contains important parameters to network operation. In Figure 10.12, we show the remarkable decline in the prediction of the probability of malicious users at time $t = 11$ and the usual behaviour of nodes. The malicious user is punished by the larger decaying factor which manifests the intentional packet drops.

   The above examples describe the adaption of trust to allow desirable CRN operation by using the learning algorithms.

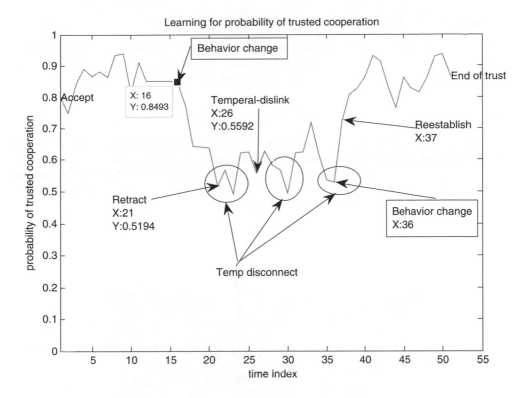

Figure 10.11 Node changes behaviour under light traffic density

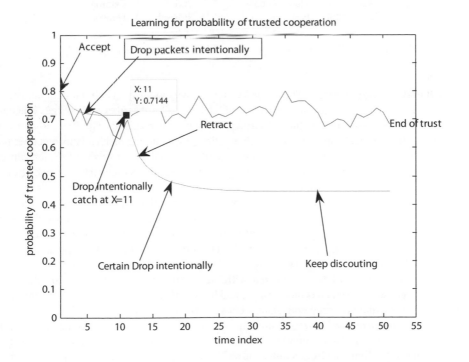

Figure 10.12 Node drops the major part of the traffic intentionally

10.4 Security in CRN

As described in Chapter 9, the CRN may be established based on different levels of trust and security. Having introduced trust, we are going to discuss security. Security for wireless networks has been widely studied and we therefore focus more on the new challenges for security in CRN.

10.4.1 Security Properties in Cellular Data Networks

We start with cellular security, and the cellular architecture is generically depicted as Figure 10.13, with UE representing user equipment, RAN representing the radio access network and CN representing the core network. The primary goal of security services in cellular data networks is to protect information and resources from attacks and misbehaviour. Compared with other wireless data networks such as WLAN, the cellular data network has the following characteristics related to security:

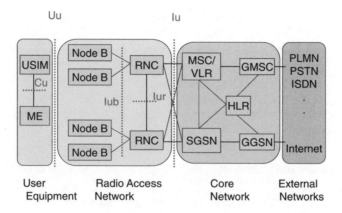

Figure 10.13   Network elements in a public land mobile network (PLMN)

- *Packet-based charging services:* This is the key reason why security in cellular data network is so important. The cellular data network provides services to each subscriber and charges it according to the accumulated packets that the subscriber has sent and received. The malicious users are stimulated to bypass legal authentication, authorisation or accounting mechanisms by either paying less or gaining more. Thus, an absolute prerequisite of the cellular data network of each subscriber (i.e., identity) is varied, since nobody wants to pay for fraudulent data transmissions made by others.
- *Centralised infrastructure with identity authentication:* The centralised infrastructure leverages a centralised subscription database named the Home Subscriber Station (HSS) to record the identity and subscriber information of all users. Typically, an International Mobile Subscriber Identity (IMSI) number is used as the primary identity for the subscriber. To prevent fraudulent use of services, the network authenticates the identity of a user through a challenge-response mechanism, provides the UE and the SN with session keys and allows the UE and SN to establish connections protected with the session keys. Compared with the existing key establishment mechanism, the challenge-response mechanism is more simple and lightweight, and benefits from information such as identity and pre-defined key being shared between the user and network sides.
- *Separation of control and data transmission:* In order to maintain the QoS for a data service, cellular data networks prevent in-band signalling by isolating data transmission and control transmission. However, due to the limited wireless link bandwidth, this separation means that the control channel

can be overloaded easily. Moreover, to transfer a similar amount of data, a lot of control messages are necessary for data bearer establishment. After data transmission is complete, lots of signalling message exchanges are introduced to release the radio resource for data transmission. Such establishment and release procedures incur heavy signalling overheads and make the control plane vulnerable to overload attacks.

- *Hierarchical-based architecture:* The hierarchical nature of current cellular data networks is considered beneficial for certain critical system functions such as location management, handover management and power control. However, it introduces larger number of signalling messages, causes longer signalling latency and thus degrades the overall performance. Furthermore, heavy signalling overheads may increase the possibility of being attacked through overloading of the system.

In dealing with the attacks in cellular data networks, we first consider the definition of attack: an attack exploits a law or vulnerability and results in a violation of security properties of a specific domain. Consequently, we may classify attacks in cellular data networks into two basic dimensions: domain and impact. We organise information about attacks in a hierarchical manner, beginning with the domain the attack located, moving lower to the impact of the attack and further narrowing down to the vulnerability that caused the attack. The domains in which are attacks located in the cellular data networks are as follows:.

- *Access domain:* Ensures that the authenticated UE connects to a valid HN and provides the UE and the SN with confidentiality and identity keys to protect following data or signalling connections.
- *Network domain:* Enables the network entities in different networks (such as HN and SN) or different systems (such as cellular and WLAN) to exchange signalling data and messaging data securely.
- *Device domain:* Prevents the UE with high computing power and various network connection capabilities from being hacked into through worms or viruses. Once the UE is infected by these malicious codes, UE becomes a zombie and is manipulated to spread these codes to other users or attack the network entities.

We now introduce security properties and list the basic kinds of attacks that violate these properties. However, an attack may have many consequences, and for simplification we only consider the immediate impact of an attack. Any well-known attack can be decomposed and the components can all be classified into one of the following attack techniques:

- *Authentication:* One of the primary advantages of the cellular network is the centralised authentication property. By leveraging identifications of subscribers, authentication determines whether people are who they declare themselves to be and ensures that legal people are using the network.
  - *Session hijacking:* A malicious user can hijack an already established session, and act as a legitimate base station.
  - *Message replay:* An attacker can intercept an encrypted message and replay it back at a later time and the user might not know that the packet received is not the right one.
  - *Man-in-the-middle attack:* An attacker can sit between a cell phone and an access station and intercept messages between them, and change them.
- *Authorisation:* Permits authenticated users to gain access to the cellular network and use it for services. Thus, some sort of role based access control is necessary.
  - *Buffer overflow:* An attacker overflows its buffer by inserting designed codes to corrupt data values in memory addresses adjacent to the allocated buffer, to execute commands only allowed to the administrator and illegally increase the access privilege.

- *Confidentiality:* Protects a message or data file from being revealed and understood by anyone other than the desired recipient who is authorised to have that information. Confidentiality is typically enabled by applying data encryption.
  - *Eavesdropping:* If the traffic on the wireless link is not encrypted, an attacker can eavesdrop and intercept sensitive communication such as confidential calls, sensitive documents, etc.
- *Integrity:* Ensures that a message or data file cannot be modified by any malicious users during its transmission. Integrity is typically ensured by applying one-way hashes.
  - *Message forgery:* If the communication channel is not secure, an attacker can intercept messages in both directions and change the content without the users ever known.
- *Non-repudiation:* Ensures that a entity cannot deny the sending of a message that it originated. This ability is much more important in cellular data networks because every packet sent or received needs to be charged. Cellular data networks provide a complex billing mechanism to achieve that aim.
- *Availability:* Ensures that the desired network services are available whenever they are expected.
  - *Channel jamming:* An attacker may jam the wireless channel to deny access to any legitimate users in the network.
  - *Denial of service (DoS):* This attack is accomplished by sending excessive data to the network to overload the network and results in users being unable to access network resources. The high signalling overhead and heavy control processing properties make cellular data networks vulnerable to this attack.
  - *Distributed denial of service (DDoS):* This attack is similar to the previous one, but in this case multiple compromised users flood the bandwidth or resources of the cellular data network. Due to the popularity of smart phones, attackers may physically co-locate with many infected victims through viruses, worms or Trojan horses, and cause such attacks.

10.4.2 Dilemma of CRN Security

CRNs face unique security problems not faced by conventional wireless networks. The current focus of the CR/Software-Defined Radio (SDR) community is on preventive security measures that secure the radio software download process and on schemes that thwart the tampering of radio software once it is installed. However, preventive security is not sufficient. CRNs are also more open to attacks such as jamming and selfish node misbehaviour. Authentication of the users can prevent use of the network by unauthorised users. The typical method for the identification of such terminals is the transmitter verification procedure, in which the location verification is done in addition to the signal energy detection.

The objective of *denial of service attack* that is likely in CRN is to prevent other nodes from accessing the channel and using the channel to increase their own throughput. Denial of service attack causes the network to become unavailable to users and disrupts the services. The nodes exchange MAC control frames to contend for the channel. Due to the absence of a base station, the channel negotiation process is done in a distributed manner or by using a centralised spectrum manager. The false control frames transmitted are responsible for the denial of service attacks and selfish misbehaviour. With the MAC-layer authentication scheme, these problems can be solved.

We now expand on some of the security concerns raised in the previous section:

- *Denial of Service:* In multi-channel environments, high traffic load may cause frequent exchange of control packets saturating the control channel. From the security point of view if attackers can saturate the control channel successfully, it can hinder the channel negotiation and allocation process and thus cause denial of service.
- *Selfish Misbehaviour:* During the channel negotiation process, a selfish node tries to take unfair advantage and improve its own performance. The channel negotiation process is done using the

results from spectrum sensing and the fairness depends on the cooperation of the contending nodes. A selfish node may conceal the available data channels from others and reserve it for its own use.

- *Licensed User Emulation Attack:* CRs use the licensed spectrum when it is free and otherwise use the unlicensed band. The attacker may jam the licensed band and emulate the primary user, thus limiting the CRN to operating in the unlicensed bands and therefore limiting CRN capacity. A solution to this problem is not yet known.
- *Common Control Channel Jamming:* In this case, the attacker transmits periodical pulses in the control channel spectrum. The jamming of one channel blocks the probable communication between all CR nodes. UWB ultra wideband as common control channel deployment may solve the problem of jamming.
- *Attacks on Spectrum Managers:* There cannot be just one spectrum manager for assigning frequency bands because it may be a single point of attack on the network. So, when the spectrum manager is not available, communication between CR nodes is not possible. Thus, the spectrum availability should be distributed and replicated in CRNs. The attack can be prevented possibly by use of a specific pilot channel in the licensed band.
- *Eavesdropping:* The transmission range of CRs is not limited to a short distance because it uses bands lower than UNII and ISM. This allows collection of data by attackers invisible to the emergency services. Therefore, a strong data encryption is needed at the physical level.
- *Privacy Compromised:* No disclosure of data should be allowed without permission of the user. The distribution and sharing of context information should be governed by rules and policies that ensure that the user's privacy is not endangered.
- *Network Management Infrastructure:* The infrastructure is often designed and deployed based on the hierarchical manager/agent architecture, with extensions for ad-hoc wireless networks and techniques for passively and actively monitoring overlays.
- *Policy Management:* Throughout the CRN, policy management will be achieved by properly propagating profile information updates by the CRN administrator. Policies are personalised according to user interaction and can be achieved by propagation of policy updates throughout the CRN.

10.4.3 Requirements and Challenges for Preserving User Privacy in CRNs

In addition to security, privacy is usually required for wireless networks, which is particularly important in heterogeneous wireless networks such as CRN. Privacy has the following aspects in the CRN:

- *Maintaining information privacy*, i.e., to prevent any disclosure of information directly related to the individual to a service or application without the user's prior approval or knowledge.
- *Maintaining context privacy,* i.e., to prevent any disclosure of information related to the context in which the user is using the service (for example his/her current device parameters) and from which indirect information about the user could be extracted.
- *Maintaining location privacy* of the user, i.e., to deny an attacker the knowledge of a device's current and past location and preventing link ability.
- *Preserving anonymity* of the users' identifiable parameters for distinct scenarios, i.e., preserving their 'state of being not identifiable within a set of subjects'. Anonymity impacts location privacy, because as long as a user is anonymous, location privacy is preserved.

Anonymity mechanisms should allow the user to use the network services while protecting the identity or other identification information from possible abuse. One way of providing anonymity is using label name/pseudonyms. For CRNs, that requires some form of identification of the user or other privacy-related user information, including the use of pseudonyms if necessary. The problem here is the managing of the pseudonyms and their time life.

One solution could be the capability-based privacy-preserving scheme. In this proposed solution, the policy manager contains descriptions of the user, context, scenario, device and service. The context-aware policy management and the privacy protection and anonymity mechanism have to be transparent to the authorised user and non-obtrusive, giving the user control in all situations. More user profiles, security and privacy policies, scenarios and rules have to be easy to create based on provided templates and easy to update by the user. For keeping the user anonymous, there should not be a possibility of linking any parameters of the user identity (names, address, social security number, bank account number, etc.) with any context-based information (location, IP address, time, presence, type of service, etc.).

The user should be given options to select different levels of privacy and to negotiate network access preserving his/her privacy and anonymity. Policy rules for preserving a user's privacy include the following:

- requesting only the minimum necessary user sensitive data for the service to operate (the data minimisation principle);
- keeping the data only for the time duration stated;
- not disclosing the data to any other party without prior notification and agreement of the user;
- clearly stating the trustworthiness of the provided location and context information to the user;
- clearly stating a transparent policy for liability of the networks.

Here are the main features of the proposed context-aware privacy preserving mechanisms:

- Preserving any type of sensitive information that the user considers private and for any level of granularity – the user decides how to protect his/her sensitive information and anonymity, and location privacy. Data abstractions over all types of low-level sensitive data are part of the mechanism and they are processed first allowing faster filtering and default setting.
- Hiding all the complexity of the system from the user and delegating privacy decisions from the user to the device by using descriptive profiles and policies, roles, scenarios and context.
- Rule-based access over the private data helps to delegate the decisions to the device and to take actions concerning the correct provision of privacy parameters to the services.
- Whenever the context attributes change, the privacy protection mechanisms evaluate the overall privacy status and act accordingly based on the predefined rules.

10.4.4 Implementation of CRN Security

Identification of the primary user is vital to grant access and prioritise resources, services, etc., and therefore the use of non-forgetable characteristics of primary user signals is a must in order to identify and authenticate primary users successfully. This requires tighter architectural integration of CR identification and network identification. In other words a single set of user credentials are desirable. The solution here is CRN Admission Control (CRNAC), which is used in the 802.1x. For user identification, identity management is required, which is a collection of tools and processes that manage the lifecycle of information elements that constitute a user's digital identity. Some of the functions that identity management addresses include the following:

- Identifying the entities that interact in the system, authenticating each entity and applying authorisation rules as specified by policies, billing and auditing.
- Providing information about an entity (e.g., device capabilities, service features, user preferences, or roles) to other system entities or third parties, according to privacy management rules.

Each CR node needs to be 802.1x-enabled. When user authentication is requested, the CR node creates two virtual ports through which traffic will flow. One port is for control traffic and the other is for data; by default, the port that carries the data is disabled; only the port for carrying the control (Extensible Authentication Protocol over Wireless (EAPOW)) traffic is opened, but this will not carry data traffic if authentication has not been completed.

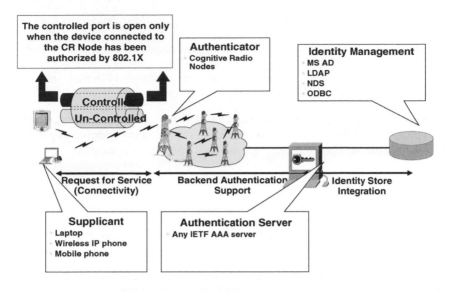

Figure 10.14   802.1x Access Control Model

Next, a security role or clearance server is required. Security roles are necessary for better management and control over the access on a security system, and they also enhance usability of the provided policy administration tools. Security roles are directly associated with the appropriate access policies as these are applied onto the access control lists of CR nodes; for example, policy decision points need to have access at all time to this association. This information is therefore necessary when availability anytime-anywhere is required. These needs, however, pose a great challenge to security.

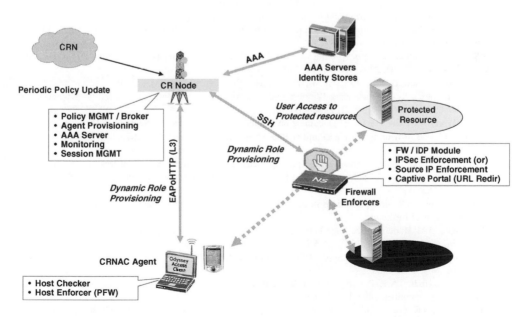

Figure 10.15   Secure cognitive network access control architecture

The integrity and secure handling of this association is mandatory in order to

- avoid false delegation of rights and granting of privileges to users, which may subsequently harm the system;
- avoid denial of service to users who should have rights and privileges when the system fails to make use of the relevant associations and credentials.

Figure 10.15 depicts the security framework architecture for CRNs through mutual authentication based on a 802.1x and AAA server working in concert with layer 3 networking security features such as firewall, virtual private network, intrusion detection and protection mechanisms.

References

[1] J. Mitola, III, G.Q. Maguire. 'Cognitive Radio: Making Software Radios More Personal', *IEEE Personal Communications*, **6**(4), 1999, 13–18.

[2] J. Mitola, III, *Cognitive Radio Architecture*, John Wiley & Sons, Inc., New Jersey, 2006.

[3] S. Haykin, 'Cognitive Radio: Brain-empowered Wireless Communications', *IEEE Journal on Selected Areas in Communications*, **23**(2), 2005, 201–220.

[4] K.C. Chen, L.H. Kung, D. Shiung, R. Prasad, S. Chen, 'Self-Organizing Terminal Architecture for Cognitive Radio Networks', *Proceeding Wireless Personal Multimedia Communications Conference, Jaipur, India, 3–6 December*, 2007.

[5] K.C. Chen, Y.C. Peng, N. Prasad, Y.C. Liang, S. Sun, 'Cognitive Radio Network Architecture: Part I – General Structure', *Proceedings of the ACM International Conference on Ubiquitous Information Management and Communication, Seoul*, 2008.

[6] K.C. Chen, Y.C. Peng, N. Prasad, Y.C. Liang, S. Sun, 'Cognitive Radio Network Architecture: Part II – Trusted Network Layer Structure', *Proceedings of the ACM International Conference on Ubiquitous Information Management and Communication, Seoul*, 2008.

[7] W. Rudin, *Functional Analysis*, McGraw-Hill, 1991.

[8] G. Theodorakopoulos, J.S. Baras, 'On Trust Models and Trust Evaluation Metrics for Ad Hoc Networks', *IEEE Journal on Selected Areas in Communications*, **24**(2), 2006, 318–328, 2006.

[9] H.C. Tijms, *Stochastic Models*, John Wiley & Sons, Inc., New Jersey, 1994.

[10] M. Puterman, *Markov Decision Process*, John Wiley & Sons, Inc., New Jersey, 1994.

[11] V. Poor, *Introduction to Detection and Estimation*, 2nd edition. Springer, 1994.

[12] Y.C. Ho, R.C.K. Lee, 'A Bayesian Approach to Problems in Stochastic Estimation and Control', *IEEE Trans. On Automatic Control*, October 1964, 333–339.

[13] R.P.S. Mahler, 'Multitarget Bayes Filtering via First-Order Multitarget Moments', *IEEE Trans. On Aerospace and Electronic Systems*, **39**(4), 2003, 1152–1178.

[14] S. Thrun, W. Burgard, D. Fox, *Probabilistic Robots*, MIT Press, Cambridge, MA, 2005.

[15] Bob Mayer, Patrick Kelly, 'Enhancing Wireless LAN Security with Cognitive Radios,' http://www.eetasia.com/ART_8800375666_590626_8e30c51f200509.HTM.

[16] Wireless Working Group, 'Technical Document on Cognitive Radio Networks', September 15, 2006.

[17] R. Chen, J.-M. Park, 'Ensuring Trust in Cognitive Radio Networks', Wireless Personal Communication Symposium, June 2006.

[18] R. Chen, J.-M. Park, 'Ensuring Trustworthy Spectrum Sensing in Cognitive Radio Networks', IEEE Workshop on Networking Technologies for Software Defined Radio Networks (held in conjunction with IEEE SECON 2006), September 2006.

[19] J. Polson, 'Cognitive Radio Applications in Software Defined Radio', SDR 04 Technical Conference and Product Exposition. www.sdrforum.org.

[20] M. Zivkovic, M.M. Buddhikot, K. Lagerberg, J. Bemmel, 'Authentication Across Heterogeneous Networks', *Bell Labs Technical Journal*, **10**(2), 2005, 39–56.

[21] A.K. Dey, 'Providing Architectural Support for Building Context-Aware Applications', PhD thesis, Georgia Inst. Tech., USA, November 2000.

[22] A. Mitseva, M. Imine, N.R. Prasad, 'Context-Aware Privacy Protection with Profile Management', WMASH'06, ACM 2006, 29 September, 2006, Los Angeles, CA.

[23] D. Kyriazanos, N.R. Prasad, G.I. Stassinopoulos, 'Ubiquitous Access Control and Policy Management in Personal Networks', IWUAC2006, IEEE Mobiquitous 2006, San Jose, CA, July 2006.

[24] V.M. Gure, R.D. Williams, 'Taxonomies of Attacks and Vulnerabilities in Computer Systems', *IEEE Commun. Surveys Tuts.*, **10**(1), 2008, 6–19.

11

Spectrum Management of Cognitive Radio Networks

CR users are capable of accessing both the licensed portions of the spectrum used by primary users and the unlicensed portions of the spectrum through certain wideband access technology. Consequently, the operation types for CRN can be classified as *licensed band operation* and *unlicensed band operation*, and hybrid *multi-radio co-existing*.

- *Licensed band operation:* The licensed band is primarily used by the primary network. Hence, CRNs are focused mainly on the detection of primary users in this case. The channel capacity depends on the interference at nearby primary users. Furthermore, if primary users appear in the spectrum band occupied by CR users, the latter should vacate that spectrum band and move to available spectrum immediately.
- *Unlicensed band operation:* In the absence of primary users, CR users have the same right to access the spectrum. Hence, sophisticated spectrum sharing methods are required for CR users to compete for the unlicensed band.
- *Multi-radio co-existing:* Within the same geographical region, through a flexible RF segment, CRs might be able to access multiple frequency bands and multiple co-existing radio systems. As described in Chapter 5, a self-organising coordinator is required to operate among licensed band(s) and unlicensed bands in a cooperative way.

In the above three types of CRN operation, CRs might access the CRN through the following possible ways:

- *CR network access:* CR users can access their own CR base station, on both licensed and unlicensed spectrum bands. Because all interactions occur inside the CRN, their spectrum sharing policy can be independent of that of the primary network.
- *CR ad-hoc access:* CR users can communicate with other CR users through an ad-hoc connection on both licensed and unlicensed spectrum bands.
- *Primary network access:* CR users can also access the primary base station through the licensed band. Unlike for other access types, CR users require an adaptive medium access control (MAC) protocol, which enables roaming over multiple primary networks with different access technologies. Please note that the primary network here may mean any available primary system (PS) or existing communication wired/wireless infrastructure or backbone.

The CRN therefore imposes unique challenges due to co-existence with primary systems/networks as well as diverse QoS requirements. Thus, CRN spectrum management functions are require with the following critical design challenges [1]:

- *Interference avoidance:* CR nodes in the CRN should avoid interference with primary systems.
- *QoS awareness:* To decide on an appropriate spectrum band, CRNs should support QoS-aware communication, considering the dynamic and heterogeneous spectrum environment.
- *Seamless communication:* CRNs should provide seamless communication regardless of the appearance of primary users.

The spectrum management process consists of four major steps [1]:

- *Spectrum sensing:* CR users can allocate only an unused portion of the spectrum. Therefore, these users should monitor the available spectrum bands, capture their information and then detect spectrum holes. Chapter 7 and Section 9.4 describe the scope of this technology.
- *Spectrum decision:* Based on the spectrum availability, CR users can allocate a channel. This allocation not only depends on spectrum availability, but is also determined based on internal (and possibly external) policies. Chapters 5 and 8 reveal some of the technology scenario.
- *Spectrum sharing:* There may be multiple CR users trying to access the spectrum, and so CRN access should be coordinated to prevent multiple users colliding in overlapping portions of the spectrum.
- *Spectrum mobility:* CR nodes are regarded as roaming mobile users to the spectrum. Hence, if the specific portion of the spectrum in use is required by a primary user, the communication must be continued in another vacant (or available under appropriate incentives) portion of the spectrum.

The entire spectrum management functions of CRN, including spectrum sensing, access, network layer functions, QoS, etc., can be summarised as shown in Figure 11.1.

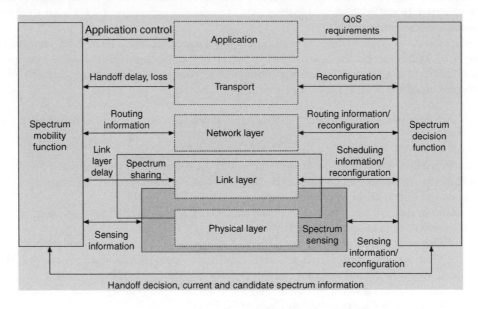

Figure 11.1   CRN management functions (from Figure 3 in [1])

Spectrum sharing and mobility, as part of spectrum management, is introduced in the following sections.

11.1 Spectrum Sharing

Effective spectrum sharing would be the first step toward overall CRN spectrum management. This is a completely new problem and we start from an homogeneous CRN, that is, an ad-hoc network consisting of totally N nodes [2]. Among these nodes, there are a set of L uni-cast communication 'sessions'. $s(l)$ and $d(l)$ denote the source node and destination node of the session $l \in L$ with rate $r(l)$.

We can further denote $\mathcal{M}_i$ as the set of available bands for node $i \in \mathbb{N}$ where $\mathbb{N}$ represents the set of nodes in CRN. $\mathcal{M} = \cup_{i \in \mathbb{N}} \mathcal{M}_i$ as the set of all available bands in CRN. $|\mathcal{M}| = M$ denotes the total number of available bands in (CRN. $\mathcal{M}_{ij} = \mathcal{M}_i \cap \mathcal{M}_j$ is the set of available bands on the link (i, j), beeen node i and node j in $\mathbb{N}$. The bandwidth of the band $m \in \mathcal{M}$ is W_m and this bandwidth might be divided into K_m sub-bands for more flexible resource allocation of bandwidth $u_{m, k} W_m$ while

$$\sum_{k=1}^{K_m} u_{m,k} = 1$$

To study interference, we use the channel gain between node i and node j in $\mathbb{N}$ with distance d_{ij} as in Chapter 6

$$h_{ij} = c_\alpha d_{ij}^{-\alpha}$$

The power spectral density (psd) from the CR-Tx is Q. After radio propagation, successful reception can be achieved if the received psd exceeds the threshold Q_T. On the other hand, if the received interference psd is larger than Q_I, it is non-negligible. The transmission range is therefore $R_T = \sqrt[\alpha]{c_\alpha Q/Q_T}$ and the interference range is similarly $R_I = \sqrt[\alpha]{c_\alpha Q/Q_I}$. In common cases, $Q_I < Q_T$, which is equivalent to $R_T < R_I$.

Our problem is now to assign frequency sub-bands at a node for transmission and reception. The feasible scheduling must ensure no interference (or with tolerable interference) among the nodes. Suppose band m is available at both node i and node j. We set up an indicator function (like a *medium occupancy indicator*):

$$\mathbf{1}_{ij}^{m,k} = \begin{cases} 1, & \text{if communication between node } i \text{ and node } j \text{ over sub-band } m_k \\ 0, & \text{otherwise} \end{cases}$$

For a node $i \in \mathbb{N}$ and a band $m \in \mathcal{M}_i$, the set of other nodes that can use band m is represented by

$$N_i^m = \{j : d_{ij} \le R_T, j \ne i, m \in \mathcal{M}_i\}$$

Since node i cannot simultaneously transmit to multiple nodes over the same frequency sub-band, it gives

$$\sum_{q \in N_i^m} \mathbf{1}_{iq}^{m,k} \le 1 \tag{11.1}$$

Also, node i cannot transmit and receive over the same frequency sub-band, which is consistent with usual radio operation to avoid self-interference and signal leakage. Mathematically, when node i is transmitting, for node j

$$\mathbf{1}_{ij}^{m,k} + \sum_{q \in N_j^m} \mathbf{1}_{jq}^{m,k} \le 1$$

In other words, if $\mathbf{1}_{ij}^{m,k} = 1$, node j cannot use the same band for transmission. If $\mathbf{1}_{ij}^{m,k} = 0$, node j can use the frequency band for one receiving node $q \in N_j^m$.

To deal with potential interference, we denote P_j^m as the set of nodes to generate interference from node j at the band m. That is,

$$P_i^m = \left\{ p : d_{pj} \leq R_I, p \neq i, N_p^m \neq \phi \right\}$$

Consequently, for $p \in P_i^m, p \neq i$,

$$\mathbf{1}_{ij}^{m,k} + \sum_{q \in N_p^m} \mathbf{1}_{pq}^{m,k} \leq 1$$

In the above equation, if $\mathbf{1}_{ij}^{m,k} = 0$, two nodes can produce interference to node j, but are far apart outside the interference range, so that it is possible to use the band for transmission. To further simplify the mathematical equation, let us denote

$$I_i^m = \left\{ p : d_{pj} \leq R_I, N_p^m \neq \phi \right\}$$

which actually means

$$I_i^m = \begin{cases} P_i^m \cup \{j\}, & \text{if} N_p^m \neq \phi \\ P_i^m, & \text{otherwise} \end{cases}$$

The resulting representation of the constraint is

$$\mathbf{1}_{ij}^{m,k} + \sum_{q \in N_p^m} \mathbf{1}_{pq}^{m,k} \leq 1 \, (p \in N_i^m, p \neq i)$$

In the CRN, a CR source node may have a few cooperative relay nodes (and thus paths) to route data to its destination node, for either more reliable transportation or more aggregated bandwidth, to result in overall multi-hop networking. The spectrum sharing can be facilitated by routing the traffic to different paths via different frequency (sub)-bands. We denote $f_{ij}(j)$ as the data rate on link (i, j) that is attributed to session l, where $i \in \mathbb{N}, j \in \cup_{m \in \mathcal{M}_i} I_i^m$, and $l \in L$. We have three possibilities as the flow balance equation:

- If node i is the source node of session l, that is $i = s(l)$, then

$$\sum_{j \in N_i^m} f_{ij}(l) = r(l)$$

- If node i is a relay node for session l, that is, $i \neq s(l)$, $i \neq d(l)$, then

$$\sum_{j \in N_i^m, j \neq s(l)} f_{ij}(l) = \sum_{p \in N_i^m, p \neq s(l)} f_{pi}(l)$$

- If node i is the destination node of session l, that is $i = d(l)$, then

$$\sum_{p \in N_i^m} f_{pi}(l) = r(l)$$

The aggregated flow rates on each radio link should not exceed the corresponding link capacity, which suggests that if node i sends packets to node j over sub-band m_k (i.e., $\mathbf{1}_{ij}^{m,k} = 1$), the capacity on link (i, j) in sub-band m_k is

$$c_{ij}^{m,k} = \mathbf{1}_{ij}^{m,k} u_{m,k} W_m \log_2 \left(1 + \frac{h_{ij} Q}{\sigma_n^2} \right)$$

The link capacity constraint becomes

$$\sum_{l \in L, i \neq d(l), j \neq s(l)} f_{ij}(l) \leq \sum_{m \in \mathcal{M}_{ij}} \sum_{k=1}^{K_m} c_{ij}^{m,k} = \sum_{m \in \mathcal{M}_{ij}} \sum_{k=1}^{K_m} \mathbf{1}_{ij}^{m,k} u_{m,k} W_m \log_2 \left(1 + \frac{h_{ij}Q}{\sigma_n^2}\right)$$

Spectrum sharing of the CRN now needs to find an optimal solution to divide the set of available frequency bands at each node, the scheduling of sub-bands for transmission and reception, and allocation of flow in cooperative relay paths, so that the total required radio bandwidth in CRN is minimised.

Proposition 11.1: CRN spectrum sharing is to find $\mathbf{1}_{ij}^{m,k}$, $u_{m,k}$, and $f_{ij}(l)$ by

$$\min \sum_{i \in \mathbb{N}} \sum_{m \in \mathcal{M}_i} \sum_{j \in N_i^m} \sum_{k=1}^{K_m} \mathbf{1}_{ij}^{m,k} u_{m,k} W_m \tag{11.2}$$

with the following constraints

(i) $\sum_{k=1}^{K_m} u_{m,k} = 1 \ (m \in \mathcal{M})$

(ii) $\sum_{q \in N_i^m} \mathbf{1}_{iq}^{m,k} \leq 1 \ (i \in \mathbb{N}, m \in \mathcal{M}_i, 1 \leq k \leq K_m)$

(iii) $\mathbf{1}_{ij}^{m,k} + \sum_{q \in N_p^m} \mathbf{1}_{pq}^{m,k} \leq 1 \ (i \in \mathbb{N}, m \in \mathcal{M}_i, 1 \leq k \leq K_m, j \in N_i^m, p \in I_l^m, p \neq i)$

(iv) $\sum_{l \in L, i \neq d(l), j \neq s(l)} f_{ij}(l) \leq \sum_{m \in \mathcal{M}_{ij}} \sum_{k=1}^{K_m} \mathbf{1}_{ij}^{m,k} u_{m,k} W_m \log_2 \left(1 + \frac{h_{ij}Q}{\sigma_n^2}\right) \ (i \in \mathbb{N}, j \in N_i^m)$

(v) $\sum_{j \in N_i^m} f_{ij}(l) = r(l) \ (l \in L, i = s(l))$

(vi) $\sum_{j \in N_i^m, j \neq s(l)} f_{ij}(l) = \sum_{p \in N_i^m, p \neq s(l)} f_{pi}(l) \ (l \in L, i \in \mathbb{N}, i \neq s(l), i \neq d(l))$

This optimisation is a *mixed-integer nonlinear programming* problem, which is NP-hard in general. Effective sub-optimal solutions with reasonable complexity are therefore open to CRN spectrum sharing. Finally, please note that the above development is based on frequency domain, but the same principle can be applied to time domain or just logical channel bands by appropriate physical layer constraints.

11.2 Spectrum Pricing

Up to this point, we note that the CRN does not provide anything good for PS(s), and PS definitely operates with some small vulnerable duration to interfere with its original operation. Therefore, appropriate incentives to PS users to allow such flexible spectrum utilisation would be a key to facilitating realistic CRN operation. For efficient dynamic spectrum sharing, an economic model would consequently be required for the spectrum/system operators and spectrum users so that the appropriate incentives (i.e., maybe revenue and hence profit) and the user satisfaction can be maximised. We may generally treat the spectrum owners as operators. Spectrum users may be primary users and secondary users (i.e., cognitive radio users). The priority of primary (PS) users is often larger than secondary (CR) users, depending on the specific policy for CRN operation.

Pricing and resource allocation are closely related. This is due to the fact that while a service provider (typically an operator) wants to maximise its revenue, the user (primary and secondary users) desires to maximise his/her satisfaction in terms of QoS performance and price. From this scenario, there are three parties involved: operators, primary users and secondary (CR) users. Please note that a regulator representing the general public interest, of maximising the entire spectrum utilisation by minimising each user's cost, could be a fourth party in the CRN, and we will discuss this possibility later. These three parties have different utility functions and different incentives to optimise their own 'benefits'. Operators want to maximise the spectrum utilisation to maximise revenue/profit. When there are spectrum bands that are not allocated, the operators can allocate and 'sell' the spectrum bands to secondary CR users in an opportunistic way. The operator can generate extra revenue by doing so and hence make profit). Primary and secondary users want to get better service (more bandwidth and better channel quality) and pay less, while PS users can enjoy more incentives to deduct their payment. Because they have their own utility functions, they may compete and/or cooperate with each other to gain more. For example, when secondary users want to get better QoS, it may make the QoS of the primary users decrease, given a fixed spectrum. We can briefly define the utility functions of each party as follows:

$$U_{Operator} = R_{Primary} + R_{Second} - \text{Cost}$$
$$U_{Primary} = V(\text{spectrum size, QoS}) - \text{Cost(payment)}$$
$$U_{Secondary} = V(\text{spectrum size, QoS}) - \text{Cost(payment)}$$

The cost of $U_{Operator}$ may come from the income of the licensed spectrum, compensation for the users, and so on. Niyato and Hossam [18] developed spectrum pricing in a CRN where multiple primary service providers (i.e., operators) compete with each other to offer spectrum access opportunities to the secondary CR users to improve spectrum utilisation. By using the pricing scheme, each of the primary service providers aims to maximise its profit under the QoS constraint for primary users. This scenario was formulated as an *oligopoly market* consisting of a few firms (operators) and a customer (secondary CR user) by using the *Bertrand game* model, which is shown in Figure 11.2.

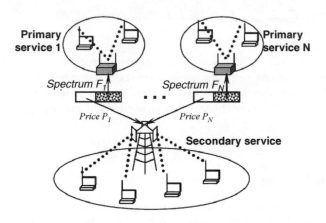

Figure 11.2   System model if spectrum sharing (from [18])

The interaction between the operators and secondary CR users is summarised as follows. Decreasing QoS for primary users by sharing spectrum with secondary users is considered to be the cost of the operators. Operators compensate for the decreasing QoS for the primary users. With the Bertrand model, they analysed the impacts of system parameters such as spectrum substitutability, which represents the ability of the secondary service to switch among the operating frequency spectra offered by the different primary services and channel quality in the *Nash equilibrium*. The distributive algorithm is adopted to obtain the solution for the game. Because the solution of the Nash equilibrium is not the maximised profit for operators, a collusion scheme between operators is proposed to maximise the profits for operators. Because the solution of collusion is not in the Nash equilibrium, the players (operators) want to improve their profits by deviating from their actions. This action decreases other profits. A punishment mechanism to prevent deviated behaviour from operators is thus introduced by forcing primary service providers to consider their long-term profits. The collusion is maintained if the long-term profit due to adopting collusion is higher than that due to deviation.

In this game theory approach, a utility function is used to qualify the spectrum demand of the secondary services. Then, for a primary service, the cost of offering spectrum access to the secondary services is formulated. The cost is based on the degradation in the QoS performance for the local connection. The following commonly used quadratic utility function is used as a secondary utility function:

$$u(\mathbf{b}) = \sum_{i=1}^{N} b_i k_i^{(s)} - \frac{1}{2} \left(\sum_{i=1}^{N} b_i^2 + 2\nu \right) - \sum_{i=1}^{N} p_i b_i$$

where $\mathbf{b}$ is the set consisting of the size of shared spectrum from all the primary services, i.e., $\mathbf{b} = \{b_1, b_2, \cdots, b_N\}$, p_i is the price offered by primary service i and $k_i^{(s)}$ is the spectral efficiency used by a secondary user. This utility also takes spectrum substitutability into account through parameter ν. To derive the demand function for spectrum f_i, they differentiate the $u(\mathbf{b})$ with respect to b_i as follows:

$$\frac{\partial u(\mathbf{b})}{\partial b_i} = 0 = k_i^{(s)} - b_i - \nu \sum_{i \neq j} b_j - p_i$$

We can obtain the demand function given the prices of all primary services by solving the above equation. The spectrum demand function can expressed as follows:

$$D_i = \frac{(k_i^{(s)} - p_i)(\nu(N-2)+1) - \nu \sum_{i \neq j}(k_i^{(s)} - p_i)}{(1-\nu)(\nu(N-1)+1)} \tag{11.3}$$

The revenue of primary user i is $R_i = c_1 M_i$ and the cost is

$$C_i(b_i) = c_2 M_i (B_i^{req} - k_i^{(p)} \frac{W_i - b_i}{M_i})^2$$

where c_1 and c_2 denote weights for the revenue and the cost function. B_i^{req} is the bandwidth requirement for a primary connection. The profit of primary user i is

$$P_i(\mathbf{p}) = b_i p_i + R_i - C_i(b_i)$$

Based on the aforementioned system model, a Bertrand game can be formulated. The players in this game are the primary services. The strategy of each player is the price unit of spectrum (denoted by p_i). The payoff of each player is the profit. The solution of the game is the Nash equilibrium. To obtain the solution, they have to solve the following equation: $\frac{\partial P_i(\mathbf{p})}{\partial p_i} = 0$ for all i. if we want to maximise the total profit of primary services. We must solve the following equation:

$$\frac{\partial \sum_{j=1}^{N} P_j(\mathbf{p})}{\partial p_i} = 0 \tag{11.4}$$

For primary service i, let P_i^o, P_i^n, and P_i^d denote the profits due to the optimal price, the price at the Nash equilibrium and the price due to the deviation, respectively. Then, for the case in which the collusion is maintained forever, the long-term profit of primary service i can be expressed as follows:

$$P_i^o + \delta_i P_i^o + \delta_i^2 P_i^o + \cdots = \frac{P_i^o}{1 - \delta_i}$$

If one primary service deviates from optimal price, that service will gain the deviated profit in the first stage, while during the rest of the stages, the primary service will experience the profit at the Nash equilibrium. Therefore, the long-term profit of primary service i can be expressed as follows:

$$P_i^d + \delta_i P_i^n + \delta_i^2 P_i^n + \cdots = P_i^d + \frac{P_i^n}{1 - \delta_i}$$

The collusion will maintained if the long-term profit due to adopting collusion is higher than that due to deviation, i.e.,

$$\frac{P_i^o}{1 - \delta_i} \geq P_i^d + \frac{P_i^n}{1 - \delta_i} \Longrightarrow \delta_i \geq \frac{P_i^d - P_i^o}{P_i^d - P_i^n}$$

The paper formulates a repeated game to analyse this situation. We illustrate the interaction of operators, primary users and secondary users in this case as shown in Figure 11.3.

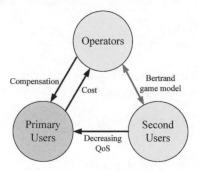

Figure 11.3   Interaction of operators, primary users and second users

This scenario considers the issue of equilibrium among multiple operators and the stability of bidding in a competitive environment, spectrum demand function of the secondary users in presence of spectrum substitutability, and the stability of strategy adaption. However, there is no guarantee on the QoS of the primary users and secondary users. It just compensates primary users for decreasing QoS and limits the users (primary and secondary) through the economic mechanism. It does not access control among all users, or consider spectrum allocation and QoS requirements among end users. The repeated game may need several iterations to create a stable solution. Variation of the number and using the time of the users (primary and secondary) based on the different service models is an important factor of the spectrum utilisation that is still required at this point.

We can further consider the problem of spectrum sharing among a primary user and multiple secondary users [19], again using an oligopoly market competition and a non-cooperative game to obtain the spectrum allocation for secondary users. The static game is developed assuming that the secondary users have the current information of the adopted strategies and the payoff of each other, to explore the case of bounded rationality in which the secondary users gradually and iteratively adjust their strategies based on the observation of previous strategies (i.e., dynamic game).

The analytical scenario depicted in Figure 11.4 considers a primary user and N secondary (CR) users. The primary user wants to share portion of its spectrum (b_i) to secondary users. In a general non-cooperative game model for an oligopoly market, all the firms compete in terms of product quality. Here the firms are secondary users and each wants to compete with spectrum size (i.e., product quality). The profit of the secondary users is based on the charge by the primary user and the benefits gained from the allocated spectrum. In the static game model, the players are secondary users, the strategy of each player is the requested/allocated spectrum size (denoted by b_i for secondary user i) and the payoff function of each secondary users represents profits. The pricing function used by the primary user is defined as

$$c(\mathbf{B}) = x + y(\sum_{b_j \in \mathbf{B}} b_j)^\tau \tag{11.5}$$

where x, y and τ are nonnegative constants, $\tau \geq 1$ (so that this pricing function is convex), and $\mathbf{B}$ denotes the set of all secondary users (i.e., $\mathbf{B} = \{b_1, \ldots, b_N\}$). The revenue of secondary users i can be obtained from $r_i \times k_i \times b_i$, where r_i is the revenue of secondary users i per unit of achievable transmission rate and k_i is the transmission rate that can be obtained by secondary user i. Hence, the profit of secondary user i can be obtained as follows: $\pi_i(\mathbf{B}) = r_i \times k_i \times b_i - b_i c(\mathbf{B})$. The best response of secondary user I, given the size of the shared spectrum by other secondary users b_j, where $j \neq i$, is defined as follows: $BR(\mathbf{B}_{-i}) = \arg\max_{b_i} \pi_i(\mathbf{B}_{-i} \cup \{b_i\})$. Mathematically, one can obtain the Nash equilibrium of this game if and only if

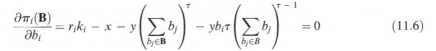

$$\frac{\partial \pi_i(\mathbf{B})}{\partial b_i} = r_i k_i - x - y \left(\sum_{b_j \in B} b_j \right)^\tau - y b_i \tau \left(\sum_{b_j \in B} b_j \right)^{\tau-1} = 0 \tag{11.6}$$

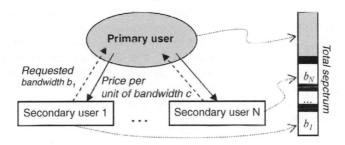

Figure 11.4   System model for spectrum sharing [19]

The dynamic game is somewhat different from static game. The adjustment of requested/allocated spectrum size can be modelled as a follows:

$$b_i(t+1) = Q(b_i(t)) = b_i(t) + \alpha_i b_i(t) \frac{\partial \pi_i(\mathbf{B})}{\partial b_i(t)}$$

where $b_i(t)$ is the allocated spectrum size at time t and α_i is the adjustment speed parameter (i.e., learning rate) of secondary users.

We can clearly see the interaction of the primary user and secondary users from this scenario shown in Figure 11.5, mainly focusing on the requirement of 'harmonic' co-existence of both licensed and unlicensed users. An important question is whether the primary user can determine the price or share the spectrum bands by itself without the control of the operator, which may invoke some interference to other primary users. Nevertheless, the scenario can make sure that the spectrum utilisation can be improved in an efficient way.

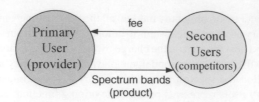

Figure 11.5 Interaction of the primary user and second users

We may observe that the collusion among selfish network users may seriously deteriorate the efficiency of the dynamic spectrum sharing. In order to fully use scarce spectrum resources by using the dynamic spectrum allocation, spectrum allocation in wireless networks with multiple selfish legacy spectrum holders and unlicensed users can be modelled as a *multi-stage dynamic game* [20] with a general network scenario in which multiple primary users (legacy spectrum holders) and secondary CR users (unlicensed users) co-exist and primary users attempt to 'sell' unused spectrum resources to secondary users for incentives or gain. For instance, primary users can be the spectrum broker connected to the core network and secondary users the base stations equipped with CR technologies; or primary users can be the access points of a mesh network and secondary users mobile devices. Such a pricing game can be modelled as an *auction* game.

One prevalent cheating behaviour is bidding collusion among primary or secondary users. To be specific, the bidders (or sellers) act collusively and engage in bid-rigging with a view to obtaining lower prices (or higher prices). The outcome deviates from *Competitive Equilibrium* (CE): it is the price at which the number of buyers willing to buy is equal to the number of sellers willing to sell. In order to combat user collusion, a pricing-based collusion-resistant dynamic spectrum allocation is proposed to optimise overall efficiency, while not only keeping the participating incentives of the selfish users but also combating possible user collusion. The method against collusion is to set an optimal reversed price. The reversed price is based on the number of collusive users. But the number of the collusive users may be unknown (i.e., incomplete information). The players need to construct briefs of other player's future possible strategies to assist their decision making.

The expected revenue of an operator from second users with reverse price ϕ_{r,p_i} by releasing a spectrum band can be written as

$$E_{V_i}[U_{p_i}(a_i^j, \phi_{r,p_i})] = (\phi_{r,p_i} - E[b_i])(F_{C_M^{-i}}(\phi_{r,p_i}) - F_{C_{M-1}^{-i}}(\phi_{r,p_i})) + \int_{\phi_{r,p_i}}^{M_b} (z - E[b_i])f_{C_M^{-i}}(z)dz \tag{11.7}$$

Assuming that an interior maximum exists for the above equation, the optimal reverse price is ϕ_{r,p_i}^* satisfying the following first-order condition of the above equation:

$$F_{C_M^{-i}}(\phi_{r,p_i}^*) - F_{C_{M-1}^{-i}}(\phi_{r,p_i}^*) - F_{C_M^{-i}}(\phi_{r,p_i}^*)f_{C_{M-1}^{-i}}(\phi_{r,p_i}^*) = 0 \tag{11.8}$$

with *a prior* knowledge about the exact distribution of C_{M-1}^{-i} and C_M^{-i}. Sometimes, the operator does know the collusion among users. Hence, how to further obtain an optimal reverse price considering the constraints is still unanswered. It is thus suggested that the secondary users' belief is the ratio of their bid being accepted at different price levels. Let y be the bid price of secondary users. The ratio of bids from secondary users at y that can have been accepted can be written as

$$\tilde{r}_s(y) = \frac{\eta_A(y)}{\eta(y)}$$

where $\eta_A(y)$ and $\eta(y)$ are the number of accepted bids at y, respectively. $r_S{}^\sim(y)$ can be usually accurately estimated if a great number of buyers are participating in the pricing at the same time. We have the following observations: if a bid $y^\sim > y$ is rejected, the bid at y will also be rejected; if a bid $y^\sim < y$ is accepted, the bid at y will also be accepted. To find the secondary users' belief, for each potential bid at y, define:

$$\overset{\cup}{r_S}(y) = \begin{cases} 1 & y = 0 \\ \dfrac{\sum\limits_{w \ge x} \eta_A(w)}{\sum\limits_{w \ge x} \eta_A(w) + \sum\limits_{w \le x} \eta_R(w)} & y \in (0, M_b) \\ 0 & y \ge w \end{cases}$$

where $\eta_R(w)$ is the number of bids at w that have been rejected and M_b is a large enough value so that the bids greater than M_b will be definitely. Therefore the belief function above can be represented as the CDF of C_M^{-i}.

The scenario only considers co-existence of operators and secondary users or co-existence of primary users and secondary users, without considering all the three parties at the same time. The goal is to improve the spectrum utilisation and maximise total utilities, from the interaction between operators and secondary users or primary users and secondary users as shown in Figure 11.6.

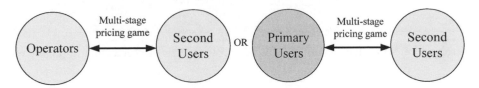

Figure 11.6   Interaction of operators, primary users and second users

Through multi-stage dynamic games, coordinating the spectrum allocation among primary and secondary users through a bilateral pricing process can be achieved to maximise the utilities of both primary and second users according to spectrum dynamics, while focusing on the countermeasures to selfish users' collusion behaviour. A distributed collusion-resistant dynamic pricing approach with optimal reserve prices is designed to achieve efficient spectrum allocation while combating user collusion. The pricing overhead can be significantly decreased by introducing a *belief function* for each user to help decision making. Moreover, the *Nash Bargain Solution* is applied for deriving the performance lower bound of the proposed scheme considering the presence of user collusion. The limitation of this approach is that the secondary users cannot bid based on their requirements of spectrums.

The key challenge to improving spectrum utilisation, by enabling CR users to access the spectrum dynamically without disturbing licensed primary users, is to implement an efficient MAC mechanism that can efficiently and adaptively allocate transmission powers and spectrum bands among CR users based on surrounding environments. A novel joint power/channel allocation scheme that improves the performance through a distributed pricing approach is now proposed and the spectrum allocation is modelled as a *non-cooperative game*. A priced-based iterative water-filling (PIWF) algorithm to enable CR users to reach a good *Nash equilibrium* (NE) is proposed and implemented in a distributed way with CRs repeatedly negotiating their best transmission powers and spectrums. A MAC protocol is also suggested to implement the radio resource management.

Focusing on an opportunistic CRN where the CR users are secondary users that co-exist with the primary users, the interaction among secondary users is taken into account. The players of game model are secondary users, with corresponding actions as transmission powers. The goal is thus to maximise the transmission rate, and spectrum pricing is used to improve the total transmission rate of secondary users. Power constraint implemented by the MAC protocol has been first introduced in spectrum management.

Each non-cooperative CR user is interested in maximising its own achievable rate. Such a greedy mechanism can be modelled as a game: $\vartheta = \{\Omega, P, \{U_i\}\}$, $\Omega = \{1, 2, \cdots, N\}$ (the links of CR users), where $P = P_1 \times P_2 \times \cdots \times P_N$ is the action space with P_i (Power) and $U_i : P \to \Re$ is the utility function of player i. Players are CR users. To maximise the transmission rate, the solution of the game is again a NE, though such a solution is not a global maximised sum transmission rate. That is, the resulting NE may be far from the Pareto optimal. In practice, we are interesting in maximising a sum of the utilities of all users, defined as $\max_P \sum_{i \in \Omega_N} U_i(P_i, P_{-i})$. To derive the NE towards the Pareto optimal boundary, pricing is used as an incentive for each selfish user to work in a cooperative manner. The new utility function with pricing is then defined: $\tilde{U}_i(P_i, P_{-i}) = \sum_{K \in \Omega_K} u_i^{\sim}(P_i(f_k))$. The game formulation is defined as follows:

$$\max_{P_i} \tilde{U}_i(P_i, P_{-i}), \forall i \in \Omega_N$$

$$s.t$$

$$C_1 : P_i(f_k) \geq 0, \forall i \in \Omega_N \ and \ \forall k \in \Omega_K$$

$$C_2 : \sum_{k \in \Omega_K} P_i(f_k) \leq P_{\max}, \forall i \in \Omega_N \quad\quad\quad (11.9)$$

$$C_3 : p_i(f_k) \leq P_{mask}(f_k), \forall i \in \Omega_N \ an$$

If there exists a NE for the game ϑ and this NE is Pareto optimal, then the linear pricing function factor is

$$\lambda_i(f_k)^{opt} = \sum_{j \in NBR_i} \frac{h_{jj}(f_k) P_j(f_k) h_{ij}(f_k)}{M_j(f_k)(M_j(f_k) + h_{jj}(f_k) P_j(f_k))} \quad\quad\quad (11.10)$$

The linear pricing function can be found by Lagrangian Multiplier. The Lagrangian function for user i can be expressed as

$$
\begin{aligned}
J_i &= \sum_{k \in \Omega_K} \tilde{u}_i(P_i(f_k)) + \sum_{k \in \Omega_K} \alpha_{i,k} P_i(f_k) - \beta_i(\sum_{k \in \Omega_K} P_i(f_k) - P_{\max}(f_k)) \\
&= \sum_{k \in \Omega_K} u_i(P_i(f_k) - \lambda_i(f_k) P_i(f_k)) + \sum_{k \in \Omega_K} \alpha_{i,k} P_i(f_k) - \beta_i(\sum_{k \in \Omega_K} P_i(f_k) - P_{\max}(f_k))
\end{aligned}
$$

The higher pricing factor $\lambda_i(f_k)$ intuitively presents user i from using a large transmission power on channel k. Two types of algorithms can be proposed for the NE solution: *sequential PIWF* and *parallel PIWF*. Each individual CR user, say i, first adjusts its pricing factor $\lambda_i(f_k)$ over all channels according to the game formulation and then determines its best action, i.e., the optimal channel/power/rate combination, by measuring the total noise plus interference M_i over all channels. The best response of user i is to maximise his/her utility function subjects to the constraints C1~C3. The same procedure is repeated for all users in the CRN. If the users are to make their best-response decisions sequentially according to a fixed order, they have the following sequential PIWF algorithm.

Algorithm 11.1 Sequential PIWF

0: Initialise $P_i(f_k) = 0, \forall i \in \Omega_N, k \in \Omega_K$; initialise iteration count $l = 0$.
1: Repeat iterations:
2: $l = l + 1$;
3: **for** $i = 1$ to N users **do**
4: **for** $k = 1$ to N channels **do**

5: Estimate the total interference plus noise level $M_i (f_k)$;
6: Compute the pricing factor $\lambda_i (f_k)$; using step 5;
7: Estimate the channel gain $h_{ii}(f_k)$ using the received signal power of the control packet.
8: **end for**
9: $P_i^{(l)} = \text{BR}_i(P_{-i}) = [P_1^{(l)}, \ldots, P_{i-1}^{(l)}, P_{i+1}^{(l-1)}, \ldots, P_N^{(l-1)}]$.
10: Transmit on selected channels using $P_i^{(l)}$.
11: **end for**
12: until $l > L_{\max}$ or $\left(\left\| P_i^{(l)} - P_i^{(l-1)} \right\| / \left\| P_i^{(l-1)} \right\| \right) \leq \varepsilon$ for all $i \in \Omega_N$.

Otherwise, if the users are to make their best-response decisions simultaneously, they have a parallel PIWF algorithm as follows.

Algorithm 11.2 : Parallel PIWF

0: Initialise $P_i(f_k) = 0, \forall i \in \Omega_N, k \in \Omega_K$; initialise iteration count $l = 0$.
1: Repeat iterations:
2: $l = l + 1$;
3: **for** $i = 1$ to N users **do**
4: **for** $k = 1$ to N channels **do**
5: Estimate the total interference plus noise level $M_i (f_k)$;
6: Compute the pricing factor $\lambda_i(f_k)$ using step 5;
7: Estimate the channel gain $h_{ij} (f_k)$ using the received signal power of the control packet.
8: **end for**
9: $P_i^{(l)} = \text{BR}_i(P_{-i}) = \left[P_1^{(l)}, \ldots, P_{i-1}^{(l)}, P_{i+1}^{(l-1)}, \ldots, P_N^{(l-1)} \right]$.
10: **end for**
11: **for** $i = 1$ to N users **do**
12: Transmit using $P_i^{(l)}$.
13: **end for**
14: until $l > L_{\max}$ or $\left(\left\| P_i^{(l)} - P_i^{(l-1)} \right\| / \left\| P_i^{(l-1)} \right\| \right) \leq \varepsilon$ for all $i \subset \Omega_N$.

Both algorithms require system parameters to be correctly estimated for each CR, which may not be practical. To overcome the problem, a 'relaxed' algorithm is proposed by requiring each CR to remember its most recent policy choices together with the choices of others. A relaxed version of the sequential PIWF and parallel PIWF algorithm can be achieved if the best response function in Algorithm 11.1 and Algorithm 11.2 are replaced by

$$P_i^{(l)} = wP_i^{l-1} + (1-w)\text{BR}_i(P_1^l, \cdots, P_{i-1}^l, P_{i+1}^{l-1}, \cdots, P_N^{l-1})$$

where w is interpreted as the memory factor.

By auction approach, 15] provides a compromise spectrum management to meet the requirement among all parties in CRN.

11.3 Mobility Management of Heterogeneous Wireless Networks

CRN, as a heterogeneous wireless network, is most likely realised as an *Internet Protocol* (IP) network. With almost no literature concerning CRN mobility, we focus on the introduction of mobile IP management in heterogeneous wireless networks.

In order to achieve the transparency to the individual network system, IP is used as the inter-connect protocol. The IP-based interconnection has the advantage of hiding the heterogeneities of the lower layer technologies from higher layers. IP is also a more general communication protocol, providing the globally successful infrastructure for the upper application in a cost effective way. It is expected to become the core backbone network associated with CRN. Ian F. Akyildiz, Shantidev Mohanty and Jiang Xie [27] proposed a method to construct the mobile IP architecture, by defining two new entities: Network Interoperating Agent (NIA) and Interworking Gateway (IG). For example, there is a mobile terminal/node, which roams to WLAN network from GPRS network. As Figure 11.7 shows, the IG is constructed in the GPRS network and connected between Gateway GPRS Support Node (GGSN) and NIA. In the WLAN network, IG is connected between an Access Router (AR) and NIA. IG is responsible for the mobility management, traffic management, authentication and accounting. NIA consists of the operator database, handoff management, authentication and accounting. Many man-agement units such as mobility, traffic, authentication and accounting are all constructed in the IG. In these cases, the operator database and handoff are in the NIA. In addition, QoS is considered in the handoff management of NIA and mobility management of IG. To implement the QoS, both NIA and IG need to take account of the scheduling function, too. These issues raise the great complexity in the implementation.

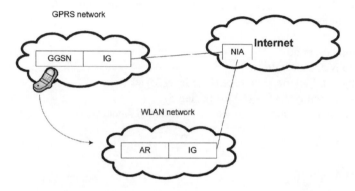

Figure 11.7   NIA-based architecture in the mobile IP network

To achieve handoff management in mobile IP, the architecture requires three new functional entities: home agent (HA), foreign agent (FA) and a mobile terminal (MT) or mobile node (MN). The relation of these entities is shown in Figure 11.8.

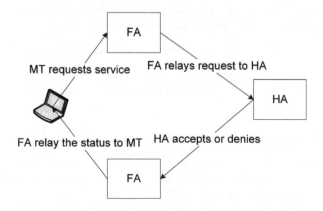

Figure 11.8   The relation between three new entities in mobile IP

The movement of mobile users between two network domains is referred to as *macro-mobility*; the movement of mobile users between two subnets within one domain is referred to as *micro-mobility*. This is the same as the *horizontal handoff* and *vertical handoff* we mentioned previously. The macro-mobility is summarised as follows:

- *Agent discovery:* A MN is able to detect whether it has moved into a new subnet by periodically receiving unsolicited Agent Advertisement messages broadcast from each FA. A MN can also send Agent Solicitation messages to learn about the presence of any prospective mobility agent.
- *Registration:* When a MN discovers it is in a foreign network, it obtains a new care-of address (CoA). This CoA can be obtained by soliciting or listening for FA advertisements (a FA CoA), or contacting the Dynamic Host Configuration Protocol (DHCP) or Point to Point Protocol (PPP) (a collocated CoA). The MN registers the new CoA with its HA. Then the HA updates the mobility binding by associating the CoA of the MN with its permanent IP address.
- *Routing and tunnelling:* Packets sent by a correspondent node (CN) to a MN are intercepted by the HA. The HA encapsulates the packets and tunnels them to the MN's CoA. With a FA CoA, the encapsulated packets reach the FA serving the MN, which 'decapsulates' the packets and forwards them to the MN, as shown in steps a, b, and c in Figure 11.9. With a collocated CoA, the encapsulated packets reach the MN, which then decapsulates them. In Figure 11.9, the *tunnelling* ends at the MN instead of at the FA.

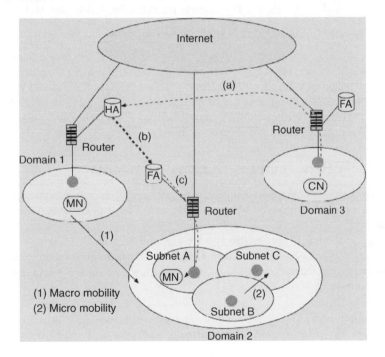

Figure 11.9   Mobile IP architecture (from [28])

When a MN moves from one subnet to another, the handoff procedure is carried out by the following steps, as in Figure 11.10. This may explain the Handoff Management in Mobile IP.

- The MN obtains a new CoA when it enters a new subnet.
- The MN registers the new CoA with its HA. The HA sets up a new tunnel to the end point of the new CoA and removes the tunnel from the old CoA.

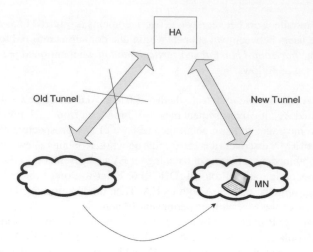

Figure 11.10   HA sets up the new tunnel when MN moves to a new subnet

- Once the new tunnel is set up, the HA tunnels packets destined to the MN using the MN's new CoA.

Some more issues should be considered in handoff, such as reduction of the service delay or interruption caused from handoff. To provide the service quality assurance, the mobile agents can request the reservation before the mobile terminal or user handoff to another (heterogeneous) wireless network. Moreover, good bandwidth reservation policy can overcome the service interruption caused by network congestion, and can reduce the blocking rate of new generated calls and handoff calls. As far as micro-handoff in CRN is concerned, it is still an open problem.

11.4 Regulatory Issues and International Standards

CRN is not only a technology but also involves the economic side as mentioned in Section 11.2. Furthermore, CRN is heavily involved with regulation, and international standardisation efforts have been on-going.

The term 'cognitive' originally referred to a device's capability to sense the surrounding environment conditions and adapt its behaviour accordingly. Thus, cognitive radio is based on the methodology of humans by understanding, learning and then adapting to the surrounding. Cognitive radio as defined by J. Mitola III as follows:

> *The term cognitive radio identifies the point at which wireless personal digital assistants (PDAs) and the related networks are sufficiently computationally intelligent about radio resources and related computer-to-computer communications to:*
> (a) *detect user communications needs as a function of use context, and*
> (b) *provide radio resources and wireless services most appropriate to those needs.*

Features such as Observe, Adapt, Reason, Learn, Plan, Decide and Act make CR a 'self-aware' radio, as shown in Figure 11.11. State-of-the-art terminal re-configurability is the sole responsibility of the user device, which manually or automatically switches and reconfigures the terminal as it moves to different geographic locations, or according to his various needs. CR is an 'intelligent' radio, which can reconfigure itself by sensing the environment and adapting its transmission accordingly under given circumstances, completely transparently to the user.

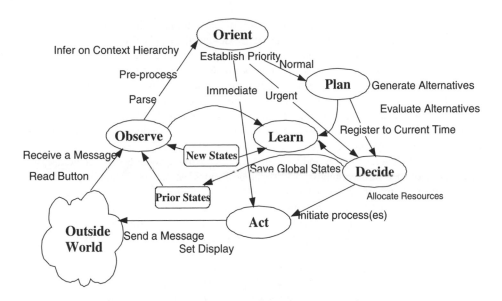

Figure 11.11 Simplified cognitive radio high-level operation

Before CR becomes a reality, a lot of issues have to be dealt with. Starting from the underlying technology to the regulatory aspects of deployment of CR, everything needs to be looked into with detailed analysis and research to provide a flexible and efficient radio platform for future use. Among the various concerns, proper and improved spectrum utilisation and interference to primary users contained within acceptable limits are of prime importance.

11.4.1 Regulatory Issues

The major applications of CR create potential regulatory issues in Radio Resource Management (RRM), spectrum management and restriction of insignificant interference. The main objective of RRM is to use efficiently the scarce available radio resources. Spectrum management deals with efficient use of available spectrum either by spectrum sharing or dynamic spectrum allocation. The large numbers of heterogeneous devices produce variable levels of interference, and the major concern for CR is to produce an acceptable level of interference to the licensed systems.

The success of reconfigurable network operation depends upon effective management of the available resources:

- more efficient utilisation of available spectrum;
- an intelligent network planning process;
- management of radio resources belonging to different RATs with fixed spectrum allocation.

RRM is a complex but necessary process. It consists of dynamically managing resources such as spectrum, as well as allocating traffic dynamically to the Radio Access Technologies (RATs) participating in a heterogeneous wireless access infrastructure. Consequently, RRM can be seen as a superset of Spectrum Management and Joint Radio Resource Management (JRRM).

In new spectrum management, specific RATs should not have fixed frequency bands applied to them. Dynamic allocation of bands to RATs can be done through intelligent management mechanisms so that the capacity of each RAT is maximised and interference is minimised. The radio resource is not only the radio spectrum, but also realised in the real radio network as access rights for individual mobile users, time period that a mobile user is active, channelisation codes, transmission power, connection mode, etc., which require the management functions being designed in different time scales. JRRM is a mechanism for controlling communication between heterogeneous networks, in which mobile terminals and radio networks consisting of a number of RATs co-exist together and cooperate with each other, but with fixed spectrum allocation. These mechanisms work simultaneously over multiple RATs with the necessary support of reconfigurable/multimode terminals.

There is a strong relationship between resource management and CR: the latter will provide the technical means for determining in real-time the best band and best frequency to provide the services desired by the user at any time.

The growth in wireless technology and ever-increasing demand for wireless multimedia services has created the issue of spectrum availability. Multimedia services require significant bandwidth that even the existing 3G technology does not seem capable of fulfilling with today's spectrum regulation. Contrary to the popular belief that there is a scarcity of radio spectrum, actual surveys show that most of the allocated spectrum is vastly underused at any specific time and location. Spectrum access and its dynamic allocation is a more significant problem than the physical scarcity of spectrum. Thus, there is a need to change the archaic spectrum management regulations.

Stepping towards efficient spectrum usage, in 2004, the Federal Communications Commission (FCC) issued a Notice of Proposed Rule Making (NPRM) raising the possibility of permitting unlicensed users/secondary users to 'borrow' spectrum temporarily from licensed holders, as long as no undue interference is seen by the primary user. Instead of having two alternatives (licensed or unlicensed spectrum) dynamic spectrum assignment can provide a more flexible way of coordinating the spectrum utilisation. CR has the capability to sense its spectrum and reconfigure its transmission ability (power levels, modulation schemes and protocols) while co-existing with licensed spectrum users.

11.4.1.1 Regulatory Certification for CR

CR needs a generalised testing procedure for certification of its regulatory aspects to ensure its proper operation in the ecosystem. A study group in the IEEE 1900 committee has recently been formed to investigate technical issues and standardisation opportunities for correct operation according to the specification for dynamic spectrum access of CR. Spectrum policies are vastly variable in nature, depending upon country, spectrum owner, time and geographical location. In any of its operational modes, to ensure compliance to spectrum regulations, CR needs these spectrum policies as an integral part of its knowledge base. Work is on-going to express the spectrum policies in a machine readable language in the Defense Advanced Research Projects Agency – neXt Generation (DARPA XG) programme. Ontology web Language (OWL) is used for this purpose. A regulatory framework and enforcement is needed for monitoring, detecting and stopping policy violations. Policies are to be updated in a secure manner utilising a secure method such as digital certificates or light weight crypto systems. The software reconfiguration should not modify the secure module and it should be sufficiently tamper-proof.

11.4.1.2 Concurrent Transmission

The concept of filling the void in the wireless spectrum was the basis of CR technology. But researchers are also looking for more flexible and spectrally efficient approaches of concurrent transmission at the same time or over the same frequency band as that of the primary user. Under this possibility the CR does not need to wait for an idle channel to start transmission. But there are fundamental limits on the communication possible over such a channel, such as the achievable data rates at which two users can transmit, and its comparison parameters when the devices are not CR, regulatory issues and engineering problems to be faced. Also measures must be taken to ensure that the concurrent secondary user does not abuse its rights and adversely affect concurrent users.

11.4.1.3 Cooperation among CRs

Cooperation between CRs operating in the same band can reduce the detection time and thus increase the overall agility. Compared to totally non-cooperative schemes, cooperation among CRs increases agility gain by manifolds.

11.4.1.4 Interference

CRs not only causes out-of-cell interference but also out-of-system interference. The large number of heterogeneous devices with varying powers, duty cycles and even propagation path losses are sources for this interference. Interference can be mitigated if the CR is able to approximate the proximity of primary users and adjust its power level accordingly. Adequate sensitivity is a fundamental requirement for a practical CR in order to detect low SNR signals in the conditions of shadowing. A consequent major concern for CR is to induce an acceptable level of interference to the licensed or legacy primary systems. The fundamental question always exists of whether cognitive systems can always operate without causing harmful interference to legacy users, because it seems that any reliable sensing may still have vulnerable situations. More precisely, a CR may not be able to detect a primary user reliably and therefore may start sending even though the primary user is using the radio resource, which makes the technology related in Chapter 7 critical for practical regulations.

Another possible angle from which to consider this challenge is that interference can also be catered for based on interference temperature limit. For a given frequency in a given geographic region, an interference temperature is defined. In 2003, the idea of interference temperature for 'quantifying and managing interference' was introduced by the FCC. By regulating the received power, CRs are capable of measuring the current interference environment and using the frequency band by adjusting their transmission characteristics without causing harmful interference to the licensees. Therefore, Interference Temperature Multiple Access (ITMA) using the interference temperature to regulate power and bandwidth in a CDMA-like network has been proposed. It is a promising new multiplexing technique for multi-user ad-hoc wireless communication networks within licensed spectrum bands without interfering with existing signals. There is on-going research to fully understand the interactions between primary users and ITMA-based secondary spectrum users.

11.4.1.5 Beacon Concept

The spectrum sharing in CRN can be divided into two types:

- *Vertical spectrum sharing:* The secondary users periodically search for the vacant band in the spectrum to transmit and receive data. Upon sensing the return of the primary users, it vacates the band and switches to another vacant spectrum band.
- *Horizontal spectrum sharing:* If a secondary user detects the presence of another secondary user in its spectrum band and there is no better chance to get another band, it may decide to share the band with the other secondary user.

Horizontal and vertical spectrum sharing, as shown in Figure 11.12, is done by CRs depending upon the regulatory status of the radio system that operates in the same spectrum. Horizontal sharing provides flexibility and a higher level of freedom for secondary radio systems. Coordination in vertical spectrum sharing involves a novel approach of using *beacons*.

The beacon concept is an operator assisted signalling mechanism that provides dynamic spectrum assignment. It is proposed to coordinate secondary spectrum utilisation with centralised decision making. Permission (grants) and denials of spectrum access are two signalling mechanisms or beacons transmitted by primary users. When used individually these beacons have limited reliability but when

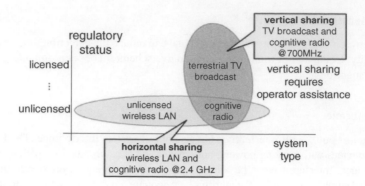

Figure 11.12   Vertical sharing and horizontal sharing: depending on the regulatory status of the other radio systems, the CRs share the spectrum with different types of systems

used simultaneously, spectrum usage can be coordinated more efficiently. Beacons can be implemented with various approaches. They are transmitted at some power and detected at some sensitivity. Detection of a grant beacon and no denial beacon only allows the CR to use the spectrum. Simultaneous detection of both beacons refrains CR from spectrum use. This beacon concept ensures high reliability and an interference free environment.

11.4.2 International Standards

There are some on-going efforts related to international standards for cognitive radios. The Wireless World Research Forum (WWRF) has a working group in cognitive radios and another special interest group in self-organising networks. IEEE 802.19 is dealing with technical advice to multi-radio co-existing systems, and cognitive radio technology is definitely within its scope.

The IEEE 802.22 Working Group was formed in November 2004 for Wireless Regional Area Networks (WRAN). This working group is dedicated to develop an air interface (i.e., MAC and PHY) based on CRs for unlicensed operations in the TV broadcast bands. This standard plays a key role in the evolution of CRs and aims to define an international standard that may regulate in any regulatory regime [5].

The IEEE P1900 Standards Group was established in the first quarter of 2005 jointly by the IEEE Communications Society and the IEEE Electromagnetic Compatibility Society. The objective is to develop supporting standards dealing with new technologies and techniques being developed for next generation radio and advanced spectrum management. Please note that this sponsor group is called SCC41 – The Standards Coordination Committee 41. The individual working groups are still called IEEEP1900.x.

There are the following working groups within IEEE P1900.

- *IEEE P1900. 1:* Standard Terms, Definitions and Concepts for Spectrum Management, Policy Defined Radio, Adaptive Radio and Software Defined Radio.
- *IEEE P1900.2:* Recommended Practice for Interference and Coexistence Analysis.
- *IEEE P1900.3:* Recommended Practice for Conformance Evaluation of Software Defined Radio (SDR) Software Modules.
- *IEEE P1900.a:* Dependability and Evaluation of Regulatory Compliance for Radio Systems with Dynamic Spectrum Access. This is at present a study group. It has submitted a PAR, however, moves are on to merge this with the P1900.3.
- *IEEE P1900.4:* Architectural Building Blocks Enabling Network-Device Distributed Decision Making for Optimised Radio Resource Usage in Heterogeneous Wireless Access Networks.

* *IEEE 1900.A:* Dependability and Evaluation of Regulatory Compliance for Radio Systems with Dynamic Spectrum Access.

Although standardisation efforts are more prevalent in the CR link level, together with the booming Internet CRNs are expected to become a reality in a few years time.

References

[1] I.F. Akyildiz, W.-Y. Lee, M.C. Vuran, S. Mohanty, 'A Survey of Spectrum Management of Cognitive Radio Networks', *IEEE Communications Magazine*, April 2008, 40–48.

[2] Y.T. Hou, Y. Shi, 'Spectrum Sharing for Multi-hop Networking with Cognitive Radio', *IEEE Journal on Selected Areas in Communications*, **26**(1), 2008, 146–155.

[3] FCC, Cognitive Radio Workshop, 19 May, 2003, [Online]. Available: http://www.fcc.gov/searchtools.html.

[4] J. Mitola, Cognitive Radio: An Integrated Agent Architecture for Software Defined Radio, Ph.D. dissertation Royal Institute of Technology (KTH), 2000.

[5] C. Cordeiro, K. Challapali, D. Birru, S.N. Shankar, 'IEEE 802.22: The First Worldwide Wireless Standard based on Cognitive Radios', *IEEE Journal*, 2005.

[6] J. Mitola III, G.Q. Maguire, Jr., 'Cognitive Radio: Making Software Radios More Personal', *IEEE Personal Communications*, **6**(4), 1999, 13–18.

[7] J. von Neumann, O. Morgenstein, *Theory of Games and Economic Behavior*, Princeton University Press, New Jersey, 1947.

[8] D. Fudenberg, D.K. Levine, *The Theory of Learning in Games*, MIT Press, Cambridge, MA, 1999.

[9] J. Mitola III, 'Signal Processing Technology Challenges of Cognitive Radio'.

[10] 'Information on IEEE P1900 Purpose, Plans, Progress and Relevance to the work of ITU-R WP8F', May, 2005.

[11] P. Martigne, 'Overview of Some Standardization Activities on Cognitive Radio', ETSI Workshop on SDR/CR, February 2007.

[12] D. Cabric, I.D. O'Donnel, M.S.-W. Chen, R.W. Brodersen, 'Spectrum Sharing Radios', *IEEE Circuits and Systems Magazine*, Second Quarter 2006.

[13] NTIA, 'U.S. frequency allocation', [Online]. http://www.ntia.doc.gov/osmhomc/allochrt.pdf

[14] First Report and Order, Federal Communication Commission Std. FCC 02–48, Feb. 2002.

[15] H.B. Chang, K.C. Chen, N. Prasad, C.W. Su, 'Auction Based Spectrum Management for Cognitive Radio Networks', IEEE Vehicular Technology Conference, Spring, 2009.

[16] Q. Zhao, B.M. Sadler, 'A Survey of Dynamic Spectrum Access', *Signal Processing Magazine IEEE*, **24**(3), 2007, 79–89.

[17] S. Haykin, 'Cognitive Radio: Brain-empowered Wireless Communications', *IEEE Journal on Selected Areas in Communications*, **23**(2), 2005, 201–220.

[18] D. Niyato, E. Hossain, 'Competitive Pricing for Spectrum Sharing in Cognitive Radio Networks: Dynamic Game, Inefficiency of Nash Equilibrium, and Collusion', *IEEE Journal on Selected Areas in Communications*, **26**(1), 2008, 192–202.

[19] D. Niyato, E. Hossain, 'Competitive Spectrum Sharing in Cognitive Radio Networks: A Dynamic Game Approach', *IEEE Transactions on Wireless Communications*, **7**,July 2008, 2651–2660.

[20] Z. Ji, K.J.R. Liu, 'Multi-Stage Pricing Game for Collusion-Resistant Dynamic Spectrum Allocation', *IEEE Journal on Selected Areas in Communications*, **26**(1), 2008, 182–191.

[21] F. Wang, M. Krunz, S. Cui, 'Price-Based Spectrum Management in Cognitive Radio Networks', *IEEE Journal of Selected Topics in Signal Processing*, **2**(1), 2008, 74–87.

[22] S. Gjerstad, J. Dickhaut, 'Price Formation in Double Auctions', *Games and Economic Behavior*, **22**,1998, 1–29.

[23] V. Krishna, *Auction Theory*, Academic Press, 2002.

[24] N. Singh, X. Vives, 'Price and Quantity Competition in a Differentiated Duopoly', *RAND J. Economics*, **15**(4), 1984, 546–554.

[25] W. Hu, D. Willkomm, M. Abusubaih, J. Gross, G. Vlantis, M. Gerla, A. Wolisz, 'Cognitive Radio for Dynamic Spectrum Access – Dynamic Frequency Hopping Communities for Efficient IEEE 802.22 Operation', *IEEE Communications Magazine*, **45**(5), 2007, 80–87.

[26] M. Sherman, A.N. Mody, R. Martinez, C. Rodriguez, R. Reddy, 'IEE Standards Supporting Cognitive Radio and Networks, Dynamic Spectrum Access, and Coexistence', *IEEE Communications Magazine*, **46**(7), 2008, 72–79.

[27] I.F. Akyildiz, S. Mohanty, J. Xie, 'A Ubiquitous Mobile Communication Architecture for Next-Generation Heterogeneous Wireless Systems', *IEEE Communications Magazine*, **43**,2005, s29–s36.

[28] C.E. Perkins, 'Mobile Networking Through Mobile IP', *IEEE Internet Computing*, January/February 1998, 58–69.

Index

1-persistent 71
A/D converter 52, 166
Access Negotiation 237, 242
Access point (AP) 116
Achievability of multiple access 103
Achievable rate/region 134
Active Probing 216, 224
Ad hoc networks 59, 160, 178
Adaptive antenna array (AAA) 25
Adaptive Frequency hopping (AFH) 128
Adaptive MMSE Receiver 38
Adaptive modulation and coding
  (AMC) 142
Adaptive Power-Rate Adjusting scheme 268
Ad-hoc on demand distance vector (AODV) 276,
  283
Aggressive sensing 216, 220, 224
ALOHA 60, 251
Amplify and forward 103, 108
Association 276, 299, 311
Auction 344
Authentication 329
Authorisation 329
Average likelihood ratio test (ALRT) 221

Backbone/core networks 159
Base station/access point (BSs/APs) 159
Beamforming 24
Bellman-Ford 83
Bellman-Ford Algorithm 85
Bertrand game 340
Block Markov Code 100
Broadcast channel 97
Broadcasting 83

Capacity Region 97
Carrier sense multiple access with collision
  detection (CSMA/CD) 79
Carrier sensing 71, 259
Channel capacity 134
Channel fading coefficient 113
Channel state information (CSI) 131
Clear channel assessment (CCA) 140
Clear channel indicator (CCI) 140, 232
Clipping 20
Clustered head 116
Code division multiple access (CDMA) 2
Cognitive cycle 167, 195
Cognitive radio (CR) 121, 123, 161, 183, 350
Cognitive Radio networks (CRN) 146, 179, 275
Cognitive radio relay network 145
Coherence bandwidth 15
Collision 61
Collision Avoidance/Resolution 237
Collision resolution period (CRP) 67
Competitive Equilibrium (CE) 344
Composite detection 198
Composite hypothesis-testing 221
Concurrent transmission 259, 352
Connectivity 118
Cooperation 114, 171
Cooperative communication 96, 146
Cooperative diversity 96
Cooperative relay 145
Cooperative relay network 151
Cooperative Spectrum Sensing 190
Counting rule 192
Cox Axiom 311
CRN on-demand (CRNO) routing 289

CSMA 71, 259
CSMA with spatial-reuse transmissions (CSMA-ST) 260
Cyclic prefix 18
Cyclic suffix 18
Cyclostationary detection 184

De Moivre-Laplace 315
Decode and forward (DF) 103, 108, 146
Decode-Forward (DF) 100, 174
Decor-relating Receiver 37
Dense network 177
Detection and avoidance (DAA) 128
Digital radio processing/processor (DRP) 48, 57
Dijkstera's Algorithm 85, 318
Doppler shifts 10
Dynamic source routing (DSR) 276, 278
Dynamic Spectrum Access(DSA) 124, 140, 232

Effective bandwidth 174
Elliot-Gilbert channel 136
Energy detection 184
Error control 292

FCFS Splitting Algorithm 68
Fixed relay 173
Flooding 83
Flow Control 89, 291
Forward-path table 289
Frequency offset 9
Frequency Synthesiser 51
Fundamental Frequency 208
Fusion centre 191

Game theory 341
Generalized likelihood ratio test (GLRT) 198, 221
3GPP long-term evolution (LTE) 22

Handover 164
Harvard architecture 43
Heterogeneous wireless networks 276
Hidden terminal problem 187, 190, 232, 242
High-order statistics 211
Hop Reservation Multiple Access 244
Hybrid Automatic-Repeat-Request (HARQ) 174, 292

IEEE 802.16e 22
IEEE 802.16m 22
IEEE 802.22 354
IEEE P1900 354
Incremental relay 174

Infrastructure 59, 160
Inter Channel Interface (ICI) 3
Inter Symbol Interface (ISI) 3
Interference 269, 353
Interference Canceller (IC) 38
Interference channel 101
Inter-system handover 300
Interweave 126, 138
ISI mitigation 16

Leaky Bucket 91
Linear programming 155
Link allocation 154
Link availability 276
Listen-before-transmission (LBT) 71
Log-normal shadowing 201

Markovian Trust Process 319
Matched filter 184
Max-flow min-cut theorem
Maximum likelihood (ML) decoding algorithm 27
Medium access control (MAC) 168, 231, 232, 345
Medium occupancy indicator (MOI) 140, 337
Mesh network 160
MIMO precoding 31
Mixed-integer nonlinear programming 339
ML decoding 26
(mobile) ad-hoc networks (MANET) 116, 276
Mobile IP 348
Mobile station (MS) 159
Modified Bayesian Learning Algorithm 319
MUD 46
Multicarrier CDMA 22
Multichannel MAC 232
Multi-commodity flow 147, 154
Multi-hop packet radio network 60
Multi-input-multi-output (MIMO) 1, 24, 136
Multipath intensity profile 14
Multiple access 60
Multiple access channel 96
Multiple-hop relay network 176
Multi-radio 121, 207, 336
Multi-radio systems 128, 205, 216
Multi-user detection (MUD) 34, 142
Multi-user OFDM 21
MUSIC algorithm 208

Nash equilibrium (NE) 340, 345
Network information theory 146
Network Tomography 296

Neyman-Person hypothesis testing   191, 314
Non-cooperative game   345
Non-persistent   71
NP-hard problem   36
Nyquist criterion   16

oligopoly market   340
Opportunistic Relay   112
Opportunity Spectrum Access   126, 130
Orthogonal Frequency Division Multiple Access
   (OFDMA)   3, 21, 176, 194
Orthogonal frequency division multiplexing
   (OFDM)   1, 2
Outage   107
Outage probability   115, 179, 225
Overlay   126

Parallel channel decoding   100
Parallel cooperative Relay Network   151
Path Discovery Phase   290
Path-loss exponents   225
Peak-to-average power ratio   19
Phase/frequency detector (PFD)   51
Phase-locked loop (PLL)   51
Power delay profile   14
p-Persistent   72
Primary System (PS)   161, 183
Primary users   340
Privacy   331
Proactive   276
Pseudo carrier   218, 222
Public key infrastructure (PKI)   307

QoS guarantee   174

Radio Access Network Selection   170
Radio Resource   201
Radio Resource Allocation   22
Radio Resource Management   303, 345, 351
Randomized decision rule   313
Rate-distance   140, 201, 260
Reactive   276
Received signal strength indicator (RSSI)   184
Re-configurable MAC   165, 168
Relay channel   98
Route Discovery   278
Routing   82, 275, 317

Secondary (CR) users   340
Security   328, 332

Selection relay   174
Self-configuration   298
Self-healing   299
Selfishly try   135
Selflessly act   135
Self-optimisation   298
Self-organisation   124, 166, 298
Sensing phase   289
Shortest Path Routing   83
Singular value decomposition (SVD)   208
Software-defined radio (SDR)   41,
   123, 165
Space-time codes   24, 110
Spatial multiplexing   24
Spectrum allocation   121
Spectrum decision   122, 336
Spectrum holes   128, 336
Spectrum mobility   122, 336
Spectrum re-farming   303
Spectrum sensing   122, 336
Spectrum sharing   122, 336, 339
Splitting Algorithms   66
Sum rate   177
Superposition coding   146, 149

Table Update Phase   290
Time division multiplexing   105
Transmission power control (TPC)   131
Trust   308, 319
Trust Processing Theorem   310
Trusted   286, 312, 317
Trusted Routing   318
Trust Path Theorem   310

Ultrawide band (UWB)   128
Underlay   126, 137
Unidirectional link   163, 276
Uniformly most powerful   198

Voltage controlled oscillator (VCO)   51
Von Neumann machine   43
Voting Rule   192

Wavelet detection   184
Welch's method   211
White space   136
Window Flow Control   89
World Research Forum (WWRF)   354

Zero-forcing   16